Foundations in Signal Processing, Communications and Networking

Series Editors: W. Utschick, H. Boche, R. Mathar

Martin Schubert · Holger Boche

Interference Calculus

A General Framework for Interference Management and Network Utility Optimization

Authors:

Martin Schubert
Heinrich Hertz Institute for
Telecommunications HHI Einsteinufer 37
10587 Berlin
Germany
E-mail: martin.schubert@hhi.fraunhofer.de

Holger Boche
TU Munich
Institute of Theoretical Information Technology
Arcisstrasse 21 80290 Munich,
Germany
E-mail: boche@tum.de

ISSN 1863-8538 e-ISSN 1863-8546

ISBN 978-3-642-24620-3 e-ISBN 978-3-642-24621-0

DOI 10.1007/978-3-642-24621-0

Springer Dordrecht Heidelberg London New York

Library of Congress Control Number: 2011941485

Typesetting: Scientific Publishing Services Pvt. Ltd., Chennai, India.

Cover design: eStudio Calamar S.L.

Printed on acid-free paper

Springer is part of Springer Science+Business Media (www.springer.com)

In memory of my grandmother Maria Amende (1908-1996)

Holger Boche

In memory of my father Roland Schubert (1939-1997) and my grandmother Sofie Schubert (1911-2003)

Martin Schubert

Preface

This book develops a mathematical framework for modeling and optimizing interference-coupled multiuser systems. At the core of this framework is the concept of *general interference functions*, which provides a simple means of characterizing interdependencies between users. The entire analysis builds on the two core axioms *scale-invariance* and *monotonicity*, which are introduced in Section 1.2.

Axiomatic approaches are frequently used in science. An example is cooperative game-theory, where interactions between users (or agents, players) are modeled on the basis of axioms (axiomatic bargaining). The proposed framework does indeed have a conceptual similarity with certain game-theoretic approaches. However, game theory originates from economics or social science, while the proposed interference calculus has its roots in power control theory and wireless comunications. It adds theoretical tools for analyzing the typical behavior of interference-coupled networks. In this way, it complements existing game-theoretic approaches (see e.g. Chapter 4).

The proposed framework should also be viewed in conjunction with optimization theory. There is a fruitful interplay between the theory of interference functions and convex optimization theory. By jointly exploiting the properties of interference functions (notably monotonicity), it is possible to design algorithms that outperform general-purpose techniques that only exploit convexity.

The title "*interference calculus*" refers to the fact that the theory of interference functions constitutes a generic theoretical framework for the analysis of interference coupled systems. Certain operations within the framework are "closed", that is, combinations of interference functions are again, interference functions. Also, certain properties are preserved under such operations. Interference calculus provides a methodology for analyzing different multiuser performance measures that can be expressed as interference functions or combinations of interference functions.

Over the last ten years, the authors have been involved in research on resource allocation for wireless comunication networks. Hence, this book is largely influenced by problems from this area. Among the most influential works is Yates' seminal paper on power control [1], from which we have adopted the term *interference function.* There are indeed close connections between Yates' *standard interference functions* [1] and our framework of *general interference functions*, although they are defined by different axioms. Both frameworks are compared in Section 2.4.

Our first results on general interference functions were published in the monograph [2]. Additional properties were shown in a series of journal articles [3–11]. These extensions provide a deeper and more complete understanding of the subject. An overview tutorial on interference functions and applications was given at ICASSP 2010 [12]. Among the many comments we received, there was the repeated requests for a comprehensive overview that summarizes the important facts and concepts of interference functions.

The present book was written in response to these requests. It provides an overview on the recent advances [3–11]. Particular emphasis is put on analyzing elementary structure properties of interference functions. Exploiting structure is essential for the design of efficient optimization algorithms.

Although the focus of this book is on wireless communication, the proposed axiomatic framework is quite general. Therefore, it is our hope that researchers from other disciplines will be encouraged to work in this area.

The targeted audience includes graduate students of engineering and applied mathematics, as well as academic and industrial researchers in the field of wireless communications, networking, control and game theory. No particular background is needed for reading this book, except for some familiarity with basic concepts from convex analysis and linear algebra. A general willingness to carry out detailed mathematical analysis is, however, important. The proofs and detailed calculations should help the reader in penetrating the subject. Reading our previous book [2] is not a prerequisite, although it might be helpful since it covers additional fundamental aspects of interference functions.

Our scientific work was motivated and influenced by many researchers. Among those who were most influential, we would like to name Tansu Alpcan, Mats Bengtsson, Michael Joham, Josef Nossek, Björn Ottersten, Arogyaswami Paulraj, Dirk Slock, Sławomir Stańczak, Sennur Ulukus, Wolfgang Utschick, and Roy Yates. We thank them and their group members for their inspiring work.

We also thank the funding agencies that made the research possible. The work was funded by the *Federal Ministry of Education and Research* (Bundesministerium für Bildung und Forschung, BMBF) within the projects EASY-C (01BU0631), TEROPP (01SF0708), Scalenet (01BU566), by the *German Research Foundation* (Deutsche Forschungsgemeinschaft, DFG) under grants BO1734/5-2, BO1734/15-1, and SCHU2107/2-1, by the European Commission within the FP6 project MASCOT (IST-26905). We further acknowledge the support of the Deutsche Telekom (SAGA project) and the Technical University Munich Start-up Fund.

Berlin,
August 2011

Martin Schubert
Holger Boche

Contents

1 **Introduction** ... 1
1.1 Notation ... 3
1.2 Basic Axiomatic Framework of Interference Functions ... 4
1.3 Convexity, Concavity, and Logarithmic Convexity ... 5
1.4 Examples – Interference in Wireless Networks ... 7

2 **Systems of Coupled Interference Functions** ... 17
2.1 Combinations of Interference Functions ... 18
2.2 Interference Coupling ... 18
2.3 Strict Monotonicity and Strict Log-Convexity ... 21
2.4 Standard Interference Functions and Power Control ... 23
2.5 Continuity ... 26
2.6 QoS Regions, Feasibility, and Fixed Point Characterization ... 29
2.7 Power-Constrained QoS Regions ... 32
2.8 The QoS Balancing Problem ... 35

3 **The Structure of Interference Functions and Comprehensive Sets** ... 39
3.1 General Interference Functions ... 40
3.2 Synthesis of General Interference Functions ... 46
3.3 Concave Interference Functions ... 50
3.4 Convex Interference Functions ... 61
3.5 Expressing Utility Sets as Sub-/Superlevel Sets of Convex/Concave Interference Functions ... 69
3.6 Log-Convex Interference Functions ... 72
3.7 Application to Standard Interference Functions ... 82
3.8 Convex and Concave Approximations ... 89

4 **Nash Bargaining and Proportional Fairness for Log-Convex Utility Sets** ... 99
4.1 Nash Bargaining for Strictly Log-Convex Utility Sets ... 100

4.2 The SIR Region of Log-Convex Interference Functions 109
4.3 Proportional Fairness – Boundedness, Existence, and Strict Log-Convexity 122
4.4 SINR Region under a Total Power Constraint 138
4.5 Individual Power Constraints – Pareto Optimality and Strict Convexity 140

5 QoS-Constrained Power Minimization 155
5.1 Matrix-Based Iteration 157
5.2 Super-Linear Convergence 164
5.3 Convergence of the Fixed Point Iteration 171
5.4 Worst-Case Interference and Robust Designs 177

6 Weighted SIR Balancing 183
6.1 The Max-Min Optimum 184
6.2 Principal Eigenvector (PEV) Iteration 188
6.3 Fixed Point Iteration 192
6.4 Convergence Behavior of the PEV Iteration 195

A Appendix 197
A.1 Irreducibility 197
A.2 Equivalence of Min-Max and Max-Min Optimization 198
A.3 Log-Convex QoS Sets 199
A.4 Derivatives of Interference Functions 201
A.5 Non-Smooth Analysis 202
A.6 Ratio of Sequences 203
A.7 Optimizing a Ratio of Linear Functions 204
A.8 Continuations of Interference Functions 205
A.9 Proofs 208

References 227

Index 235

1

Introduction

A fundamental problem in the analysis and optimization of multi-user communication networks is that of modeling and optimizing performance tradeoffs. Tradeoffs occur when users share a limited resource or if they are coupled by mutual interference. In both cases, the users cannot act independently. If one user increases the performance by using more resources, then this generally comes at the cost of reducing the available performance margin of other users.

A typical example is a wireless multi-user system, where the signal transmitted by one user causes interference to other users. Interference is not restricted to the physical layer of the communication system, it also affects routing, scheduling, resource allocation, admission control and other higher-layer functionalities. In fact, interference is one of the main reasons why a cross-layer approach is often advocated for wireless systems [13].

Interference may also be understood in a more general way, as the competition for resources in a coupled multi-user system. Interference is not limited to wireless communication scenarios. It is also observed in wireline networks. For example, interference occurs between twisted-pair copper wires used for DSL transmission. There are many other forms of interference in different contexts.

The modeling and optimization of coupled multi-user systems can be difficult. Adaptive techniques for interference mitigation can cause the interference to depend on the underlying resources in a complicated nonlinear fashion. In order to keep the complexity of the resource allocation manageable, interference is often avoided by allocating resources to users in an orthogonal manner, and residual interference is treated as noise. Then the system becomes a collection of quasi-independent communication links. This practical approach greatly simplifies the analysis of multi-user systems. However, assigning each user a separate resource is not always an efficient way of organizing the system. If the number of users is high, then each user only gets a small fraction of the overall resource. Shortages are likely to occur when users have high capacity requirements. This will become even more problematic for future wireless networks, which are expected to provide high-rate services for densely populated

M. Schubert, H. Boche, *Interference Calculus*, Foundations in Signal Processing, Communications and Networking 7,

user environments. The system then might be better utilized by allowing users to share resources.

This development drives the demand for new design principles based on the dynamic reuse of the system resources *frequency*, *power*, and *space* (i.e. the distribution and usage of transmitting and receiving antennas over the service area). Interference is no longer "just an important issue", but rather emerges as the key performance-limiting factor. The classical design paradigm of independent point-to-point communication links is gradually being replaced by a new network-centric point of view, where users avoid or mitigate interference in a flexible way by dynamically adjusting the resources allocated to each user. Interference modeling is an important problem in this context, because the quality of any optimization strategy can only be as good as the underlying interference model.

The development of sophisticated resource sharing strategies requires a thorough understanding of interference-coupled systems. It is important to have theoretical tools which enable us to model and optimize the nonlinear dependencies within the network. The interdependencies caused by interference are not confined to the lower layers of the communication system. For example, it was shown in [14] how the manipulability of certain resource allocation strategies depends on the interference coupling.

This book proposes an abstract theory for the analysis and optimization of interference-coupled multi-user systems. At the core of this theory lies the concept of an *interference function*, which is defined by a framework of axioms (positivity, scale-invariance, monotonicity), as introduced in Section 1.2. This axiomatic approach has the advantage of being quite general. It is applicable to various kinds of interference-coupled systems.

The proposed axiomatic framework was strongly influenced and motivated by power control theory. It generalizes some known concepts and results.

For example, linear interference functions are included as special cases. It will be shown later that certain key properties of a system with linear interference functions extend to logarithmically convex (log-convex) interference functions. In many respects, log-convex interference functions can be regarded as a natural extension of linear interference functions. This area of research is closely linked with the Perron-Frobenius theory, which has numerous important applications [15].

The proposed framework is also useful for the analysis of SIR and SINR regions, with and without power constraints. This includes the problem of finding a suitable operating point within the region. This typically involves a compromise between *fairness* and *efficiency* among the users of the system. Often, these are conflicting goals, and there is more than one definition of fairness. This will be studied from a game-theoretic perspective in Chapter 4.

Finally, the results of this book also contribute to a deeper understanding of *standard interference functions* [1]. The framework of standard interference functions is conceptually similar to the one used here. Both follow an axiomatic approach. The difference will be discussed in Section 2.4. It will be shown that standard interference functions are included in the theory presented here.

After introducing some notational conventions, we will introduce the basic axiomatic framework in Section 1.2. These axioms are the basis for all following derivations. Additional properties, like convexity and logarithmic convexity will be introduced in Section 1.3. In Section 1.4 we will discuss examples of interference functions.

1.1 Notation

We begin with some notational conventions.

- The sets of non-negative reals and positive reals are denoted by $\mathbb{R}_+$ and $\mathbb{R}_{++}$, respectively. $\mathbb{R}^K$ denotes the Kg-dimensional Euclidean vector space.
- Matrices and vectors are denoted by bold capital letters and bold lowercase letters, respectively.
- Let $\boldsymbol{y}$ be a vector, then $y_l = [\boldsymbol{y}]_l$ is the lth component. Likewise, $A_{mn} = [\boldsymbol{A}]_{mn}$ is a component of the matrix $\boldsymbol{A}$.
- A vector inequality $\boldsymbol{x} > \boldsymbol{y}$ means $x_k > y_k$, for all k. The same holds for the reverse directions.
- $\boldsymbol{y} > 0$ means component-wise greater zero.
- $\boldsymbol{y} \geq \boldsymbol{x}$ means $y_l \geq x_l$ for all components
- $\boldsymbol{y} \gneq \boldsymbol{x}$ means $\boldsymbol{y} \geq \boldsymbol{x}$ and there is at least one component l such that $y_l > x_l$.
- $\boldsymbol{y} \neq \boldsymbol{x}$ means that inequality holds for at least one component.
- $\exp(\boldsymbol{y})$ and $\log(\boldsymbol{y})$ means component-wise exponential and logarithm, respectively.

Some often used variables and quantities are as follows.

$\boldsymbol{r}$, $\boldsymbol{p}$ Throughout this book, $\boldsymbol{r}$ is a K-dimensional non-negative vector of "system resources", which are not further specified. A special case is $\boldsymbol{r} = \boldsymbol{p}$, where $\boldsymbol{p}$ is a K_u-dimensional vector containing the transmission powers of K_u users.

$\mathcal{I}(\boldsymbol{r})$ General interference function (see Section 1.2)

$\mathcal{J}(\boldsymbol{p})$ Standard interference function (see Section 2.4)

$\underline{\boldsymbol{p}}$ Extended power vector $\underline{\boldsymbol{p}} = [p_1, \dots, p_{K_u}, \sigma_n^2]^T$, where σ_n^2 is the noise power. Sometimes, we normalize $\sigma_n^2 = 1$.

$\mathcal{K}$, $\mathcal{K}_u$ Index sets of cardinality K and K_u, respectively.

SIR Signal-to-interference ratio

SINR Signal-to-interference-plus-noise ratio

QoS Quality of Service, defined as a strictly monotone and continuous function of the SIR or SINR

$\boldsymbol{\gamma}$ Vector of SIR or SINR targets

$\boldsymbol{\Gamma}$ Diagonal matrix $\boldsymbol{\Gamma} = \text{diag}\{\boldsymbol{\gamma}\}$

$\boldsymbol{V}$ Coupling matrix, $\boldsymbol{V} = [\boldsymbol{v}_1, \dots, \boldsymbol{v}_{K_u}]^T$, where $\boldsymbol{v}_k$ contains the coupling coefficients of user k

$\boldsymbol{W}$ Coupling matrix for log-convex interference functions

1.2 Basic Axiomatic Framework of Interference Functions

Axiomatic characterizations have a long-standing tradition in science. Well-known examples include the *axiomatic bargaining theory* introduced by Nash [16, 17] (see also [18, 19]) and the axiomatic characterization of the Shannon entropy by Khinchin [20] and Faddeev [21] (see also [22]). Analyzing the basic building blocks of a theoretical model often provides valuable new insight into its underlying structure.

In this book, interference is defined as a monotone scale-invariant (homogeneous) function.

Definition 1.1. *Let* $\mathcal{I} : \mathbb{R}_+^K \mapsto \mathbb{R}_+$. *We say that* $\mathcal{I}$ *is a general interference function (or simply* interference function*) if the following axioms are fulfilled:*

A1 *(positivity)* *There exists an* $\boldsymbol{r} > 0$ *such that* $\mathcal{I}(\boldsymbol{r}) > 0$

A2 *(scale invariance)* $\mathcal{I}(\alpha \boldsymbol{r}) = \alpha \mathcal{I}(\boldsymbol{r})$ *for all* $\alpha \geq 0$

A3 *(monotonicity)* $\mathcal{I}(\boldsymbol{r}) \geq \mathcal{I}(\boldsymbol{r}')$ *if* $\boldsymbol{r} \geq \boldsymbol{r}'$

The framework A1, A2, A3 is related to the concept of *standard interference functions* introduced by Yates [1], where *scalability* was required instead of *scale invariance*. Scalability was motivated by a specific power control problem. It will be shown in Section 2.4 that standard interference functions can be comprehended within the framework A1, A2, A3.

Concrete examples of interference functions will be discussed in Section 1.4. Most of these examples focus on multi-user communication systems, where $\boldsymbol{r}$

is a vector of transmission powers, and $\mathcal{I}(\boldsymbol{r})$ is the resulting interference at some receiver. For example, $\mathcal{I}(\boldsymbol{r})$ can measure the impact of some system variables that are collected in the vector $\boldsymbol{r}$.

If one component of $\boldsymbol{r}$ is increased then axiom A3 states that the resulting interference increases or remains constant. This property is closely related to the game-theoretic concept of *comprehensiveness*, which will be discussed in Chapter 3.

Scale-invariance (A2) is best understood by studying the examples in Section 1.4.

An immediate consequence of A2 and A3 is *non-negativity*, i.e., $\mathcal{I}(\boldsymbol{r}) \geq 0$ for all $\boldsymbol{r} \geq 0$. This follows from A2 and A3 by contradiction. Suppose $\mathcal{I}(\boldsymbol{r}) < 0$. For $0 < \lambda \leq 1$ we have $0 > \mathcal{I}(\boldsymbol{r}) \geq \mathcal{I}(\lambda\boldsymbol{r}) = \lambda\mathcal{I}(\boldsymbol{r})$. Letting $\lambda \to 0$ leads to a contradiction, thus proving non-negativity.

Sometimes positivity is needed instead of non-negativity. Axiom A1 (positivity) states that there exists at least one $\boldsymbol{r} > 0$ such that $\mathcal{I}(\boldsymbol{r}) > 0$. It was shown in [2] that A1 is equivalent to the statement $\mathcal{I}(\boldsymbol{r}) > 0$ for *all* $\boldsymbol{r} > 0$. If this is not fulfilled, then we have the trivial interference function $\mathcal{I}(\boldsymbol{r}) = 0$ for all $\boldsymbol{r} > 0$. Hence, the only purpose of A1 is to rule out this trivial case.

The axiomatic framework A1, A2, A3 is analytically appealing. Some basic results were already shown in [2]. But the case of real interest is when the framework is extended by additional properties. It will be shown in the following that under the assumption of certain monotonicity and convexity properties, interference functions offer enough structure to enable efficient algorithmic solutions for different kinds of resource allocation problems.

1.3 Convexity, Concavity, and Logarithmic Convexity

Convexity plays an important role in engineering, economics, and other scientific disciplines [23]. When investigating a problem, a common approach is to first look whether it is convex or not. Theoretical advances have given us new tools that are successfully applied to the optimization of multi-user communication systems [24]. Many examples can be found, for example, in the context of multi-user MIMO and robust optimization [25–30].

1.3.1 Convex and Concave Interference Functions

Standard convex optimization strategies are applicable to any kind of convex problem. However, standard approaches typically ignore the particular analytical structure of the problem at hand. Thus, they are not necessarily a good choice when convergence speed and complexity matters.

In this book we are interested in convex problems that arise from *convex interference functions*, which are not just convex, but which also fulfill the basic axioms A1, A2, A3.

Definition 1.2. *A function* $\mathcal{I} : \mathbb{R}_+^K \mapsto \mathbb{R}_+$ *is said to be a* convex interference function *if A1, A2, A3 are fulfilled and in addition* $\mathcal{I}$ *is convex on* $\mathbb{R}_+^K$. *Likewise, a function* $\mathcal{I} : \mathbb{R}_+^K \mapsto \mathbb{R}_+$ *is said to be a* concave interference function *if A1, A2, A3 are fulfilled and in addition* $\mathcal{I}$ *is concave on* $\mathbb{R}_+^K$.

Examples of nonlinear convex and concave interference functions will be discussed in Section 1.4. Linear interference functions are both convex and concave. nonlinear concave interference functions typically occur when interference is minimized. This includes adaptive receive or transmit strategies, e.g. beamforming [2, 26, 27, 31–33], CDMA [34, 35], or base station assignment [36,37]. Convex interference functions typically occur when interference is maximized. Such "worst-case" strategies are known from robust optimization [29,30].

One of the important goals of this book is to show that convex interference functions have a rich mathematical structure that can be exploited to yield efficient algorithmic solutions. Examples are the SI(N)R-balancing algorithms that will be discussed in Chapters 5 and 6.

1.3.2 Log-Convex Interference Functions

Sometimes, a problem is not convex but there exists an equivalent convex problem formulation. Then the original non-convex problem can be solved indirectly by solving the equivalent problem instead. This is sometimes referred to as "hidden convexity".

Here we are interested in a particular case of hidden convexity, namely *logarithmic convexity* (log-convexity).

Definition 1.3. *A function* $f(\boldsymbol{s})$, *with* $\boldsymbol{s} \in \mathbb{R}^K$ *is said to be log-convex if* $\log f(\boldsymbol{s})$ *is convex. An equivalent condition is [23]*

$$f\big((1-\lambda)\hat{\boldsymbol{s}} + \lambda\check{\boldsymbol{s}}\big) \leq \big(f(\hat{\boldsymbol{s}})\big)^{1-\lambda} \cdot \big(f(\check{\boldsymbol{s}})\big)^{\lambda}, \qquad \textit{for all } \hat{\boldsymbol{s}}, \check{\boldsymbol{s}} \in \mathbb{R}^K .$$

In this book we will investigate log-convexity as a property of interference functions. To this end, we introduce a change of variable

$$\boldsymbol{r} = \mathrm{e}^{\boldsymbol{s}} \qquad \text{(component-wise exponential)} \tag{1.1}$$

where $\boldsymbol{r}$ is the argument of the interference function. This approach was already used in the context of linear interference functions in [38] and later in [39–44].

Definition 1.4. *A function* $\mathcal{I} : \mathbb{R}_+^K \mapsto \mathbb{R}_+$ *is said to be a* log-convex interference function *if* $\mathcal{I}(\boldsymbol{r})$ *fulfills A1, A2, A3 and in addition* $\mathcal{I}(\exp\{\boldsymbol{s}\})$ *is log-convex on* $\mathbb{R}^K$. *Log-concave interference functions are defined accordingly.*

Let $f(\boldsymbol{s}) := \mathcal{I}(\exp\{\boldsymbol{s}\})$. Then a necessary and sufficient condition for log-convexity is [23]

$$f\big(\boldsymbol{s}(\lambda)\big) \le f(\hat{\boldsymbol{s}})^{1-\lambda} f(\check{\boldsymbol{s}})^{\lambda}, \quad \forall \lambda \in (0,1);\ \hat{\boldsymbol{s}}, \check{\boldsymbol{s}} \in \mathbb{R}^K , \tag{1.2}$$

where

$$\boldsymbol{s}(\lambda) = (1-\lambda)\hat{\boldsymbol{s}} + \lambda\check{\boldsymbol{s}} , \quad \lambda \in (0,1) . \tag{1.3}$$

The corresponding vector $\boldsymbol{r}(\lambda) = \exp \boldsymbol{s}(\lambda)$ is

$$\boldsymbol{r}(\lambda) = \hat{\boldsymbol{r}}^{(1-\lambda)} \cdot \check{\boldsymbol{r}}^{\lambda} \tag{1.4}$$

The change of variable $\boldsymbol{r} = \exp\{\boldsymbol{s}\}$ was already used by Sung [38] in the context of linear interference functions (see the following example), and later in [39–43].

With (1.2) it is clear that $\mathcal{I}_k(\mathrm{e}^{\boldsymbol{s}})$ is log-convex if and only if

$$\mathcal{I}_k\big(\boldsymbol{r}(\lambda)\big) \le \big(\mathcal{I}_k(\hat{\boldsymbol{r}})\big)^{1-\lambda} \cdot \big(\mathcal{I}_k(\check{\boldsymbol{r}})\big)^{\lambda} , \quad \lambda \in (0,1) . \tag{1.5}$$

Later, in Subsection 3.6.5, it will be shown that if $\mathcal{I}(\boldsymbol{r})$ is convex, then $\mathcal{I}(\mathrm{e}^{\boldsymbol{s}})$ is log-convex. That means that every convex interference function is log-convex in the sense of Definition 1.4. The converse is not true, however. Therefore, the class of log-convex interference functions is broader than the class of convex interference functions. Log-convex interference functions include convex interference functions as special case. Therefore, the requirement of log-convexity is relatively weak.

Log-convex interference functions offer interesting analytical possibilities similar to the convex case, while being less restrictive. In remainder of this book, we will discuss the properties of log-convex interference functions in detail. It will turn out that log-convex interference functions preserve many of the properties that are known for the linear case. An example is the SIR region studied in Chapter 4.

For completeness, we also discuss the class of *log-concave interference functions*. They were not studied in the literature so far. This is because log-concave interference functions do not have the same advantageous properties as the log-convex interference functions. For example, it is not true that every concave interference function is a log-concave interference function. A simple example is the linear interference function which is log-convex but not log-concave. There are further differences, e.g. the sum of log-convex interference functions is a log-convex interference function, however the same is not true for log-concave interference functions.

1.4 Examples – Interference in Wireless Networks

In this section we will discuss examples of interference functions satisfying the axioms A1, A2, A3. These examples originate mainly from research in wireless communication, especially *power control* theory. However, the analysis of coupled multiuser systems is a broad and diverse field (see e.g. [15]), therefore, more application examples certainly exist.

Consider a wireless communication system with K_u users sharing the same resource, thus mutual interference occurs. The users' transmission powers are collected in a vector

$$\boldsymbol{p} = [p_1, \dots, p_{K_u}]^T \in \mathbb{R}_+^{K_u} . \tag{1.6}$$

The goal is to control the powers $\boldsymbol{p}$ in such a way that a good system performance is achieved. The performance of user k is measured in terms of its *signal-to-interference ratio* (SIR)

$$\mathrm{SIR}_k(\boldsymbol{p}) = \frac{p_k}{\mathcal{I}_k(\boldsymbol{p})} , \quad k \in \mathcal{K}_u . \tag{1.7}$$

Here, $\mathcal{I}_k(\boldsymbol{p})$ is the interference (power cross-talk) observed at user k, for given transmission powers $\boldsymbol{p}$. The functions $\mathcal{I}_1, \dots, \mathcal{I}_{K_u}$ determine how the users are coupled by mutual interference (see Fig. 1.1).

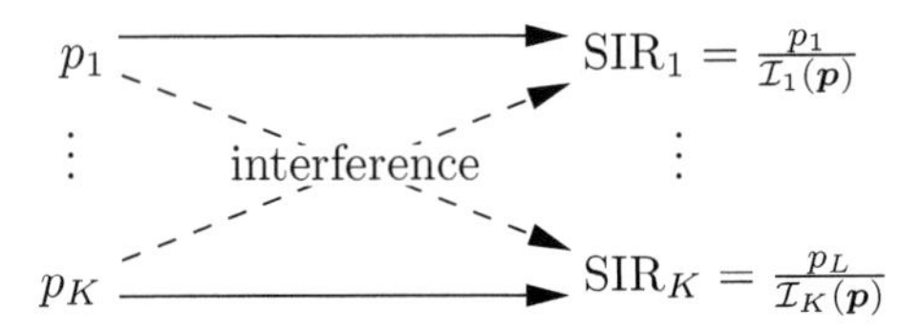

Fig. 1.1. Illustration of an interference-coupled system consisting of K_u transmitter-receiver pairs.

A simple approach to interference modeling is by means of linear functions. This is the basic model in power control theory (see e.g. [42, 45–47] and the references therein).

1.4.1 Linear Interference Function

The interference of user k is defined as

$$\mathcal{I}_k(\boldsymbol{p}) = \boldsymbol{p}^T \boldsymbol{v}_k , \quad k \in \mathcal{K}_u , \tag{1.8}$$

where $\boldsymbol{v}_k \in \mathbb{R}_+^{K_u}$ is a vector of *coupling coefficients*. By collecting all K_u coupling vectors in a *coupling matrix* or *link gain matrix*

$$\boldsymbol{V} = [\boldsymbol{v}_1, \dots, \boldsymbol{v}_{K_u}]^T , \tag{1.9}$$

we can rewrite (1.8) as

$$\mathcal{I}_k(\boldsymbol{p}) = [\boldsymbol{V}\boldsymbol{p}]_k, \quad k \in \mathcal{K}_u . \tag{1.10}$$

The popularity of the linear model is due to its simplicity, but also to its close connection with the rich mathematical theory of non-negative matrices (Perron-Frobenius theory). In the past, this has led to many theoretical results and power control algorithms, e.g. [48–52]. The applicability of the Perron-Frobenius theory is not limited to power control. There are many further examples of systems characterized by a non-negative irreducible matrix. For an overview we refer to [15].

For practical applications, the *signal-to-interference-plus-noise ratio* (SINR) is a typical performance measure. The SINR is also defined as (1.7), where $\mathcal{I}_k$ also depends on noise power σ_n^2. To this end, we introduce the *extended power vector*

$$\underline{\boldsymbol{p}} = [p_1, \dots, p_{K_u}, \sigma_n^2]^T \in \mathbb{R}_+^{K_u+1} \tag{1.11}$$

The resulting interference-plus-noise power is

$$\mathcal{I}_k(\underline{\boldsymbol{p}}) = \underline{\boldsymbol{p}}^T \cdot \left[\begin{smallmatrix} \boldsymbol{v}_k \\ 1 \end{smallmatrix}\right] = \boldsymbol{p}^T \boldsymbol{v}_k + \sigma_n^2 \,. \tag{1.12}$$

Note, that the interference function (1.12) has the same structure as (1.8). The only difference is the dependence on the power vector $\underline{\boldsymbol{p}}$ which is extended by one dimension. This notation will allow us later to investigate different problems within a single unifying framework. Some properties are shared by both models, no matter whether there is noise or not. For example, most structure results from Chapter 3 readily extend to the case where there is additional noise.

However, noise clearly makes a difference when investigating resource allocation algorithms in a power-constrained multi-user system. Then it is important to consider the special properties resulting from the assumption of a constant noise component. This will be studied in detail in Section 2.4.

Linear interference functions are concave, convex, and also log-convex after a change of variable (see Section 1.3.2). Hence, all results in this book hold for linear interference functions.

1.4.2 Beamforming

The linear model is well understood and there is a wealth of interesting results and applications, not limited to communication scenarios (see e.g. [15]). However, interference often depends on the transmission powers in a *nonlinear* way, e.g., if adaptive receive and transmit strategies are employed to avoid or mitigate interference. Using a linear model may oversimplify the real situation. Therefore, it is desirable to extend the linear model.

An example is the following *nonlinear* interference function resulting from multi-user beamforming. This scenario was studied, e.g., in [26–28, 31–33].

Consider an uplink system with K_u single-antenna transmitters and an M-element antenna array at the receiver. Independent signals $S_1, \dots, S_{K_u}$ with powers $p_k = \mathrm{E}[|S_k|^2]$ are transmitted over vector-valued channels $\boldsymbol{h}_1, \dots, \boldsymbol{h}_{K_u} \in \mathbb{C}^M$, with spatial covariance matrices $\boldsymbol{R}_k = \mathrm{E}[\boldsymbol{h}_k \boldsymbol{h}_k^H]$. The superimposed signals at the array output are received by a bank of linear filters $\boldsymbol{u}_1, \dots, \boldsymbol{u}_{K_u}$ (the 'beamformers'). The output of the kth beamformer is

$$y_k = \boldsymbol{u}_k^H \Big(\sum_{l \in \mathcal{K}_u} \boldsymbol{h}_l S_l + \boldsymbol{n} \Big) , \tag{1.13}$$

where $\boldsymbol{n} \in \mathbb{C}^M$ is an AWGN vector, with $\mathrm{E}[\boldsymbol{n}\boldsymbol{n}^H] = \sigma_n^2 \boldsymbol{I}$. The SINR of user k is

$$\mathrm{SINR}_k(\boldsymbol{p}, \boldsymbol{u}_k) = \frac{\mathrm{E}[|\boldsymbol{u}_k^H \boldsymbol{h}_k S_k|^2]}{\mathrm{E}[| \sum_{l \in k} \boldsymbol{u}_k^H \boldsymbol{h}_l S_l + \boldsymbol{u}_k^H \boldsymbol{n}|^2]} = \frac{p_k \boldsymbol{u}_k^H \boldsymbol{R}_k \boldsymbol{u}_k}{\boldsymbol{u}_k^H \big(\sum_{l \neq k} p_l \boldsymbol{R}_l + \sigma_n^2 \boldsymbol{I} \big) \boldsymbol{u}_k} .$$

With the common normalization $\|\boldsymbol{u}_k\|_2 = 1$, the interference function for the beamforming case is

$$\begin{aligned} \mathcal{I}_k(\underline{\boldsymbol{p}}) &= \min_{\|\boldsymbol{u}_k\|=1} \frac{p_k}{\mathrm{SINR}(\boldsymbol{p}, \boldsymbol{w}_k)} = \min_{\|\boldsymbol{u}_k\|_2=1} \frac{\boldsymbol{u}_k^H \big(\sum_{l \neq k} p_l \boldsymbol{R}_l + \sigma_n^2 \boldsymbol{I} \big) \boldsymbol{u}_k}{\boldsymbol{u}_k^H \boldsymbol{R}_k \boldsymbol{u}_k} \\ &= \min_{\|\boldsymbol{u}_k\|_2=1} \frac{\sum_{l \neq k} p_l \boldsymbol{u}_k^H \boldsymbol{R}_l \boldsymbol{u}_k + \sigma_n^2 \|\boldsymbol{u}_k\|^2}{\boldsymbol{u}_k^H \boldsymbol{R}_k \boldsymbol{u}_k} = \min_{\|\boldsymbol{u}_k\|_2=1} \underline{\boldsymbol{p}}^T \boldsymbol{v}_k(\boldsymbol{u}_k) , \end{aligned} \tag{1.14}$$

where $\boldsymbol{v}_k(\boldsymbol{u}_k)$ is a vector of coupling coefficients defined as follows.

$$[\boldsymbol{v}_k(\boldsymbol{u}_k)]_l = \begin{cases} \frac{\boldsymbol{u}_k^H \boldsymbol{R}_l \boldsymbol{u}_k}{\boldsymbol{u}_k^H \boldsymbol{R}_k \boldsymbol{u}_k} & 1 \leq l \leq K_u,\ l \neq k \\ \frac{\|\boldsymbol{u}_k\|^2}{\boldsymbol{u}_k^H \boldsymbol{R}_k \boldsymbol{u}_k} & l = K_u + 1, \\ 0 & l = k . \end{cases} \tag{1.15}$$

It can be observed that the interference coupling is not constant. For any power vector $\boldsymbol{p} > 0$, the beamformer $\boldsymbol{u}_k$ adapts to the interference in such a way that the signal-to-interference-plus-noise ratio (SINR) is maximized. This optimization can be solved efficiently via an eigenvalue decomposition [53].

A special case occurs if the channels $\boldsymbol{h}_1, \ldots, \boldsymbol{h}_{K_u}$ are deterministic, then $\boldsymbol{R}_l = \boldsymbol{h}_l \boldsymbol{h}_l^H$. In this case, the interference resulting from optimum beamformers can be written in closed form

$$\mathcal{I}_k(\underline{\boldsymbol{p}}) = \frac{1}{\boldsymbol{h}_k^H \big(\sigma_n^2 \boldsymbol{I} + \sum_{l \neq k} p_l \boldsymbol{h}_l \boldsymbol{h}_l^H \big)^{-1} \boldsymbol{h}_k} . \tag{1.16}$$

Although the interference function (1.16) is more complicated than the linear one, it has an analytical structure that allows for efficient algorithmic solutions. The interference function (1.14) is concave, as the minimum of linear functions. Thus (1.16) is concave as well. It will be shown later (Theorem 3.23) that all concave interference functions have a structure that enables efficient algorithmic solutions. Examples are the interference balancing algorithms from Chapters 5 and 6.

1.4.3 Receive Strategies

The next example shows that the interference functions (1.14) and (1.16) can be understood within a more general and abstract framework of *adaptive receive strategies*.

For every user k, we define an abstract *receive strategy* z_k from a non-empty compact set $\mathcal{Z}_k$. The receive strategy z_k leads to coupling coefficients $\boldsymbol{v}_k(z_k) \in \mathbb{R}_+^{K_u}$. Since we aim for generality, we do not specify the nature of the parameter z_k or how the interference coupling $\boldsymbol{v}_k$ depends on z_k. The name "receive strategy" refers to the typical behavior of a receiver which maximizes the SINR, or equivalently, minimizes the interference. That is is, for any given power vector $\boldsymbol{p}$ we choose z_k such that the interference $\boldsymbol{p}^T\boldsymbol{v}_k(z_k)$ becomes minimal. The resulting interference functions are

$$\mathcal{I}_k(\boldsymbol{p}) = \min_{z_k \in \mathcal{Z}_k} \boldsymbol{p}^T \boldsymbol{v}_k(z_k) , \quad \forall k \in \mathcal{K}_u . \tag{1.17}$$

Noise can be included by using the extended power vector (1.11) and the extended coupling vector, as in (1.12).

$$\mathcal{I}_k(\underline{\boldsymbol{p}}) = \min_{z_k \in \mathcal{Z}_k} \underline{\boldsymbol{p}}^T \underline{\boldsymbol{v}}_k(z_k) , \quad \forall k \in \mathcal{K}_u . \tag{1.18}$$

A special case is the previous example (1.14), where beamformers $\boldsymbol{u}_k$ were used as receive strategies. The beamformers were chosen from the unit sphere, i.e., $\|\boldsymbol{u}_k\|_2 = 1$. Note, that the model (1.17) allows for arbitrary other constraints. For example, beamformers with shaping constraints were studied in [27]. This is included in the generic model (1.17), where we only require that the set $\mathcal{Z}_k$ is compact in order to ensure the existence of the minimum. The set $\mathcal{Z}_k$ can also be discrete, for example when there is a choice between several receivers. A special case is the problem of joint beamforming and base station assignment [36, 37].

As in the previous example, the resulting interference function is concave.

1.4.4 The Spectral Radius – Indicator of Feasibility

Consider again the example of linear interference functions from Subsection 1.4.1. The function $\mathcal{I}_k(\boldsymbol{p}) = [\boldsymbol{V}\boldsymbol{p}]_k$ is based on a non-negative and irreducible coupling matrix $\boldsymbol{V}$. Irreducibility means that each user depends on the transmission power of any other user, either directly or indirectly (see Appendix A.1 for a formal definition). The concept of irreducibility is fundamental for the analysis of interference-coupled systems. It will be used at several points throughout this book.

A fundamental question is, under what conditions can certain SIR values $\mathrm{SIR}_1(\boldsymbol{p}), \ldots, \mathrm{SIR}_{K_u}(\boldsymbol{p})$ be achieved jointly? This depends on how the users are coupled by interference. Let γ_k be the target SIR of user k. The targets of all K_u targets are collected in a vector

$$\boldsymbol{\gamma} = [\gamma_1, \gamma_2, \ldots, \gamma_{K_u}]^T \in \mathbb{R}_{++}^{K_u} . \tag{1.19}$$

If all SIR targets $\boldsymbol{\gamma}$ can be achieved then we say that $\boldsymbol{\gamma}$ is *feasible*. It was already observed in early work [48, 54] that the feasibility depends on the spectral radius

$$\rho_V(\gamma) = \rho(\boldsymbol{\Gamma V}) , \quad \text{where } \boldsymbol{\Gamma} = \operatorname{diag}\{\gamma\} . \tag{1.20}$$

In the context of non-negative irreducible matrices, ρ_V is also referred to as the *Perron root.*

If $\rho_V(\gamma) \leq 1$, then there exists a $\boldsymbol{p} > 0$ such that $\mathrm{SIR}_k(\boldsymbol{p}) \geq \gamma_k$ for all $k \in \mathcal{K}_u$. The feasible SIR region is defined as follows.

$$\mathcal{S} = \{\gamma > 0 : \rho_V(\gamma) \leq 1\} . \tag{1.21}$$

The function $\rho_V(\gamma)$ is an indicator for the feasibility of an SIR vector γ. It provides a single measure for the system load caused by the K_u users. It is observed that $\rho_V(\gamma)$ fulfills the axioms A1, A2, A3. Thus, the SIR region (1.21) is a sublevel set of an interference function.

The structure of the region $\mathcal{S}$ is directly connected with the properties of $\rho_V(\gamma)$. This was already exploited in [38], where it was shown that the SIR region $\mathcal{S}$ is convex on a logarithmic scale. Additional properties were shown in [39–41]. The log-convexity of $\mathcal{S}$ can also be understood as a special case of [5], where the SIR region was studied within the framework of log-convex interference functions. In fact, $\rho_V(\gamma)$ is a log-convex interference function, in the sense of Definition 1.4. That is, $\rho_V(\exp \boldsymbol{q})$ is log-convex on $\mathbb{R}^{K_u}$, where we use the change of variable $\gamma = \exp \boldsymbol{q}$. The region $\mathcal{S}$ is a sub-level set of the log-convex (thus convex) indicator function $\rho_V(\exp \boldsymbol{q})$. Consequently, the log-SIR region is convex.

The spectral radius is an indicator function resulting from linear interference functions. This is a special case of the min-max function $C(\gamma)$ that will be discussed in the next subsection.

1.4.5 Min-Max Balancing and Feasible Sets

Consider arbitrary interference functions $\mathcal{I}_1, \ldots, \mathcal{I}_{K_u}$. In contrast to the previous example we only require the basic axioms A1, A2, A3. We wish to know whether there exists a $\boldsymbol{p} > 0$ such that $\mathrm{SIR}_k(\boldsymbol{p}) \geq \gamma_k$ for all $k \in K_u$, or equivalently

$$\max_{k \in K_u} \frac{\gamma_k}{\mathrm{SIR}_k(\boldsymbol{p})} = \max_{k \in K_u} \frac{\gamma_k \cdot \mathcal{I}_k(\boldsymbol{p})}{p_k} \leq 1$$

Thus, the following function $C(\gamma)$ is an indicator for the feasibility of γ.

$$C(\gamma) = \inf_{\boldsymbol{p} > 0} \left(\max_{k \in K_u} \frac{\gamma_k \cdot \mathcal{I}_k(\boldsymbol{p})}{p_k} \right) . \tag{1.22}$$

The optimizer of this problem (if existent) maximizes the minimum SIR (see Appendix A.2).

Some SIR vector $\gamma > 0$ is feasible if and only if $C(\gamma) \leq 1$. If $C(\gamma) = 1$ and the infimum (1.22) is not attained, then this means that γ is a boundary point that can only be achieved asymptotically. Our definition of feasibility includes this asymptotic case, but for most practical scenarios, 'inf' can be replaced by 'min', which means that γ is actually attained by some $\boldsymbol{p} > 0$.

The feasible SIR region is defined as

$$\mathcal{S} = \{\boldsymbol{\gamma} > 0 : C(\boldsymbol{\gamma}) \leq 1\} . \tag{1.23}$$

If the interference functions are linear, i.e., $\mathcal{I}_k(\boldsymbol{p}) = [\boldsymbol{V}\boldsymbol{p}]_k$ as in the previous example (1.10), then $C(\boldsymbol{\gamma})$ is simply the spectral radius (1.20) of the weighted coupling matrix $\boldsymbol{\Gamma V}$. This can be seen from the Collatz-Wielandt type characterization [55] (see also [56–58]).

$$C(\boldsymbol{\gamma}) = \inf_{\boldsymbol{p}>0} \Big(\max_{k \in \mathcal{K}_u} \frac{[\boldsymbol{\Gamma V p}]_k}{p_k} \Big) = \rho(\boldsymbol{\Gamma V}) . \tag{1.24}$$

The indicator function $C(\boldsymbol{\gamma})$ fulfills the axioms A1, A2, A3. If the underlying interference functions are log-convex, then $C(\boldsymbol{\gamma})$ is a log-convex interference function. the function $C(\exp \boldsymbol{q})$ is log-convex after a change of variable $\boldsymbol{\gamma} = \exp \boldsymbol{q}$. Every log-convex function is convex [23]. Hence, the log-SIR region $\{\boldsymbol{q} \in \mathbb{R}^{K_u} : C(\exp \boldsymbol{q}) \leq 1\}$ is a convex set. This can be generalized to arbitrary log-convex functions $\boldsymbol{\gamma}(\boldsymbol{q})$), as shown in [5] (see also Section 2.6).

In this example, interference functions occur on different levels. The physical interference is modeled by $\mathcal{I}_1, \dots, \mathcal{I}_{K_u}$. On a higher level, the interference function $C(\boldsymbol{\gamma})$ provides a measure for the system load. The properties of the resulting SIR region depends on the properties of $C(\boldsymbol{\gamma})$, which depends on the properties of $\mathcal{I}_1, \dots, \mathcal{I}_{K_u}$. These aspects will be studied in more detail in Section 2.6. In Section 2.7 we will discuss how to incorporate power constraints.

1.4.6 Transmit Strategies and Duality

Consider a system of K_u users with an irreducible coupling matrix $\boldsymbol{G} \in \mathbb{R}_+^{K_u \times K_u}$. Assume that $\boldsymbol{G}$ depends on parameters $z = (z_1, \dots, z_{K_u})$ in a *column-wise* fashion. That is, the kth column of $\boldsymbol{G}$ only depends on $z_k \in \mathcal{Z}_k$. As a consequence, the interference $[\boldsymbol{G}(z)\boldsymbol{p}]_k$ of user k depends on *all* parameters $z = (z_1, \dots, z_{K_u})$. This is typical for "transmit strategies" that optimize the communication links at the transmitter side (e.g. transmit beamforming). Thus, we refer to z_k as a *transmit strategy*, in contrast to the receive strategy discussed in the previous example.

However, the resulting interference values $[\boldsymbol{G}(z)\boldsymbol{p}]_k$ are difficult to handle since each of them not only depends on $\boldsymbol{p}$, but also on *all* transmit strategies $z_1, \dots, z_{K_u}$. The choice of any transmitter influences the interference received by all other users. Thus, we cannot write the interference in terms of K_u separate interference functions depending only on $\boldsymbol{p}$, as in the previous examples. When optimizing the system jointly with respect to $\boldsymbol{p}$ and z, then a joint optimization approach is required. An example is the problem of joint power control and downlink beamforming, for which suboptimal heuristics were proposed in early work [31, 59].

Fortunately, there is a simple way of getting around the problem of coupled transmit strategies. We can exploit that the columns of $\boldsymbol{G}(z)$ are independent with respect to $z_1, \dots, z_{K_u}$. The key idea is to optimize the transpose system $\boldsymbol{V}(z) = \boldsymbol{G}^T(z)$ instead of the original system $\boldsymbol{G}(z)$. Similar to (1.9) we define

$$\boldsymbol{V}(z) = [\boldsymbol{v}_1(z_1), \dots, \boldsymbol{v}_{K_u}(z_{K_u})]^T \,,$$

The kth row of this "virtual" system $\boldsymbol{V}(z)$ only depends on the parameter z_k. Hence, the resulting interference can be expressed in terms of the interference functions (1.17). Introducing an auxiliary variable $\boldsymbol{q} \in \mathbb{R}_+^{K_u}$, we have

$$\mathcal{I}_k(\boldsymbol{q}) = \min_{z_k \in \mathcal{Z}_k} [\boldsymbol{V}(z)\boldsymbol{q}]_k = \min_{z_k \in \mathcal{Z}_k} \boldsymbol{q}^T \boldsymbol{v}_k(z_k) \,, \quad \forall k \in \mathcal{K}_u \,. \tag{1.25}$$

The variable $\boldsymbol{q}$ can be regarded as the power vector of the virtual system. Because the kth row of $\boldsymbol{V}(z)$ only depends on the parameter z_k we obtain K_u interference functions which can be optimized independently with respect to the parameters z_k. The transmit strategy z becomes a "virtual receive strategy".

It remains to show that the optimization of the virtual system $\boldsymbol{V}(z)$ leads to the optimum of the original system $\boldsymbol{G}(z)$. Whether such a "duality" exists depends on the optimization problem under consideration. An example for which duality holds is the problem of *SIR balancing*, where the aim is to maximize the worst SIR among all users. This problem will be studied in detail in Chapter 6. For the special case of transmit beamforming, the problem was studied in [8, 59, 60].

This duality between transpose systems was already observed in [61] in a power control context. Duality was also observed in the context of the aforementioned downlink beamforming problem [62, 63]. In this work, the matrix $\boldsymbol{V}$ characterizes a downlink point-to-multipoint channel, whereas the transpose $\boldsymbol{V}^T$ has an interpretation as an uplink multipoint-to-point channel. Thus, the term "uplink-downlink duality" was introduced to refer to this reciprocity between both channels.

Examples in the context of multi-antenna signal processing include [32, 33, 64–66]. A recent extension of this line of work is [67], where per-antenna power constraints were studied. There is also an interesting relationship with the MAC/BC duality observed in information theory [65, 66, 68, 69].

1.4.7 Robust Designs

Linear interference functions (1.8) can be generalized by introducing parameter-dependent coupling coefficients $\boldsymbol{v}_k(c_k)$. Assume that the parameter c_k stands for some *uncertainty* chosen from a compact *uncertainty region* $\mathcal{C}_k$. A typical source of uncertainty are channel estimation errors or other system imperfections. Then, the worst-case interference is given by

$$\mathcal{I}_k(\boldsymbol{p}) = \max_{c_k \in \mathcal{C}_k} \boldsymbol{p}^T \boldsymbol{v}_k(c_k) \,, \quad k \in K_u \,. \tag{1.26}$$

Performing power allocation with respect to the interference functions (1.26) guarantees a certain degree of robustness. Robust power allocation was studied, e.g., in [9, 29, 30, 43].

As an example, consider again the downlink beamforming scenario discussed in the previous section. In the presence of imperfect channel estimation, the spatial covariance matrices can be modeled as $\boldsymbol{R}_k = \hat{\boldsymbol{R}}_k + \boldsymbol{\Delta}_k$, where $\hat{\boldsymbol{R}}_k$ is the estimated covariance, and $\boldsymbol{\Delta}_k \in \mathcal{Z}_k$ is the estimation error from a compact uncertainty region $\mathcal{Z}_k$. In order to improve the robustness, the system can be optimized with respect to the worst case interference functions

$$\mathcal{I}_k(\underline{\boldsymbol{p}}) = \max_{\boldsymbol{\Delta}_k \in \mathcal{Z}_k} \frac{\sum_{l \neq k} p_l \boldsymbol{u}_l^H (\hat{\boldsymbol{R}}_k + \boldsymbol{\Delta}_k) \boldsymbol{u}_l + \sigma_n^2}{\boldsymbol{u}_k^H (\hat{\boldsymbol{R}}_k + \boldsymbol{\Delta}_k) \boldsymbol{u}_k} \tag{1.27}$$

Other types of uncertainties, like *noise uncertainty* are straightforward extensions of this model.

The examples (1.26) and (1.27) are convex interference functions. Every convex interference function is a log-convex interference function in the sense of Definition 1.4. That is, (1.26) and (1.27) are log-convex interference functions, they are log-convex after a change of variable.

In related work [70, 71], an additional optimization with respect to the beamformers $\boldsymbol{u}_1, \ldots, \boldsymbol{u}_{K_u}$ is performed. This leads to the min-max interference function

$$\mathcal{I}_k(\underline{\boldsymbol{p}}) = \min_{\|\boldsymbol{u}_l\|=1} \left(\max_{\boldsymbol{\Delta}_k \in \mathcal{Z}_k} \frac{\sum_{l \neq k} p_l \boldsymbol{u}_l^H (\hat{\boldsymbol{R}}_k + \boldsymbol{\Delta}_k) \boldsymbol{u}_l + \sigma_n^2}{\boldsymbol{u}_k^H (\hat{\boldsymbol{R}}_k + \boldsymbol{\Delta}_k) \boldsymbol{u}_k} \right) \tag{1.28}$$

The interference function (1.28) is neither convex nor concave in general, but it also fulfills the basic properties A1, A2, A3.

1.4.8 Interference Functions in Other Contexts

The previous list of examples is by no means exhaustive. It shows that interference is often nonlinear, and interference functions appear in different contexts, not limited to power control.

For example, a generic performance measure is as follows.

$$\mathcal{I}_{w,\alpha}(\boldsymbol{p}) = \Big(\sum_{k \in \mathcal{K}} w_k \cdot (p_k)^\alpha \Big)^{1/\alpha} \tag{1.29}$$

where $\alpha > 0$ and $w_k > 0$, $\forall k \in \mathcal{K}$. For $w_k = 1$, this is the p-norm on $\mathbb{R}^K$. The function (1.29) fulfills the axioms A1, A2, A3, and thus falls within the framework of interference functions.

Another example is the weighted sum utility

$$U_{\mathrm{sum}}(\boldsymbol{w}) = \max_{\boldsymbol{u} \in \mathcal{U}} \sum_{k \in \mathcal{K}} w_k u_k \,, \tag{1.30}$$

where $\boldsymbol{w} \in \mathbb{R}^K_+$, with $\|\boldsymbol{w}\|_1 = 1$, are weighting factors and the utility vector $\boldsymbol{u}$ is chosen from a compact set $\mathcal{U} \subset \mathbb{R}^K_{++}$. For example, in a time-scheduled system u_k could stand for a user rate and w_k could be the queue backlog.

The function $U_{\text{sum}}(\boldsymbol{w})$ is a convex interference function. Consequently, it is a log-convex interference functions after a change of variable (see Section 1.3.2).

More properties of interference functions will be studied in the remainder of this book. The analysis of interference functions is closely tied to the analysis of feasible sets (see e.g. the example from Section 1.4.5). The properties of the feasible SIR sets are determined by the properties of the underlying interference functions. Thus, parts of the book are devoted to a detailed study of the interdependencies between interference functions and feasible sets.

Furthermore, the interference calculus is very closely connected with the theory of *monotone optimization* (see e.g. [72]), which is based on *increasing positively homogeneous functions*. This theory has been applied to the study of models in Mathematical Economics [73]. The differences and similiarities between both theories have not yet been fully explored.

2

Systems of Coupled Interference Functions

Consider a multi-user system characterized by K interference functions

$$\mathcal{I}_1(\boldsymbol{r}), \mathcal{I}_2(\boldsymbol{r}), \ldots, \mathcal{I}_K(\boldsymbol{r}),$$

which all depend on the same resource vector $\boldsymbol{r} \in \mathbb{R}_+^K$. We begin with the most general case where the interference functions are only characterized by the axioms A1, A2, A3 (see p. 4 in Section 1.2). The overall performance of the system is typically a function of all interference values, which depend on the same underlying resource vector $\boldsymbol{r}$. Specific examples were already discussed in Section 1.4.

The analysis and optimization of such a system is complicated by the fact that the interference functions can be mutually *coupled*. That is, the interference value $\mathcal{I}_k(\boldsymbol{r})$ of some user k can depend on other users' resources r_l, $l \neq k$. The users can also be coupled by sharing a common budget. This leads to joint optimization problems that are often difficult to handle. It is therefore important to have a thorough understanding of the properties of interference, and the structure of the optimization problems that result from combinations of interference functions.

In this chapter we discuss some fundamental properties of interference-coupled systems, and we show the connections with existing work in power control theory. In the context of power control, Yates [1] introduced the axiomatic framework of *standard interference functions* for modeling interference as a function of transmission powers. The theory was further analyzed and extended in [11, 34, 74, 75]. It will be shown in Section 2.4 that the axiomatic framework A1, A2, A3 with additional strict monotonicity provides an equivalent way of modeling standard interference functions. Hence, standard interference functions can be regarded as a special case, and most results derived in this book immediately transfer to standard interference functions.

M. Schubert, H. Boche, *Interference Calculus*, Foundations in Signal Processing, Communications and Networking 7,

2.1 Combinations of Interference Functions

Section 1.4.5 introduced the indicator function $C(\gamma)$, which is an example of an interference function being constructed as a combination of other interference functions. Other possible combinations exist. Consider interference functions $\mathcal{I}_1, \dots, \mathcal{I}_K$, which fulfill the axioms A1, A2, A3, then these properties are preserved by the following combinations.

- The maximum of interference functions is again an interference function.

$$\mathcal{I}(\boldsymbol{r}) = \max_{k \in \mathcal{K}} \mathcal{I}_k(\boldsymbol{r}) \,. \tag{2.1}$$

  This remains valid when the maximum is replaced by the minimum.
- Any linear combination of interference functions is an interference function.

$$\mathcal{I}(\boldsymbol{r}) = \sum_{k \in \mathcal{K}} \alpha_k \mathcal{I}_k(\boldsymbol{r}) \quad \text{where } \alpha_k \in \mathbb{R}_+ \,. \tag{2.2}$$

- Let $\tilde{\mathcal{I}}$ be an interference function, depending on other interference functions $\mathcal{I}_1, \dots, \mathcal{I}_K$, then

$$\mathcal{I}(\boldsymbol{r}) = \tilde{\mathcal{I}}(\mathcal{I}_1(\boldsymbol{r}), \mathcal{I}_2(\boldsymbol{r}), \dots, \mathcal{I}_K(\boldsymbol{r})) \tag{2.3}$$

  is an interference function.

For log-convex interference functions, the following properties hold:

- The sum of log-convex interference functions is a log-convex interference function.
- Let $\mathcal{I}^{(1)}$ and $\mathcal{I}^{(2)}$ be log-convex interference functions, then

$$\mathcal{I}(\boldsymbol{r}) = \big(\mathcal{I}^{(1)}(\boldsymbol{r})\big)^{1-\alpha} \cdot \big(\mathcal{I}^{(2)}(\boldsymbol{r})\big)^{\alpha}, \quad 0 \le \alpha \le 1 \,,$$

  is also a log-convex interference function.
- Let $\mathcal{I}^{(n)}(\boldsymbol{r})$ be a sequence of log-convex interference functions, which converges to a limit $\lim_{n \to \infty} \mathcal{I}^{(n)}(\boldsymbol{r}) = \hat{\mathcal{I}}(\boldsymbol{r}) > 0$ for all $\boldsymbol{r} > 0$, then $\hat{\mathcal{I}}$ is also a log-convex interference function.

2.2 Interference Coupling

Interference coupling was well-defined for the specific examples of Section 1.4. For linear interference functions, the coupling between the users is characterized by a *link gain* matrix $\boldsymbol{V} \geq 0$, as defined by (1.9). This is a common approach in power control theory (see, e.g., [45] and references therein).

However, the axiomatic framework A1, A2, A3 does not include the notion of a coupling matrix. It is a priori not clear whether the functions are coupled or not. For example, interference can be removed by interference cancellation strategies, or it can be avoided by allocating users to different resources.

It is desirable to have a general way of modeling interference coupling, which can be applied to arbitrary interference functions satisfying A1, A2, A3. The following notion of "interference coupling" defines whether a user causes interference to another user or not.

2.2.1 Asymptotic Coupling Matrix

Independent of the choice of $\boldsymbol{r}$, the interference coupling can be characterized by an asymptotic approach. To this end, we introduce $\boldsymbol{e}_l$, which is the all-zero vector with the l-th component set to one.

$$[\boldsymbol{e}_l]_n = \begin{cases} 1 & n = l \\ 0 & n \neq l \,. \end{cases} \tag{2.4}$$

We have the following result.

Lemma 2.1. *Assume there exists a* $\hat{\boldsymbol{r}} > 0$ *such that* $\lim_{\delta\to\infty} \mathcal{I}_k(\hat{\boldsymbol{r}} + \delta \boldsymbol{e}_l) = +\infty$, *then*

$$\lim_{\delta\to\infty} \mathcal{I}_k(\boldsymbol{r} + \delta \boldsymbol{e}_l) = +\infty \quad \text{for all } \boldsymbol{r} > 0. \tag{2.5}$$

Proof. Let $\boldsymbol{r} > 0$ be arbitrary. There exists a $\lambda > 0$ such that $\lambda \boldsymbol{r} \geq \hat{\boldsymbol{r}}$. Thus, A3 implies

$$\lim_{\delta\to\infty} \mathcal{I}_k(\lambda \boldsymbol{r} + \delta \boldsymbol{e}_l) \geq \lim_{\delta\to\infty} \mathcal{I}_k(\hat{\boldsymbol{r}} + \delta \boldsymbol{e}_l) = +\infty \,. \tag{2.6}$$

With A2 we have $\mathcal{I}_k(\lambda \boldsymbol{r} + \delta \boldsymbol{e}_l) = \lambda \mathcal{I}_k(\boldsymbol{r} + \frac{\delta}{\lambda} \boldsymbol{e}_l)$. This implies $\lim_{\delta\to\infty} \mathcal{I}_k(\boldsymbol{r} + \frac{\delta}{\lambda} \boldsymbol{e}_l) = +\infty$, from which (2.5) follows. The interference function $\mathcal{I}_k$ is unbounded and monotone increasing (axiom A3), thus the existence of the limits is guaranteed. □

For arbitrary interference functions satisfying A1-A3, condition (2.5) formalizes the notion of "user l causing interference to user k". This enables us to define interference coupling by means of a matrix.

Definition 2.2. *The* asymptotic coupling matrix *is*

$$[\boldsymbol{A}_{\mathcal{I}}]_{kl} = \begin{cases} 1 & \text{if there exists a } \boldsymbol{r} > 0 \text{ such that} \\ & \lim_{\delta\to\infty} \mathcal{I}_k(\boldsymbol{r} + \delta \boldsymbol{e}_l) = +\infty, \\ 0 & \text{otherwise.} \end{cases} \tag{2.7}$$

The matrix $\boldsymbol{A}_{\mathcal{I}}$ characterizes the way users are connected by interference. The 1-entries in the kth row of $\boldsymbol{A}_{\mathcal{I}}$ mark the positions of the power components on which $\mathcal{I}_k$ depends. Notice that because of Lemma 2.1, the condition in (2.7) does not depend on the choice of $\boldsymbol{r}$. That is, $\boldsymbol{A}_{\mathcal{I}}$ provides a general characterization of interference coupling for interference functions fulfilling

A1, A2, A3. The matrix $\boldsymbol{A}_\mathcal{I}$ can be regarded as a generalization of the link gain matrix (1.9) commonly used in power control theory. In particular, $[\boldsymbol{A}_\mathcal{I}]_{kl} = 1 \Leftrightarrow [\boldsymbol{V}]_{kl} > 0$ and $[\boldsymbol{A}_\mathcal{I}]_{kl} = 0 \Leftrightarrow [\boldsymbol{V}]_{kl} = 0$.

With $\boldsymbol{A}_\mathcal{I}$ we define the *dependency set* as follows.

Definition 2.3 (dependency set). *The* dependency set $\mathsf{L}(k)$ *is the index set of transmitters on which user k depends, i.e.,*

$$\mathsf{L}(k) = \{l \in \mathcal{K} : [\boldsymbol{A}_\mathcal{I}]_{kl} = 1\} \,. \tag{2.8}$$

The set is always non-empty because we have ruled out the trivial case $\mathcal{I}(\boldsymbol{r}) = 0$, $\forall \boldsymbol{r}$, in our axiomatic interference model (see Section 1.2). Axiom A1 implies that each interference function depends on at least one transmitter, i.e. the dependency set is non-empty and there is at least one non-zero entry in each row of $\boldsymbol{A}_\mathcal{I}$. For some of the following results we need the additional assumption that every column has at least one non-zero empty off the main diagonal. This rather natural assumption means that every user causes interference to at least one other user.

2.2.2 The Dependency Matrix

The asymptotic coupling matrix $\boldsymbol{A}_\mathcal{I}$ is a general way of characterizing interference coupling. It is applicable to arbitrary interference functions. In this section we will introduce another concept, namely the *global dependency matrix* $\boldsymbol{D}_\mathcal{I}$. It will turn out (Theorem 2.6) that $\boldsymbol{D}_\mathcal{I} = \boldsymbol{A}_\mathcal{I}$ for the special case of log-convex interference functions.

We begin with a local definition of dependency that depends on the choice of $\boldsymbol{r}$.

Definition 2.4. *For any* $\boldsymbol{r} \geq 0$, *the* local dependency matrix $\boldsymbol{D}_\mathcal{I}(\boldsymbol{r})$ *is defined as*

$$[\boldsymbol{D}_\mathcal{I}(\boldsymbol{r})]_{kl} = \begin{cases} 1 & \text{if there exists a } \delta_l(\boldsymbol{r}) > 0 \text{ such that} \\ & \text{the function } f_l(\delta, \boldsymbol{r}) = \mathcal{I}_k(\boldsymbol{r} - \delta \boldsymbol{e}_l) \\ & \text{is strictly monotone decreasing for} \\ & 0 \leq \delta \leq \delta_l(\boldsymbol{r}). \\ 0 & \text{otherwise.} \end{cases} \tag{2.9}$$

This definition can be weakened. Instead of requiring this property for a specific $\boldsymbol{r}$, we next define the system as "coupled" if there is some arbitrary $\boldsymbol{r}$ such that $[\boldsymbol{D}_\mathcal{I}(\boldsymbol{r})]_{kl} = 1$. This leads to the following definition of a *global dependency matrix*, which is independent of the choice of $\boldsymbol{r}$.

Definition 2.5. *The* global dependency matrix $\boldsymbol{D}_\mathcal{I}$ *is defined as*

$$[\boldsymbol{D}_\mathcal{I}]_{kl} = \begin{cases} 1 & \text{if there exists a } \boldsymbol{r} > 0 \text{ such that} \\ & \mathcal{I}_k(\boldsymbol{r} + \delta \boldsymbol{e}_l) \text{ is not constant for some} \\ & \text{values } \delta > 0, \\ 0 & \text{otherwise.} \end{cases} \tag{2.10}$$

Later, we will use $\boldsymbol{D}_{\mathcal{I}}$ in order to analyze how the interference coupling affects the structure of the boundary.

The following theorem connects $\boldsymbol{A}_{\mathcal{I}}$ and $\boldsymbol{D}_{\mathcal{I}}$ for the special case of log-convex interference functions. Evidently, $[\boldsymbol{A}_{\mathcal{I}}]_{kl} = 1$ implies $[\boldsymbol{D}_{\mathcal{I}}]_{kl} = 1$, but the converse is generally not true. However, both characterizations are indeed equivalent if the underlying interference functions are log-convex.

Theorem 2.6. *Let $\mathcal{I}_1, \ldots, \mathcal{I}_K$ be log-convex interference functions, then both characterizations are equivalent, i.e., $\boldsymbol{A}_{\mathcal{I}} = \boldsymbol{D}_{\mathcal{I}}$.*

Proof. The proof is given in the Appendix A.9 □

2.3 Strict Monotonicity and Strict Log-Convexity

Consider an interference function $\mathcal{I}_k(\boldsymbol{r})$ with dependency set $\mathsf{L}(k)$. The function depends on all r_l with $l \in \mathsf{L}(k)$. However, this does not necessarily mean that $\mathcal{I}_k(\boldsymbol{r})$ is strictly monotone in these components. *Strict monotonicity on the dependency set* is a fundamental property, which is often needed to ensure unique solutions to certain optimization problems.

Definition 2.7 (strict monotonicity). *$\mathcal{I}_k(\boldsymbol{r})$ is said to be strictly monotone (on its dependency set $\mathsf{L}(k)$) if for arbitrary $\boldsymbol{r}^{(1)}$, $\boldsymbol{r}^{(2)}$, the inequality $\boldsymbol{r}^{(1)} \geq \boldsymbol{r}^{(2)}$, with $r_l^{(1)} > r_l^{(2)}$ for some $l \in \mathsf{L}(k)$, implies $\mathcal{I}_k(\boldsymbol{r}^{(1)}) > \mathcal{I}_k(\boldsymbol{r}^{(2)})$.*

In other words, $\mathcal{I}_k(\boldsymbol{r})$ is strictly monotone increasing in at least one power component. Strict monotonicity plays a central role in this book, especially for the result on power control.

Whenever we address the problem of SINR optimiation in the presence of power constraints, we can use an interference model $\mathcal{I}(\underline{\boldsymbol{p}})$ that is based on an extended power vector $\underline{\boldsymbol{p}} \in \mathbb{R}_+^{K_u+1}$. An example was already given in Section 1.4.1. The component $\underline{p}_{K_u+1}$ stands for the noise power, which is assumed to be equal for all users. It will be shown in Section 2.4 that strict monotonicity with respect to $\underline{p}_{K_u+1}$ yields a framework which is equivalent to Yates' framework of standard interference functions [1]. This way, standard interference functions can be comprehended within the framework A1, A2, A3.

Next, we define strictly log-convex interference functions.

Definition 2.8 (strict log-convexity). *A log-convex interference function $\mathcal{I}_k$ is said to be* strictly log-convex *if for all $\hat{\boldsymbol{p}}$, $\check{\boldsymbol{p}}$ for which there is some $l \in \mathsf{L}(k)$ with $\hat{p}_l \neq \check{p}_l$, the following inequality holds.*

$$\mathcal{I}_k\big(\underline{\boldsymbol{p}}(\lambda)\big) < \big(\mathcal{I}_k(\hat{\underline{\boldsymbol{p}}})\big)^{1-\lambda} \cdot \big(\mathcal{I}_k(\check{\underline{\boldsymbol{p}}})\big)^{\lambda}, \quad \lambda \in (0,1) \tag{2.11}$$

where $\boldsymbol{p}(\lambda) = \hat{\boldsymbol{p}}^{1-\lambda} \cdot \check{\boldsymbol{p}}^{\lambda}$.

The following lemma shows that strict log-convexity implies strict monotonicity.

Lemma 2.9. *Every strictly log-convex interference function $\mathcal{I}_k$ is* strictly monotone *on its dependency set (see Definition 2.7).*

Proof. Consider an arbitrary fixed vector $\boldsymbol{p} \in \mathbb{R}^K_{++}$, and an arbitrary $l \in \mathsf{L}(k)$. We define

$$\boldsymbol{p}^{(l)}(x) = \boldsymbol{p} + x\boldsymbol{e}_l, \quad x > 0 \tag{2.12}$$

and

$$\boldsymbol{p}(\lambda) = (\boldsymbol{p})^{1-\lambda} \cdot (\boldsymbol{p}^{(l)}(x))^{\lambda}, \quad \lambda \in (0,1) . \tag{2.13}$$

Since $l \in \mathsf{L}(k)$, strict log-convexity implies

$$\mathcal{I}_k(\underline{\boldsymbol{p}}(\lambda)) < (\mathcal{I}_k(\underline{\boldsymbol{p}}))^{1-\lambda} \cdot (\mathcal{I}_k(\underline{\boldsymbol{p}}^{(l)}(x)))^{\lambda} . \tag{2.14}$$

By definition (2.13) we have

$$p_v(\lambda) = p_v \text{ for all } v \neq l . \tag{2.15}$$

Also, $x > 0$ implies

$$p_l(\lambda) = (p_l)^{1-\lambda} \cdot (p_l + x)^{\lambda} > p_l . \tag{2.16}$$

With A3 (monotonicity) we know that $\boldsymbol{p} \leq \boldsymbol{p}(\lambda)$ implies $\mathcal{I}_k(\underline{\boldsymbol{p}}) \leq \mathcal{I}_k(\underline{\boldsymbol{p}}(\lambda))$. With (2.14) we have

$$\mathcal{I}_k(\underline{\boldsymbol{p}}) < (\mathcal{I}_k(\underline{\boldsymbol{p}}))^{1-\lambda} \cdot (\mathcal{I}_k(\underline{\boldsymbol{p}}^{(l)}(x)))^{\lambda},$$

thus

$$(\mathcal{I}_k(\underline{\boldsymbol{p}}))^{\lambda} < (\mathcal{I}_k(\underline{\boldsymbol{p}}^{(l)}(x)))^{\lambda} , \tag{2.17}$$

which shows strict monotonicity. □

Note that the converse of Lemma 2.9 is not true. The following example shows a strictly monotone interference function which is not strictly log-convex. That is, strict monotonicity is weaker than strict log-convexity.

Example 2.10. Consider the interference function

$$\mathcal{I}(\boldsymbol{p}) = C \cdot \prod_{k \in \mathcal{K}} (p_k)^{w_k}, \quad \text{with } \sum_{l \in \mathcal{K}} w_l = 1 \text{ and } \min_{l \in \mathcal{K}} w_l > 0. \tag{2.18}$$

Using the same notation as in Definition 2.8 we have $\mathcal{I}(\boldsymbol{p}(\lambda)) = \mathcal{I}(\hat{\boldsymbol{p}})^{1-\lambda} \cdot \mathcal{I}(\check{\boldsymbol{p}})^{\lambda}$. Thus, (2.18) is log-convex but not strictly log-convex. However, (2.18) is strictly monotone.

2.4 Standard Interference Functions and Power Control

A principal goal of power control is the selection of K_u transmit powers $\boldsymbol{p} \in \mathbb{R}^{K_u}_{++}$ to achieve a good system performance. Optimization strategies are mostly based on the SIR or the SINR, depending on whether noise is part of the model or not. Good overviews on classical results are given in [45, 46].

Power control in the presence of noise and power constraints is an important special case of the axiomatic framework A1, A2, A3. The linear interference function (1.11) in Section 1.4.1 is an example that shows how noise can be included by means of an extended power vector

$$\underline{\boldsymbol{p}} = [p_1, \ldots, p_{K_u}, \sigma_n^2]^T \in \mathbb{R}^{K_u+1}_{+} . \tag{2.19}$$

While the impact of noise is easy to model in the case of linear interference functions, it is less obvious for the axiomatic framework A1, A2, A3.

In this section we discuss how noise can be included in the axiomatic framework. This is closely connected with the concept of *standard interference functions*. The results apppeared in [11].

2.4.1 Standard Interference Functions

Yates [1] introduced an axiomatic framework of *standard interference functions*.

Definition 2.11. *A function* $\mathcal{J} : \mathbb{R}^{K_u}_{+} \mapsto \mathbb{R}_{++}$ *is said to be a* standard interference function *if the following axioms are fulfilled:*

Y1 *(positivity)*	$\mathcal{J}(\boldsymbol{p}) > 0$ *for all* $\boldsymbol{p} \in \mathbb{R}^{K_u}_{+}$
Y2 *(scalability)*	$\alpha\mathcal{J}(\boldsymbol{p}) > \mathcal{J}(\alpha\boldsymbol{p})$ *for all* $\alpha > 1$
Y3 *(monotonicity)*	$\mathcal{J}(\boldsymbol{p}) \geq \mathcal{J}(\boldsymbol{p}')$ *if* $\boldsymbol{p} \geq \boldsymbol{p}'$.

A simple example for a standard interference function is the linear interference model (1.12), which can be written as $\mathcal{J}(\boldsymbol{p}) = \boldsymbol{v}_k^T\boldsymbol{p} + \sigma_n^2$. Other examples can be found in the context of beamforming [2, 26, 27, 31–33], CDMA [34, 35], base station assignment [36, 37], robust designs [29, 30], and other areas [7, 76–79].

In [1] and related work, the following power control problem is addressed.

$$\min_{\boldsymbol{p}\geq 0} \sum_{l\in K_u} p_l \quad \text{s.t.} \quad \frac{p_k}{\mathcal{J}_k(\boldsymbol{p})} \geq \gamma_k , \quad \forall k \in K_u . \tag{2.20}$$

The goal is to minimize the sum of transmission powers while satisfying SINR targets $\gamma_1, \ldots, \gamma_{K_u}$. The problem will be studied in detail later in Section 2.8. If these targets are feasible, then the following fixed point iteration converges globally to the unique optimizer of the power minimization problem (2.20).

$$p_k^{(n+1)} = \gamma_k \mathcal{J}_k(\boldsymbol{p}^{(n)}) , \quad \forall k \in \mathcal{K}_u , \quad \boldsymbol{p}^{(0)} \in \mathbb{R}^{K_u}_{+} \tag{2.21}$$

Properties of this iteration were investigated in [1, 7, 74, 75]. If a feasible solution exists, then the axioms Y1–Y3 ensure global convergence for any initialization $\boldsymbol{p}^{(0)}$.

2.4.2 Comparison between Standard Interference Functions and General Interference Functions

It was shown in [11] that standard interference functions can be understood as a special case of the axiomatic framework A1, A2, A3. This framework is based on the extended power vector (2.19), and the assumption that $\mathcal{I}(\underline{p})$ is strictly monotone in the noise component $\underline{p}_{K_u+1}$.

Definition 2.12 (Strict monotonicity w.r.t. noise). *An interference function $\mathcal{I}(\underline{p})$ is said to be strictly monotone with respect to $\underline{p}_{K_u+1} > 0$, if for arbirary given vectors $\underline{p}$ and $\underline{p}'$, with $\underline{p} \geq \underline{p}'$, we have*

$$\underline{p}_{K_u+1} > \underline{p}'_{K_u+1} \quad \Rightarrow \quad \mathcal{I}(\underline{p}) > \mathcal{I}(\underline{p}') \, . \tag{2.22}$$

When comparing the axiomatic framework A1, A2, A3 (cf. Section 1.2) with the framework Y1, Y2, Y3, it is observed that the only difference is between A2 (scale invariance) and Y2 (scalability). In order to establish a link between both frameworks, we introduce the following definition.

Definition 2.13. *A function $\mathcal{J} : \mathbb{R}_+^{K_u} \mapsto \mathbb{R}_{++}$ is said to be a* weakly standard interference function *if the following axiom Y2' is fulfilled together with Y1 (positivity) and Y3 (monotonicity).*

***Y2'** (weak scalability) $\alpha \mathcal{J}(\boldsymbol{p}) \geq \mathcal{J}(\alpha \boldsymbol{p})$ for all $\alpha \geq 1$.*

The following theorem shows how general interference functions $\mathcal{I}$ and standard interference functions $\mathcal{J}$ are related. To this end, we introduce the power set

$$\underline{\mathcal{P}} = \left\{ \underline{p} = \left[\begin{smallmatrix} \boldsymbol{p} \\ \underline{p}_{K_u+1} \end{smallmatrix} \right] : \boldsymbol{p} \in \mathbb{R}_+^{K_u}, \, \underline{p}_{K_u+1} \in \mathbb{R}_{++} \right\} . \tag{2.23}$$

In a power control context, $\boldsymbol{p}$ is a vector of transmission powers and $\underline{p}_{K_u+1}$ is the noise power. For notational convenience, we define $\mathcal{I}(\underline{p}) = \mathcal{I}(\boldsymbol{p}, \underline{p}_{K_u+1})$.

The following theorem [11] shows the connection between general and standard interference functions.

Theorem 2.14. *1) Let $\mathcal{J} : \mathbb{R}_+^{K_u} \mapsto \mathbb{R}_{++}$ be a weakly standard interference function, then the extended function*

$$\mathcal{I}_{\mathcal{J}}(\underline{p}) := \mathcal{I}_{\mathcal{J}}(\boldsymbol{p}, \underline{p}_{K_u+1}) = \underline{p}_{K_u+1} \cdot \mathcal{J}\left(\frac{p_1}{\underline{p}_{K_u+1}}, \ldots, \frac{p_{K_u}}{\underline{p}_{K_u+1}} \right) \tag{2.24}$$

is a general interference function on $\underline{\mathcal{P}}$. We have

$$\mathcal{J}(\boldsymbol{p}) = \mathcal{I}_{\mathcal{J}}(\boldsymbol{p}, 1) \quad \text{for all } \boldsymbol{p} \geq 0 \, . \tag{2.25}$$

2) Let $\mathcal{I} : \mathbb{R}_+^{K_u+1} \mapsto \mathbb{R}_+$ be a general interference function, then for any given $\underline{p}_{K_u+1} > 0$, the reduced function

$$\mathcal{J}_{\mathcal{I}}(\boldsymbol{p}) := \mathcal{I}(p_1, \ldots, p_{K_u}, \underline{p}_{K_u+1}) \tag{2.26}$$

is a weakly standard interference function on $\mathbb{R}_{++}^{K_u}$.

3) *Let $\mathcal{I}_{\mathcal{J}}$ be defined as in (2.24). Then $\mathcal{J}$ is a standard interference function if and only if $\mathcal{I}_{\mathcal{J}}$ fulfills A1, A2, A3, and for all $\boldsymbol{p} \in \mathbb{R}_+^{K_u}$, the function $\mathcal{I}_{\mathcal{J}}(\boldsymbol{p}, \underline{p}_{K_u+1})$ is strictly monotone in the sense of Definition 2.12.*

Proof. We begin by proving 1). Axiom A1 is fulfilled because $\mathcal{I}_{\mathcal{J}}(\underline{\boldsymbol{p}}) > 0$ for all $\underline{\boldsymbol{p}} \in \underline{\mathcal{P}}$. Axiom A2 (scale invariance) is fulfilled because for all $\lambda > 0$

$$\mathcal{I}_{\mathcal{J}}(\lambda \underline{\boldsymbol{p}}) = \lambda \cdot p_{K_u+1} \cdot \mathcal{J}\Big(\frac{\lambda p_1}{\lambda p_{K_u+1}}, \ldots, \frac{\lambda p_{K_u}}{\lambda p_{K_u+1}}\Big) = \lambda \mathcal{I}_{\mathcal{J}}(\underline{\boldsymbol{p}}) \,.$$

It remains to show A3 (monotonicity). Consider two arbitrary vectors $\underline{\boldsymbol{p}}^{(1)}, \underline{\boldsymbol{p}}^{(2)} \in \underline{\mathcal{P}}$ such that $\underline{\boldsymbol{p}}^{(1)} \geq \underline{\boldsymbol{p}}^{(2)}$. With $\tilde{\lambda} = p^{(1)}_{K_u+1}/p^{(2)}_{K_u+1} \geq 1$, we have

$$\begin{aligned}
\mathcal{I}_{\mathcal{J}}(\underline{\boldsymbol{p}}^{(2)}) &= p^{(2)}_{K_u+1} \cdot \mathcal{J}\Big(\frac{p_1^{(2)}}{p^{(2)}_{K_u+1}}, \ldots, \frac{p^{(2)}_{K_u}}{p^{(2)}_{K_u+1}}\Big) \\
&= p^{(2)}_{K_u+1} \cdot \mathcal{J}\Big(\tilde{\lambda}\frac{p_1^{(2)}}{p^{(1)}_{K_u+1}}, \ldots, \tilde{\lambda}\frac{p^{(2)}_{K_u}}{p^{(1)}_{K_u+1}}\Big) \\
&\leq p^{(2)}_{K_u+1} \cdot \mathcal{J}\Big(\tilde{\lambda}\frac{p_1^{(1)}}{p^{(1)}_{K_u+1}}, \ldots, \tilde{\lambda}\frac{p^{(1)}_{K_u}}{p^{(1)}_{K_u+1}}\Big) \\
&\leq p^{(2)}_{K_u+1} \cdot \tilde{\lambda} \cdot \mathcal{J}\Big(\frac{p_1^{(1)}}{p^{(1)}_{K_u+1}}, \ldots, \frac{p^{(1)}_{K_u}}{p^{(1)}_{K_u+1}}\Big) \\
&= p^{(1)}_{K_u+1} \cdot \mathcal{J}\Big(\frac{p_1^{(1)}}{p^{(1)}_{K_u+1}}, \ldots, \frac{p^{(1)}_{K_u}}{p^{(1)}_{K_u+1}}\Big) = \mathcal{I}_{\mathcal{J}}(\underline{\boldsymbol{p}}^{(1)}) \,.
\end{aligned} \tag{2.27}$$

The first inequality follows from $Y3$ (monotonicity) and the second from $Y2'$ (weak scalability).

We now prove 2). Axiom Y3 follows directly from A3. Axiom Y1 holds on $\mathbb{R}^{K_u}_{++}$ because $\mathcal{I}(\underline{\boldsymbol{p}}) > 0$ for all $\underline{\boldsymbol{p}} > 0$. This is a consequence of A1, as shown in [2]. Axiom Y2' follows from

$$\begin{aligned}
\mathcal{J}(\alpha \boldsymbol{p}) &= \mathcal{I}(\alpha \boldsymbol{p},\, \underline{p}_{K_u+1}) \\
&\leq \mathcal{I}(\alpha \boldsymbol{p},\, \alpha \underline{p}_{K_u+1}) = \alpha \mathcal{I}(\boldsymbol{p},\, \underline{p}_{K_u+1}) = \alpha \mathcal{J}(\boldsymbol{p}) \,.
\end{aligned}$$

Note that this inequality need not be strict because we did not made any assumption on whether $\mathcal{I}$ depends on $\underline{p}_{K_u+1}$ or not.

We now prove 3). Let $\mathcal{J}$ be standard. From 1) we know that $\mathcal{I}_{\mathcal{J}}(\underline{\boldsymbol{p}})$ fulfills A1, A2, A3. We now show strict monotonicity. For arbitrary $\underline{\boldsymbol{p}}^{(1)}, \underline{\boldsymbol{p}}^{(2)} \in \underline{\mathcal{P}}$, with $\boldsymbol{p}^{(1)} = \boldsymbol{p}^{(2)}$ and $\tilde{\lambda} = p^{(1)}_{K_u+1}/p^{(2)}_{K_u+1} > 1$ the second inequality (2.27) is strict. This follows from Y2 (which holds for $\alpha > 1$ because of continuity). Thus, $\mathcal{I}_{\mathcal{J}}(\underline{\boldsymbol{p}})$ is strictly monotone with respect to the component p_{K_u+1}. Conversely, let $\mathcal{I}_{\mathcal{J}}$ be strictly monotone and axioms A1, A2, A3 are assumed to be fulfilled. Then,

$$\mathcal{J}(\lambda \boldsymbol{p}) = \mathcal{I}_{\mathcal{J}}(\lambda \boldsymbol{p}, 1) = \lambda \mathcal{I}(\boldsymbol{p}, \tfrac{1}{\lambda}) < \lambda \mathcal{I}_{\mathcal{J}}(\boldsymbol{p}, 1) = \lambda \mathcal{J}(\boldsymbol{p}) \tag{2.28}$$

for all $\alpha > 0$, thus Y2 holds. Property Y3 follows directly from A3. Finally, we show Y1 by contradiction. Suppose that there exists a $\underline{\boldsymbol{p}} \in \underline{\mathcal{P}}$ such that $\mathcal{J}(\boldsymbol{p}) = 0$. Strict monotonicity of $\mathcal{I}$ implies

$$0 = \mathcal{J}(\boldsymbol{p}) = \mathcal{I}(\underline{\boldsymbol{p}}) > \mathcal{I}(\alpha\underline{\boldsymbol{p}}) = \alpha\mathcal{I}(\underline{\boldsymbol{p}}), \quad 0 < \alpha < 1 .$$

Letting $\alpha \to 0$ we obtain a contradiction, thus proving Y1. □

Theorem 2.14 shows that any standard interference $\mathcal{J}(\boldsymbol{p})$ can be expressed alternatively as $\mathcal{I}(\underline{\boldsymbol{p}})$ with $\underline{p}_{K_u+1} = 1$. Both frameworks can be used interchangeably. Results that were derived for standard interference functions can as well be derived by the axioms A1, A2, A3 plus the additional assumption of strict monotonicity (2.22).

An example is the fixed point iteration (2.21), which can be rewritten in terms of $\mathcal{I}(\underline{\boldsymbol{p}})$. An alternative convergence proof based on A1, A2, A3 was given in [2].

Another example is the positivity of the functions $\mathcal{J}(\boldsymbol{p})$. It was observed in [75] that Axiom Y1 (strict positivity) is actually redundant. Strict positivity $\mathcal{J}(\boldsymbol{p}) > 0$ already follows from Y2 and Y3. The same result can be shown for $\mathcal{I}(\underline{\boldsymbol{p}})$ on the basis of axioms A2 and A3. It was shown in Section 1.2 that A2 and A3 imply $\mathcal{I}(\underline{\boldsymbol{p}}) \geq 0$ for all $\underline{\boldsymbol{p}} \geq 0$. Axiom A1 is only needed to rule out the trivial case $\mathcal{I}(\underline{\boldsymbol{p}}) = 0$ for all $\underline{\boldsymbol{p}}$. With strict monotonicity we do not need A1 anymore. Then there always exists a $\underline{\boldsymbol{p}} > 0$ such that $\mathcal{I}(\underline{\boldsymbol{p}}) > 0$, and hence $\mathcal{I}(\underline{\boldsymbol{p}}) > 0$ for *all* $\underline{\boldsymbol{p}} > 0$. It is actually sufficient that the strictly monotone component is positive, i.e., $\underline{p}_{K_u+1} > 0$. Assume an arbitrary $\underline{\boldsymbol{p}} \geq 0$ with $\underline{p}_{K_u+1} > 0$. The proof is by contradiction: Suppose that $\mathcal{I}(\underline{\boldsymbol{p}}) = 0$, then for any α with $0 < \alpha < 1$,

$$0 = \mathcal{I}(\underline{\boldsymbol{p}}) > \mathcal{I}(\alpha\underline{\boldsymbol{p}}) = \alpha\mathcal{I}(\underline{\boldsymbol{p}}) .$$

This would lead to the contradiction $0 = \lim_{\alpha\to 0} \alpha\mathcal{I}(\underline{\boldsymbol{p}}) < 0$. Hence, strict monotonicity (2.22) and positive noise $\underline{p}_{K_u+1} > 0$ ensures that $\mathcal{I}(\underline{\boldsymbol{p}}) > 0$ for arbitrary $\underline{\boldsymbol{p}} \geq 0$.

With Theorem 2.14 the result carries over to arbitrary standard interference functions. Furthermore, the proof extends to arbitrary strictly monotone interference functions as introduced in Section 2.3. Strictly monotone interference functions are positive whenever $p_k > 0$, where p_k is the component on which the function depends in a strictly monotone way.

2.5 Continuity

Continuity is another fundamental property that will be needed throughout this book. The following result was shown in [2].

Lemma 2.15. *All interference functions $\mathcal{I}(\boldsymbol{r})$ satisfying A1, A2, A3 are continuous on $\mathbb{R}^K_{++}$. That is, for an arbitrary $\boldsymbol{p}^* \in \mathbb{R}^K_{++}$, and an arbitrary sequence $\boldsymbol{p}^{(n)} \in \mathbb{R}^K_{++}$ such that $\lim_{n\to\infty} \boldsymbol{p}^{(n)} = \boldsymbol{p}^*$, the following holds.*

$$\lim_{n\to\infty} \mathcal{I}(\boldsymbol{p}^{(n)}) = \mathcal{I}(\boldsymbol{p}^*) . \tag{2.29}$$

Lemma 2.15 shows continuity only on a restricted domain $\mathbb{R}^K_{++}$ instead of $\mathbb{R}^K_{+}$. That is, we exclude the zeros on the boundary of the set. In many cases, this is sufficient. For example, when dealing with general signal-to-interference ratios $p_k/\mathcal{I}_k(\boldsymbol{p})$, we need to avoid possible singularities. By restricting the domain to $\mathbb{R}^K_{++}$, we ensure that $\mathcal{I}_k(\boldsymbol{p}) > 0$. This technical assumption is the price we pay for generality of the interference model. An example is (1.22), where the infimum is taken over all $\boldsymbol{p} > 0$.

However, interference functions are defined on $\mathbb{R}^K_{+}$. The case $p_k = 0$ can be interpreted as user k being inactive. The ability to model inactive users is an important prerequisite for many resource allocation problems. It is therefore desirable to extend some of the results to $\mathbb{R}^K_{+}$. This motivates the *continuation* that is introduced in the following subsection.

2.5.1 Continuation on the Boundary

Certain key properties of interference functions are preserved on the boundary [11].

Assume that $\mathcal{I}(\boldsymbol{p})$ is defined on $\mathbb{R}^K_{++}$. Let $\boldsymbol{p}^{(n)} \in \mathbb{R}^K_{++}$ be an arbitrary sequence with limit $\lim_{n\to\infty} \boldsymbol{p}^{(n)} = \boldsymbol{p} \in \mathbb{R}^K_{+}$. The interference function $\mathcal{I}$ has a continuation $\mathcal{I}^c$ on the boundary, defined on $\mathbb{R}^K_{+}$.

$$\mathcal{I}^c(\boldsymbol{p}) = \lim_{n\to\infty} \mathcal{I}(\boldsymbol{p}^{(n)}) . \tag{2.30}$$

Certain properties of $\mathcal{I}$ are preserved when one or more coordinates p_k tend to zero. This result is quite useful because it means that certain results shown for $\mathbb{R}^K_{++}$ immediately extend to $\mathbb{R}^K_{+}$.

The following theorem states that for any interference function, the properties A1, A2, A3 are preserved on the boundary.

Theorem 2.16. *Let $\mathcal{I}$ be an arbitrary interference function defined on $\mathbb{R}^K_{++}$. Then, the continuation $\mathcal{I}^c(\boldsymbol{p})$ defined on $\mathbb{R}^K_{+}$ fulfills the axioms A1, A2, A3.*

Proof. We need the lemmas shown in Appendix A.8. Axiom A3 (monotonicity) follows from Lemma A.17. Axiom A2 (scale invariance) follows from Lemma A.14. Axiom A1 is also fulfilled since $\mathcal{I}(\boldsymbol{p}) = \mathcal{I}^c(\boldsymbol{p})$ for all $\boldsymbol{p} \in \mathbb{R}^K_{++}$. □

We can use this continuation to extend results that were previously shown for $\mathbb{R}^K_{++}$ to the non-negative domain $\mathbb{R}^K_{+}$. As an example, consider Lemma 2.15, which states continuity on $\mathbb{R}^K_{++}$. This is now extended to $\mathbb{R}^K_{+}$ by the following theorem. More examples will follow.

Theorem 2.17. *$\mathcal{I}^c$ is continuous on $\mathbb{R}^K_+$. For an arbitrary sequence $\boldsymbol{p}^{(n)} \in \mathbb{R}^K_+$ with $\lim_{n\to\infty} \boldsymbol{p}^{(n)} = \boldsymbol{p}^*$ we have*

$$\lim_{n\to\infty} \mathcal{I}^c(\boldsymbol{p}^{(n)}) = \mathcal{I}^c(\boldsymbol{p}^*) . \tag{2.31}$$

Proof. The proof builds on the results from Appendix A.8. We need to show that $\mathcal{I} : \mathbb{R}^K_+ \mapsto \mathbb{R}_+$ is a continuous function, i.e., (2.31) holds for any sequence $\boldsymbol{p}^{(n)} \in \mathbb{R}^K_+$ with $\lim_{n\to\infty} \boldsymbol{p}^{(n)} = \boldsymbol{p}^*$. To this end, consider $\overline{\delta}^{(n)}$ and $\overline{\boldsymbol{p}}^{(n)}$, defined as in the proof of Lemma A.16, with the inequalities (A.36) and (A.37). Combining $\lim_{n\to\infty} \mathcal{I}(\overline{\boldsymbol{p}}^{(n)}) = \mathcal{I}(\boldsymbol{p}^*)$ and (A.36) we have

$$\limsup_{n\to\infty} \mathcal{I}(\boldsymbol{p}^{(n)}) \le \mathcal{I}(\boldsymbol{p}^*) . \tag{2.32}$$

We introduce the vector $\underline{\boldsymbol{p}}^{(n)}$ with

$$[\underline{\boldsymbol{p}}^{(n)}]_k = \underline{p}_k^{(n)} = \begin{cases} p_k^{(n)} & \text{if } p_k > 0 \\ 0 & \text{if } p_k = 0 . \end{cases} \tag{2.33}$$

Since $\underline{\boldsymbol{p}}^{(n)} \le \boldsymbol{p}^{(n)}$ we have

$$\lim_{n\to\infty} \mathcal{I}(\underline{\boldsymbol{p}}^{(n)}) \le \lim_{n\to\infty} \mathcal{I}(\boldsymbol{p}^{(n)}) . \tag{2.34}$$

From Lemma A.17 we know that the right-hand side limit of (2.34) exists, thus

$$\lim_{n\to\infty} \mathcal{I}(\underline{\boldsymbol{p}}^{(n)}) = \mathcal{I}(\boldsymbol{p}^*) . \tag{2.35}$$

Combining (2.32), (2.34), and (2.35) we have

$$\mathcal{I}(\boldsymbol{p}^*) \le \liminf_{n\to\infty} \mathcal{I}(\boldsymbol{p}^{(n)}) \le \limsup_{n\to\infty} \mathcal{I}(\boldsymbol{p}^{(n)}) \le \mathcal{I}(\boldsymbol{p}^*) .$$

Thus, (2.31) is fulfilled. □

2.5.2 Continuity of Standard Interference Functions

Theorem 2.14 shows that standard interference functions can be modeled as a special case of general interference functions characterized by A1, A2, A3. This is useful, because many results that were previously shown for general interference functions immediately transfer to standard interference function. The next corollary provides an example.

Corollary 2.18. *Any weakly standard interference function $\mathcal{J}$ is continuous on $\mathbb{R}^{K_u}_{++}$, and there exists a continuation*

$$\mathcal{J}^c(\boldsymbol{p}) = \lim_{n\to\infty} \mathcal{J}(\boldsymbol{p}^{(n)}) ,$$

which is weakly standard and continuous on $\mathbb{R}^{K_u}_+$. This also holds for standard interference functions, which are a subclass of weakly standard interference functions.

Proof. This is a direct consequence of Theorem 2.14, which states that, for any $\mathcal{J}$ there exists a general interference function $\mathcal{I}_{\mathcal{J}}$ such that $\mathcal{J}(\boldsymbol{p}) = \mathcal{I}_{\mathcal{J}}(\boldsymbol{p}, 1)$ for all $\boldsymbol{p}$. Any general interference function is continuous on $\mathbb{R}^K_{++}$, as shown in [2]. The existence of a continuous continuation on the boundary follows from Theorem 2.17. □

Continuity was implicitly assumed in [1] for proving convergence of the fixed point iteration. This proof is only rigorous under the assumption of continuity. Corollary 2.18 justifies this assumption in hindsight.

2.6 QoS Regions, Feasibility, and Fixed Point Characterization

In Subsection 1.4.5 we have introduced the SIR region $\mathcal{S}$, which is a sub-level set of the min-max indicator function $C(\boldsymbol{\gamma})$. SIR values $\boldsymbol{\gamma} = [\gamma_1, \ldots, \gamma_K]^T$ are feasible if and only if $\boldsymbol{\gamma} \in \mathcal{S}$. Some additional aspects will be discussed in this section.

2.6.1 SIR-Based QoS Sets

In this book, the *quality-of-service* (QoS) is defined as an arbitrary performance measure that depends on the SIR (or SINR) by a strictly monotone and continuous function ϕ on $\mathbb{R}_+$. The QoS of user k is

$$q_k(\boldsymbol{p}) = \phi_k\big(\mathrm{SIR}_k(\boldsymbol{p})\big) = \phi_k\Big(\frac{p_k}{\mathcal{I}_k(\boldsymbol{p})}\Big) , \quad k \in \mathcal{K} . \tag{2.36}$$

The function ϕ_k is either monotone increasing or decreasing. Examples are

- MMSE: $\phi(x) = 1/(1+x)$
- BER: $\phi(x) = Q(\sqrt{x})$
- BER approximation in the high SNR regime: $\phi(\mathrm{SIR}) \approx (G_c \cdot \mathrm{SIR})^{-G_d}$, with coding gain G_c and diversity order G_d.
- capacity: $\phi(x) = \log(1+x)$.

The QoS region $\mathcal{Q} \subset \mathbb{R}^K$ is the set of QoS values that are jointly achievable by all K users. Points from $\mathcal{Q}$ are said to be *feasible*. A thorough understanding of the structure of the QoS region is fundamental for advancing research in areas like game theory, resource allocation, or network optimization.

Let γ_k be the inverse function of ϕ_k, then $\gamma_k(q_k)$ is the minimum SINR level needed by the kth user to satisfy the QoS target q_k. Assume that the QoS is defined on $\mathbb{Q}$, and the K-dimensional domain is denoted by $\mathbb{Q}^K$. Let $\boldsymbol{q} \in \mathbb{Q}^K$ be a vector of QoS values, then the associated SIR vector is

$$\boldsymbol{\gamma}(\boldsymbol{q}) = [\gamma_1(q_1), \ldots, \gamma_K(q_K)]^T . \tag{2.37}$$

QoS values $\boldsymbol{q} \in \mathbb{Q}^K$ are feasible if and only if $C\big(\boldsymbol{\gamma}(\boldsymbol{q})\big) \leq 1$, where $C(\boldsymbol{\gamma})$ is the min-max optimum as defined by (1.22). The QoS feasible set is the sublevel set

$$\mathcal{Q} = \{\boldsymbol{q} \in \mathbb{Q}^K : C(\boldsymbol{\gamma}(\boldsymbol{q})) \leq 1\} . \tag{2.38}$$

The structure of $\mathcal{Q}$ depends on the properties of the indicator function $C(\boldsymbol{\gamma}(\boldsymbol{q}))$.

An interesting special case is that of log-convex interference functions (see Subsection 1.3.2). Assume that $\gamma(\text{QoS})$ is the inverse function of the QoS $\phi(\text{SIR})$. If $\gamma(\text{QoS})$ is log-convex then $C\big(\boldsymbol{\gamma}(\boldsymbol{q})\big)$ is log-convex on $\mathbb{Q}^K$ (see Appendix A.3). Since every log-convex function is convex [23], the QoS region $\mathcal{Q}$, as defined by (2.38), is a sublevel set of a convex function. Hence, $\mathcal{Q}$ is a convex set [2].

2.6.2 Comprehensiveness

QoS regions of the form (2.38) are *comprehensive*. An illustration is given in Figure 2.1.

Definition 2.19. *A set* $\mathcal{Q} \subset \mathbb{R}^K$ *is said to be* upward-comprehensive *if for all* $\boldsymbol{q} \in \mathcal{Q}$ *and* $\boldsymbol{q}' \in \mathbb{R}^K$

$$\boldsymbol{q}' \geq \boldsymbol{q} \quad \Longrightarrow \quad \boldsymbol{q}' \in \mathcal{Q} . \tag{2.39}$$

It is said to be downward-comprehensive *if for all* $\boldsymbol{q} \in \mathcal{Q}$ *and* $\boldsymbol{q}' \in \mathbb{R}^K$

$$\boldsymbol{q}' \leq \boldsymbol{q} \quad \Longrightarrow \quad \boldsymbol{q}' \in \mathcal{Q} . \tag{2.40}$$

In the context of monotonic optmization, comprehensive sets are also referred to as *normal sets* [80]. The QoS region (2.38) is comprehensive because

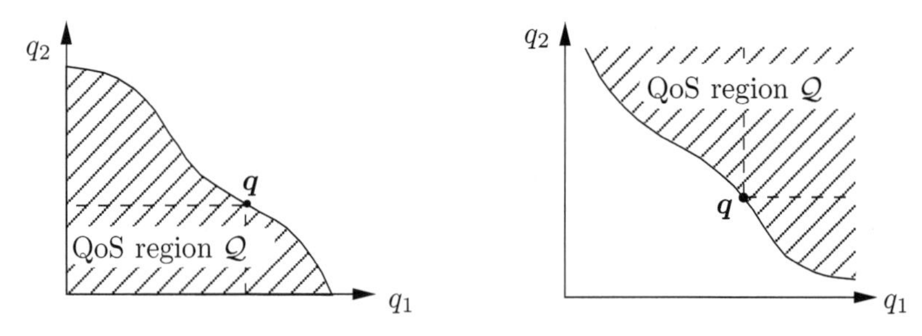

Fig. 2.1. Illustration of comprehensive sets. The left hand set is downward-comprehensive, the right-hand set is upward-comprehensive.

$C(\boldsymbol{\gamma})$ is an interference function (see Subsection 1.4.5). If $\gamma_k(q_k)$ is increasing, then the set is downward-comprehensive. This is a direct consequence of A3 (monotonicity). If $\gamma_k(q_k)$ is decreasing, then the set is upward-comprehensive.

Comprehensiveness can be interpreted as "free disposability of utility" [18]). If certain QoS values are jointly feasible for all users, then any user can reduce its QoS and the resulting point is still feasible. This is a very basic property which is fulfilled for many interference-coupled systems. Comprehensive regions are often assumed in the context of cooperative game theory (see e.g. [18]) and optimization theory [80].

Later, in Chapter 3, it will be shown that every downward-comprehensive set from $\mathbb{R}_{++}$ can be expressed as a sublevel set of an interference function. There is a close connection between comprehensive QoS sets from from $\mathbb{R}_{++}$ and interference functions. Hence by studying the properties of interference functions we can gain insight into the structure of QoS regions.

2.6.3 Fixed Point Characterization of Boundary Points

The QoS is a strictly monotone and continuous function of the SIR. Therefore, the analysis of the QoS region often reduces to the analysis of the SIR region. Properties of the SIR region carry over to the QoS region and vice versa.

The boundary $\partial\mathcal{S}$ of the SIR region (1.21) is of particular interest. The boundary structure typically determines whether the computation of an optimal power vector can be performed in an efficient manner or not. For example, if the region is convex, then efficient algorithms are readily available. The boundary is defined as follows.

$$\partial\mathcal{S} = \{\boldsymbol{\gamma} \in \mathbb{R}^K_{++} : C(\boldsymbol{\gamma}) = 1\} . \tag{2.41}$$

By definition, $\boldsymbol{\gamma} \in \partial\mathcal{S}$ is achievable, at least in an asymptotic sense. That is, for any $\epsilon > 0$ there exists a $\boldsymbol{p}_\epsilon > 0$ such that $\mathrm{SIR}_k(\boldsymbol{p}_\epsilon) \geq \gamma_k - \epsilon$ for all $k \in \mathcal{K}$.

A general characterization of achievability can be complicated. The interference framework A1, A2, A3 is quite general, and the resulting QoS region can have a complicated structure, depending on the assumed properties of the interference functions [81].

In the following we will focus on the practically relevant case when the boundary is achievable with equality.

Definition 2.20. *We say that the boundary $\partial\mathcal{S}$ is achievable with equality (or simply achievable), if for any $\boldsymbol{\gamma} \in \partial\mathcal{S}$ there exists a $\boldsymbol{p} > 0$ such that*

$$\mathrm{SIR}_k(\boldsymbol{p}) = \gamma_k, \quad \textit{for all } k \in \mathcal{K}. \tag{2.42}$$

If (2.42) is fulfilled, then $\boldsymbol{p}$ is the optimizer of the min-max balancing problem (1.22), with an optimum $C(\boldsymbol{\gamma}) = 1$. This is observed by rewriting (2.42) as follows.

$$C(\boldsymbol{\gamma}) = \frac{\gamma_k \mathcal{I}_k(\boldsymbol{p})}{p_k} = \frac{\gamma_k}{\mathrm{SIR}_k(\boldsymbol{p})}, \quad \text{for all } k \in \mathcal{K} . \tag{2.43}$$

Introducing the vector notation $\boldsymbol{\mathcal{I}}(\boldsymbol{p}) = [\mathcal{I}_1(\boldsymbol{p}), \ldots, \mathcal{I}_K(\boldsymbol{p})]^T$ and $\boldsymbol{\Gamma} := \mathrm{diag}\{\boldsymbol{\gamma}\}$, the system of equations (2.43) can be rewritten as

$$p = \frac{1}{C(\gamma)} \Gamma \mathcal{I}(p) . \quad (2.44)$$

In the following, a positive power vector $p > 0$ is said to be a *fixed point* if it satisfies (2.44). For any boundary point $\gamma \in \partial \mathcal{S}$ we have $C(\gamma) = 1$, in which case (2.44) is equivalent to (2.42).

For arbitrary $\gamma > 0$ with $C(\gamma) \neq 1$, the existence of a fixed point $p^* > 0$ implies that the infimum (1.22) is attained, and scaled SIR values $\gamma_k / C(\gamma)$ are achieved for all $k \in \mathcal{K}$. In the context of the min-max balancing problem (1.22), the values γ can be regarded as weighting factors. A uniform scaling of γ results in a scaling of $C(\gamma)$ by the same amount. If a fixed point exists for some arbitrary $\gamma > 0$, then it also exists for the boundary point $\gamma' = \beta\gamma$, where $\beta > 0$ and $C(\gamma') = 1$.

The existence of a fixed point $p^* > 0$ depends on the properties of the interference functions. For general functions characterized by A1, A2, A3, only a few basic properties are known [2].

Lemma 2.21. *Let $\mathcal{I}_1, \ldots, \mathcal{I}_K$ be interference functions characterized by A1, A2, A3, then*

1. *there always exists a $\mathbf{p}^* \geq 0$, $\mathbf{p}^* \neq 0$, such that (2.44) is fulfilled.*
2. *If $\Gamma \mathcal{I}(p^*) = \mu p^*$ for some $p^* > 0$ and $\mu > 0$, then $\mu = C(\gamma)$ and $\mathbf{p}^*$ is an optimizer of the min-max problem (1.22).*

The existence of a strictly positive fixed point $p^* \geq 0$ is important, e.g., to ensure numerical stability for certain resource allocation algorithms that operate on the boundary of the SIR set. Algorithms are usually derived under the premise that the boundary is achievable.

The existence of a positive fixed point is best understood for linear interference functions (1.10). In [81] conditions were derived based on the theory of non-negative matrices [57]. Also in [81], this was extended to the more general class of interference functions with adaptive receiver designs. Both models have in common that the interference is characterized by means of a coupling matrix.

Later, we will provide conditions under which such a fixed point exists for certain nonlinear interference functions. An example is the log-convex SIR region that will be studied in Chapter 4. Also, the boundary is typically achievable for certain SINR regions under power constraints.

2.7 Power-Constrained QoS Regions

The QoS region (2.38) is defined on the basis of general interference functions. General interference functions are scale-invariant (A2), thus the SIR $p_k / \mathcal{I}_k(p)$ is invariant with respect to a scaling of p. This means that power constraints do not have any effect on the achievable SIR region. Power constraints are only meaningful if we incorporate noise in our model. This is done as described in Section 2.4, by using the (K_u+1)-dimensional extended power vector

$$\underline{\boldsymbol{p}} = \begin{bmatrix} \boldsymbol{p} \\ 1 \end{bmatrix} .$$

We assume that strict monotonicity (2.22) holds. The component $\underline{p}_{K_u+1}$ stands for the normalized noise power, and $\mathcal{I}(\underline{\boldsymbol{p}})$ is the interference-plus-noise power. The assumption $\underline{p}_{K_u+1} = 1$ is made without loss of generality, because any other noise power can be realized by scaling $\underline{\boldsymbol{p}}$ appropriately. Such a scaling does not affect the SINR. Because of A2, the following holds for any $\alpha > 0$.

$$\frac{\alpha p_k}{\mathcal{I}(\alpha \underline{\boldsymbol{p}})} = \frac{p_k}{\mathcal{I}(\underline{\boldsymbol{p}})} .$$

Theorem 2.14 in Section 2.4.1 shows that there exists a standard interference function $\mathcal{J}_k(\boldsymbol{p})$ such that

$$\mathcal{J}_k(\boldsymbol{p}) = \mathcal{I}_k(\underline{\boldsymbol{p}}) . \tag{2.45}$$

We define

$$\mathrm{SINR}_k(\boldsymbol{p}) = \frac{p_k}{\mathcal{J}_k(\boldsymbol{p})} . \tag{2.46}$$

Although $\mathcal{J}_k$ does not fulfill the axioms A1, A2, A3, it can nevertheless be considered as a special case of this framework. Every standard interference function $\mathcal{J}_k$ is related to a general interference function $\mathcal{I}_k$ via the identity (2.45). Thus, many properties of general interference functions $\mathcal{I}_k$ transfer directly to $\mathcal{J}_k$ (see Section 2.4.1).

2.7.1 Sum-Power Constraint

Consider a sum-power constraint $\|\boldsymbol{p}\|_1 \leq P_{\max}$. The SINR region is defined as

$$\mathcal{S}(P_{\max}) = \{\boldsymbol{\gamma} \in \mathbb{R}_+^{K_u} : C(\boldsymbol{\gamma}, P_{\max}) \leq 1\} \tag{2.47}$$

where

$$C(\boldsymbol{\gamma}, P_{\max}) = \inf_{\boldsymbol{p}>0} \left(\max_{k \in \mathcal{K}_u} \frac{\gamma_k \mathcal{J}_k(\boldsymbol{p})}{p_k} \right) \quad \text{s.t. } \|\boldsymbol{p}\|_1 \leq P_{\max} . \tag{2.48}$$

The min-max optimum $C(\boldsymbol{\gamma}, P_{\max})$ is an indicator for the feasibility of SINR targets $\boldsymbol{\gamma}$.

The SINR region $\mathcal{S}(P_{\max})$ is a straightforward extension of the SIR region $\mathcal{S}$, introduced in Subsection 1.4.5. The latter one corresponds to a system without noise and power constraints. The region $\mathcal{S}(P_{\max})$ is a sublevel set of $C(\boldsymbol{\gamma}, P_{\max})$, while $\mathcal{S}$ is a sublevel set of $C(\boldsymbol{\gamma})$. Because of the properties of standard interference functions, the following inequality holds for all $\alpha > 1$.

$$\frac{\alpha p_k}{\mathcal{J}_k(\alpha \boldsymbol{p})} > \frac{p_k}{\mathcal{J}_k(\boldsymbol{p})} .$$

That is, the SINR increases when the transmission powers are increased. The unconstrained min-max optimum $C(\gamma)$ is obtained as the asymptotic limit of the power-constrained case.

$$C(\gamma) = \lim_{P_{\max} \to \infty} C(\gamma, P_{\max}) \, . \tag{2.49}$$

That is, $\mathcal{S}(P_{\max})$ is contained in $\mathcal{S}$. The SIR region $\mathcal{S}$ is approached as the total power $P_{\max}$ tends to infinity. In this regime, noise can be neglected.

Consider the QoS region introduced in Subsection 2.6.1. Throughout this book, the QoS is defined as a strictly monotone and continuous function of the SIR or SINR. Given a QoS target vector $\boldsymbol{q} \in \mathbb{Q}^{K_u}$, the corresponding SINR values are $\gamma(\boldsymbol{q})$. Hence, the sum-power constrained QoS region is defined as follows.

$$\mathcal{Q}(P_{\max}) = \{\boldsymbol{q} \in \mathbb{Q}^{K_u} : C(\gamma(\boldsymbol{q}), P_{\max}) \leq 1\} \, . \tag{2.50}$$

Alternatively, we can define $\mathcal{Q}(P_{\max})$ via the set of feasible power vectors. Without power constraints, the following set $\mathcal{P}(\boldsymbol{q})$ is the set of power vectors that achieve the targets $\boldsymbol{q}$.

$$\mathcal{P}(\boldsymbol{q}) = \{\boldsymbol{p} > 0 : \text{SINR}_k(\boldsymbol{p}) \geq \gamma_k(q_k), \; \forall k \in \mathcal{K}_u\} \, . \tag{2.51}$$

The set $\mathcal{P}(\boldsymbol{q})$ is non-empty if and only if $C\big(\gamma(\boldsymbol{q})\big) < 1$. That is, $\boldsymbol{q}$ lies in the interior of $\mathcal{Q}$ (denoted as int $\mathcal{Q}$). If $\mathcal{P}(\boldsymbol{q})$ is non-empty, then there is a unique vector

$$\boldsymbol{p}^{min}(\boldsymbol{q}) = \arg\min_{\boldsymbol{p} \in \mathcal{P}(\boldsymbol{q})} \|\boldsymbol{p}\|_1 \, , \tag{2.52}$$

which achieves $\boldsymbol{q}$ with minimum total power. This is a consequence of $J_k(\boldsymbol{p})$ being standard [2], so the results [1] can be applied. The QoS region under a total power constraint is

$$\mathcal{Q}(P_{\max}) = \{\boldsymbol{q} \in \mathbb{Q}^{K_u} : \mathcal{P}(\boldsymbol{q}) \neq \emptyset, \; \sum_{k \in \mathcal{K}_u} p_k^{min}(\boldsymbol{q}) \leq P_{\max}\} \, . \tag{2.53}$$

Hence, the sum-power constrained QoS region can be equivalently characterized either in terms of the min-max function $C(\gamma, P_{\max})$, or in terms of the power minimum $\boldsymbol{p}^{min}(\boldsymbol{q})$.

2.7.2 Individual Power Constraints

Next, consider individual power constraints

$$\boldsymbol{p} \leq \boldsymbol{p}^{\max} = [p_1^{\max}, \ldots, p_{K_u}^{\max}]^T \, .$$

As in the previous subsection, the feasibility of some SINR target vector γ depends on the min-max function

$$C(\gamma, \boldsymbol{p}^{\max}) = \inf_{0 < \boldsymbol{p} \leq \boldsymbol{p}^{\max}} \Big(\max_{k \in \mathcal{K}_u} \frac{\gamma_k \mathcal{J}_k(\boldsymbol{p})}{p_k} \Big) \, . \tag{2.54}$$

It can be observed that $C(\boldsymbol{\gamma}, \boldsymbol{p}^{\max})$ fulfills the axioms A1, A2, A3. This is another example of an interference function.

The *feasible SINR region* is the sub-level set of $C(\boldsymbol{\gamma}, \boldsymbol{p}^{\max})$.

$$\mathcal{S}(\boldsymbol{p}^{\max}) = \{\boldsymbol{\gamma} \in \mathbb{R}_+^{K_u} : C(\boldsymbol{\gamma}, \boldsymbol{p}^{\max}) \leq 1\} \tag{2.55}$$

The structure of the SINR set $\mathcal{S}(\boldsymbol{p}^{\max})$ depends on the properties of the indicator function $C(\boldsymbol{\gamma}, \boldsymbol{p}^{\max})$, which in turn depends on the properties of the underlying interference functions $\mathcal{I}_1, \ldots, \mathcal{I}_{K_u}$, as well as on the chosen power constraints $\boldsymbol{p}^{\max}$. Sub-level sets of convex functions are convex. If $C(\boldsymbol{\gamma}, \boldsymbol{p}^{\max})$ is *convex*, then $\mathcal{S}(\boldsymbol{p}^{\max})$ is a closed convex set from $\mathbb{R}_+^{K_u}$. However, convexity of $C(\boldsymbol{\gamma}, \boldsymbol{p}^{\max})$ does generally not hold, thus SINR regions are typically non-convex.

The resulting QoS region is

$$\mathcal{Q}(\boldsymbol{p}^{\max}) = \{\boldsymbol{q} \in \mathbb{Q}^{K_u} : C(\boldsymbol{\gamma}(\boldsymbol{q}), \boldsymbol{p}^{\max}) \leq 1\} \, . \tag{2.56}$$

This is one possible way of characterizing the QoS region. Another one is given by the function $\boldsymbol{p}^{min}(\boldsymbol{q})$, defined by (2.52). The vector $\boldsymbol{p}^{min}(\boldsymbol{q})$ achieves the targets $\boldsymbol{q}$ not only with minimum total power, it is also component-wise optimal. That is, for any power vector $\boldsymbol{p}$ that achieves the SINR targets, we have $\boldsymbol{p}^{min}(\boldsymbol{q}) \leq \boldsymbol{p}$ (component-wise), as shown in [1]. Thus, the function $\boldsymbol{p}^{min}(\boldsymbol{q})$ may also be used to characterize the QoS region under individual power limits.

$$\mathcal{Q}(\boldsymbol{p}_{max}) = \{\boldsymbol{q} \in \mathbb{Q}^{K_u} : \mathcal{P}(\boldsymbol{q}) \neq \emptyset, \; p_k^{min}(\boldsymbol{q}) \leq p_k^{max}, \; \forall k \in \mathcal{K}_u\} \, . \tag{2.57}$$

It is observed that the definition of QoS regions under power constraints (sum or individual) is closely tied to the problems of "QoS balancing". This problem will be studied in more detail in the following subsection.

Theorem A.4 in Appendix A.3 shows that $p_k^{min}(\boldsymbol{q})$ is log-convex if the underlying interference functions are log-convex. Since every log-convex function is convex, it follows that $\mathcal{Q}(\boldsymbol{p}_{max})$ is a convex set.

2.8 The QoS Balancing Problem

In Subsection 2.7.1 it was shown that the power-constrained QoS region $\mathcal{Q}(P_{\max})$ can be characterized in different ways, either by the *min-max solution* (2.48) or the *min-power solution* (2.52). Both strategies are closely related. They both play a central role for the analysis and optimization of power-constrained systems.

Strategy (2.48) corresponds to the problem of maximizing the worst-case SIR subject to a total power constraint (see Appendix A.2), whereas (2.52) minimizes the total power subject to a constraint on the worst-case SIR. Both problems are equivalent in a sense that the solution of one problem can be found indirectly via a bisection strategy based on solutions of the other problem, as observed in [28]. If the QoS targets are on the boundary of the feasible set, then both problems yield the same optimizer. In this sense, both problems are equivalent[1] and can be comprehended under the name "QoS balancing".

The power minimization problem can be written as follows.

$$\min_{\boldsymbol{p}\in\mathcal{P}} \sum_{l\in\mathcal{K}_u} p_l \quad \text{s.t.} \quad \frac{p_k}{\mathcal{J}_k(\boldsymbol{p})} \geq \gamma_k, \text{ for all } k \in \mathcal{K}_u \,. \tag{2.58}$$

If the targets $\boldsymbol{\gamma}$ are feasible, then it is known from [1] that the unique solution of (2.52) is the fixed point $\boldsymbol{p}^* > 0$, satisfying

$$\boldsymbol{p}^* = \boldsymbol{\Gamma}\boldsymbol{\mathcal{J}}(\boldsymbol{p}^*) \tag{2.59}$$

where $\boldsymbol{\mathcal{J}}(\boldsymbol{p}^*) = [\mathcal{J}_1(\boldsymbol{p}^*), \ldots, \mathcal{J}_{K_u}(\boldsymbol{p}^*)]$ and $\boldsymbol{\Gamma} = \text{diag}\{\boldsymbol{\gamma}\}$. The fixed point equation (2.59) can be rewritten as $\gamma_k = \text{SINR}_k(\boldsymbol{p}^*)$, for all $k \in \mathcal{K}_u$. The optimizer $\boldsymbol{p}^*$ fulfills the SINR targets $\boldsymbol{\gamma}$ with minimum power (componentwise).

2.8.1 Equivalent Convex Reformulations

Under the assumption of strict monotonicity (2.22) and $\mathcal{P} = \mathbb{R}_+^{K_u}$, and if the targets are feasible, then problem (2.58) can be solved by a globally convergent fixed point iteration. The iteration has geometric convergence [7, 74], regardless of the actual choice of $\mathcal{J}_k$ (see also Section 5.3).

More efficient solutions are available if the interference functions $\mathcal{J}_k$ are convex (see also Chapter 5 for more details). We can rewrite (2.58) in equivalent form

$$\min_{\boldsymbol{p}\in\mathcal{P}} \sum_{l\in\mathcal{K}_u} p_l \quad \text{s.t.} \quad \gamma_k \mathcal{J}_k(\boldsymbol{p}) - p_k \leq 0, \text{ for all } k \in \mathcal{K}_u \,. \tag{2.60}$$

If the power set $\mathcal{P} \subseteq \mathbb{R}_{++}^{K_u}$ is convex, which is typically fulfilled, then (2.60) is a convex optimization problem. Strict monotonicity (2.22) ensures the existence of a non-trivial solution, provided that the targets γ_k are feasible.

Next, consider the case where $\mathcal{J}_k$ is strictly monotone and *concave*. An example is the beamforming problem (see Section 1.4.2), with either individual power constraints or a total power constraint. Then, problem (2.60) is nonconvex because the constraints are *concave*, but not convex.

This observation is in line with the literature on multi-user beamforming [26–28], which is a special case of the problem at hand. In this work it was

[1] In the following we will use the notion of equivalence of optimization problems in an informal way. Two problems are called equivalent if from a solution of one, a solution of the other is readily found, and vice versa.

observed that the problem is non-convex in its direct form, but equivalent convex reformulations exist. Thus an interesting question is: does an equivalent convex reformulation also exist for the more general problem (2.60), which is only based on the axiomatic framwork with the additional assumptions of strict monotonicity and concavity? This is answered by the following theorem.

Theorem 2.22. *Let $\mathcal{J}_1, \ldots, \mathcal{J}_{K_u}$ be concave and strictly monotone interference functions, then the optimizer of problem (2.60) is equivalently obtained by the convex problem*

$$\max_{\boldsymbol{p} \in \mathcal{P}} \sum_{l \in \mathcal{K}_u} p_l \quad s.t. \quad p_k - \gamma_k \mathcal{J}_k(\boldsymbol{p}) \leq 0, \quad \forall k \in \mathcal{K}_u \ . \tag{2.61}$$

Proof. First, we observe that problem (2.60) is feasible if and only if (2.61) is feasible. Assume that (2.61) is feasible. Because of strict monotonicity (2.22) there must exist a vector $\boldsymbol{p}^* > 0$ such that all inequalities in (2.61) are fulfilled with equality. This implies feasibility of (2.60). The converse is shown likewise.

Let $\boldsymbol{\mathcal{J}}(\boldsymbol{p}) = [\mathcal{J}_1(\boldsymbol{p}), \ldots, \mathcal{J}_{K_u}(\boldsymbol{p})]^T$. The vector $\boldsymbol{p}^*$ is the unique fixed point that satisfies $\boldsymbol{p}^* = \text{diag}(\gamma)\boldsymbol{\mathcal{J}}(\boldsymbol{p}^*)$. This is the optimizer of (2.60), as shown in [1]. The same fixed point $\boldsymbol{p}^*$ is achieved by (2.61). This can easily be shown by contradiction. If there would exist a k_0 such that the optimizer $\boldsymbol{p}^*$ fulfills $p^*_{k_0} - \gamma_k \mathcal{J}_{k_0}(\boldsymbol{p}^*) < 0$, then we could increase $p^*_{k_0}$ without violating the constraints. This would mean that we could achieve a point larger than the global maximum. Therefore, (2.61) yields the fixed point $\boldsymbol{p}^*$ which also solves (2.60). □

Problem (2.61) is convex and can be solved by applying standard solutions from convex optimization theory. This also sheds some new light on the problem of multi-user beamforming [26–28, 32, 33], which is contained as a special case. It turns out that this problem has a generic convex form (2.61). The solution can be found by standard convex optmization strategies. However, general purpose solvers can be inefficient. A better performance is typically achieved by exploiting the analytical structure of the problem at hand. The structure of interference functions will be discussed in Chapter3. Later, in Chapter 5 we will discuss how these structure properties can be exploited for the design of an algorithm with superlinear convergence.

2.8.2 Equivalent Log-Convex Reformulation

Next, consider the class of log-convex interference functions (see Definition 1.4). Examples are worst-case interference designs used in the context of robust optimization (see e.g. [29, 30]). Every convex interference function is a log-convex interference function, as mentioned before. Thus, the following result also applies to convex interference functions. We introduce the following notation. If $\mathcal{P} \subseteq \mathbb{R}^n_+$, then $\log \mathcal{P} = \{\boldsymbol{s} = \log(\boldsymbol{p}) : \boldsymbol{p} \in \mathcal{P} \cap \mathbb{R}^n_{++}\}$.

Theorem 2.23. *Let $\mathcal{J}_1, \ldots, \mathcal{J}_{K_u}$ be log-convex and strictly monotone interference functions, then the optimizer* $\boldsymbol{p}^*$ *of problem (2.60) is obtained as* $\boldsymbol{p}^* = \exp \boldsymbol{s}^*$, *where* $\boldsymbol{s}^*$ *is the optimizer of*

$$\min_{\boldsymbol{s} \in \log \mathcal{P}} \sum_{l \in \mathcal{K}_u} s_l \quad s.t. \quad \log \gamma_k + \log \mathcal{J}_k(\exp \boldsymbol{s}) - s_k \leq 0 \quad \forall k \in \mathcal{K}_u \,. \tag{2.62}$$

Proof. Exploiting the strict monotonicity of the logarithm, we can rewrite the constraints in (2.60) as

$$\log \gamma_k \mathcal{J}_k(\boldsymbol{p}) - \log p_k \leq 0 \,.$$

Introducing the change of variable $\boldsymbol{s} = \log \boldsymbol{p}$, this can be rewritten as

$$\log \gamma_k + \log \mathcal{J}_k(\exp \boldsymbol{s}) - s_k \leq 0 \,.$$

Using the same argumentation as in the proof of Theorem 2.22, it follows from strict monotonicity (2.22) that the constraints in (2.62) are fulfilled with equality in the optimum, so the optimizer $\boldsymbol{s}^*$ is the unique fixed point in the transformed domain. □

The constraints in (2.62) are convex because $\log \mathcal{J}_k(\exp \boldsymbol{s})$ is convex by definition. Also, the domain $\log \mathcal{P}$ is convex if $\mathcal{P} \subseteq \mathbb{R}^{K_u}_{++}$ is a downward-comprehensive convex set (cf. Definition 2.19). Comprehensiveness is fulfilled for many cases of interest (e.g. unconstrained powers, per-user power constraints, sum-power constraint).

3

The Structure of Interference Functions and Comprehensive Sets

In the previous chapters we have introduced and motivated the analysis of interference-coupled systems by means of SIR and QoS regions. Since the QoS is defined as a strictly monotone and continuous function (2.36), both QoS and SIR regions are bijective, i.e., they can be mapped into each other in such a way that the mapping can be inverted without loss of information. Thus, we can learn about the structure of QoS regions by studying the SIR region instead. Some properties of SIR regions have a direct relationship to properties of the QoS region. We will make use of this connection many times throughout this book. Examples of such properties are *comprehensiveness* (Subsection 2.6.2) and *Pareto optimality* (Subsection 4.5.3).

The SIR region, in its basic form (1.23), is defined as the sublevel set $\mathcal{S} = \{\gamma > 0 : C(\gamma) \leq 1\}$, characterized by the min-max indicator function $C(\gamma)$. This definition holds for arbitrary systems of interference functions, and it can be extended to power-constrained systems, as shown in Section 2.7.

Since $C(\gamma)$ fulfills the axioms $\mathrm{A1}, \mathrm{A2}, \mathrm{A3}$, the SIR region $\mathcal{S}$ is a sub-level set of an interference function. This observation points to an interesting relationship between interference functions and SIR regions. The structure of the indicator function $C(\gamma)$ determines the structure of the resulting region. This means that, we can obtain valuable insight into the structure of SIR regions by analyzing properties of interference functions. A thorough understanding of interference functions and the resulting QoS regions is important for the development of efficient resource allocation algorithms.

In this chapter we discuss fundamental properties of interference functions and QoS regions, which were shown in [3–5]. In particular, every interference function can be expressed as an optimum over *elementary interference functions*. This has some very practical consequences.

For example, it was shown in [4] that every convex or concave interference function can be expressed as an optimum over linear elementary interference functions. This structure can be exploited in various ways. It allows the application of the duality concept previously discussed in Subsection 1.4.6. It

M. Schubert, H. Boche, *Interference Calculus*, Foundations in Signal Processing, Communications and Networking 7,

enables the subgradient-based algorithm from Chapter 5 and the max-min approach from Chapter 6.

The results of this chapter are based on the general axiomatic framework A1, A2, A3. This includes the special case where the interference functions also depend on some constant noise power (see Section 2.4). Most of the results can be readily extended to this case.

Parts of the chapter are based on the assumption that $\boldsymbol{p} \in \mathbb{R}^K_{++}$, i.e. we focus on the interior of the set, where p_k is strictly greater than zero. This technical assumption rules out the possible occurence of singularities (e.g. SIR tending to infinity). However, this is not much of a restriction from a practical point of view. We can deal with zero powers in an asymptotic way, by taking the infimum or supremum. That is, we can approach the boundary of the set arbitrarily close. Fundamental properties of interference functions are preserved for the continuation (2.30), as discussed in Subsection 2.5.1, thus some of the results extend to $\mathbb{R}^K_+$. For example, singularities never occur in a power-constrained system including noise, where the SINR is always well-defined.

3.1 General Interference Functions

In this section we will study the structure of interference functions defined by A1, A2, A3. Other properties are optional. For example, it will be shown later in Section 3.7 that the result can be transferred to standard interference functions, by exploiting strict monotonicity (2.22).

3.1.1 Max-Min and Min-Max Representations

We begin by showing some some fundamental properties. Consider an arbitrary interference function $\mathcal{I}$ on $\mathbb{R}^K_+$, characterized by A1, A2, A3. Here, K is an arbitrary finite dimension. We have the following result.

Lemma 3.1. *Let $\mathcal{I}$ be an arbitrary interference function characterized by A1, A2, A3. For arbitrary $\mathbf{p}$, $\hat{\mathbf{p}} > 0$, we have*

$$\Big(\min_{k\in\mathcal{K}} \frac{p_k}{\hat{p}_k}\Big) \cdot \mathcal{I}(\hat{\boldsymbol{p}}) \leq \mathcal{I}(\boldsymbol{p}) \leq \Big(\max_{k\in\mathcal{K}} \frac{p_k}{\hat{p}_k}\Big) \cdot \mathcal{I}(\hat{\boldsymbol{p}}) . \tag{3.1}$$

Proof. Defining $\bar{\lambda} = \max_k(p_k/\hat{p}_k)$, we have $\boldsymbol{p} \leq \bar{\lambda}\hat{\boldsymbol{p}}$. With A3, we have $\mathcal{I}(\boldsymbol{p}) \leq \bar{\lambda}\mathcal{I}(\hat{\boldsymbol{p}})$, which proves the right-hand inequality (3.1). The left-hand inequality is shown in a similar way. □

With Lemma 3.1, the following inequalities hold for arbitrary $\boldsymbol{p}, \boldsymbol{q} > 0$.

$$\mathcal{I}(\boldsymbol{p}) \leq \mathcal{I}(\boldsymbol{q}) \cdot \max_{k\in\mathcal{K}} \frac{p_k}{q_k} \tag{3.2}$$

$$\mathcal{I}(\boldsymbol{p}) \geq \mathcal{I}(\boldsymbol{q}) \cdot \min_{k\in\mathcal{K}} \frac{p_k}{q_k} . \tag{3.3}$$

These inequalities are fulfilled with equality if $\boldsymbol{p} = \boldsymbol{q}$. Thus,

$$\mathcal{I}(\boldsymbol{p}) = \inf_{\boldsymbol{q}>0}\Big(\mathcal{I}(\boldsymbol{q}) \max_{k\in\mathcal{K}} \frac{p_k}{q_k}\Big) = \min_{\boldsymbol{q}>0}\Big(\mathcal{I}(\boldsymbol{q}) \max_{k\in\mathcal{K}} \frac{p_k}{q_k}\Big) \tag{3.4}$$

$$\mathcal{I}(\boldsymbol{p}) = \sup_{\boldsymbol{q}>0}\Big(\mathcal{I}(\boldsymbol{q}) \min_{k\in\mathcal{K}} \frac{p_k}{q_k}\Big) = \max_{\boldsymbol{q}>0}\Big(\mathcal{I}(\boldsymbol{q}) \min_{k\in\mathcal{K}} \frac{p_k}{q_k}\Big) . \tag{3.5}$$

We can further exploit the following identities [82] (see also Appendix A.7).

$$\sup_{\boldsymbol{v}>0} \frac{\sum_{k\in\mathcal{K}} v_k p_k}{\sum_{k\in\mathcal{K}} v_k q_k} = \sup_{\boldsymbol{w}>0, \|\boldsymbol{w}\|_1=1} \frac{\prod_{k\in\mathcal{K}} (p_k)^{w_k}}{\prod_{k\in\mathcal{K}} (q_k)^{w_k}} = \max_{k\in\mathcal{K}} \frac{p_k}{q_k} \tag{3.6}$$

$$\inf_{\boldsymbol{v}>0} \frac{\sum_{k\in\mathcal{K}} v_k p_k}{\sum_{k\in\mathcal{K}} v_k q_k} = \inf_{\boldsymbol{w}>0, \|\boldsymbol{w}\|_1=1} \frac{\prod_{k\in\mathcal{K}} (p_k)^{w_k}}{\prod_{k\in\mathcal{K}} (q_k)^{w_k}} = \min_{k\in\mathcal{K}} \frac{p_k}{q_k} . \tag{3.7}$$

For $\boldsymbol{w}, \boldsymbol{v} > 0$, $\|\boldsymbol{w}\|_1 = 1$, we introduce functions

$$G_{\mathcal{I}}(\boldsymbol{q}, \boldsymbol{p}, \boldsymbol{v}) := \left(\frac{\mathcal{I}(\boldsymbol{q})}{\sum_{l\in\mathcal{K}} v_l q_l}\right) \cdot \sum_{k\in\mathcal{K}} v_k p_k \tag{3.8}$$

$$F_{\mathcal{I}}(\boldsymbol{q}, \boldsymbol{p}, \boldsymbol{w}) := \left(\frac{\mathcal{I}(\boldsymbol{q})}{\prod_{l\in\mathcal{K}} (q_l)^{w_l}}\right) \cdot \prod_{k\in\mathcal{K}} (p_k)^{w_k} . \tag{3.9}$$

The next theorem is a direct consequence of (3.4) and (3.5).

Theorem 3.2. *Consider an arbitrary interference function $\mathcal{I}$. For all $\boldsymbol{p} > 0$ we have*

$$\mathcal{I}(\boldsymbol{p}) = \inf_{\boldsymbol{q}>0} \sup_{\boldsymbol{v}>0} G_{\mathcal{I}}(\boldsymbol{q}, \boldsymbol{p}, \boldsymbol{v}) = \sup_{\boldsymbol{q}>0} \inf_{\boldsymbol{v}>0} G_{\mathcal{I}}(\boldsymbol{q}, \boldsymbol{p}, \boldsymbol{v}) \tag{3.10}$$

$$\mathcal{I}(\boldsymbol{p}) = \inf_{\boldsymbol{q}>0} \sup_{\substack{\boldsymbol{w}>0 \\ \|\boldsymbol{w}\|_1=1}} F_{\mathcal{I}}(\boldsymbol{q}, \boldsymbol{p}, \boldsymbol{w}) = \sup_{\boldsymbol{q}>0} \inf_{\substack{\boldsymbol{w}>0 \\ \|\boldsymbol{w}\|_1=1}} F_{\mathcal{I}}(\boldsymbol{q}, \boldsymbol{p}, \boldsymbol{w}) . \tag{3.11}$$

Theorem 3.2 shows that any interference function has a sup-inf and inf-sup characterization, involving functions $G_{\mathcal{I}}(\boldsymbol{q}, \boldsymbol{p}, \boldsymbol{v})$ and $F_{\mathcal{I}}(\boldsymbol{q}, \boldsymbol{p}, \boldsymbol{w})$. These functions fulfill the axioms A1, A2, A3 (with respect to the variable $\boldsymbol{p}$), so they can be regarded as *elementary interference functions*.

Note, that (3.10) and (3.11) are not saddle point characterizations, because we do not only interchange the optimization order, but also the domain. Representation (3.10) will be used in the following Sections 3.4.4 and 3.3.4, where we analyze convex and concave interference functions will be analyzed. Representation (3.11) will be needed later in Section 3.6.6, where log-convex interference functions will be analyzed.

3.1.2 Majorants and Minorants

Sometimes it is desirable to approximate an interference function by another interference function with a more favorable structure. In the following we will make use of the following definitions.

Definition 3.3. *An interference function* $\underline{\mathcal{I}}(\boldsymbol{p})$ *is said to be a* minorant *of* $\mathcal{I}(\boldsymbol{p})$ *if* $\underline{\mathcal{I}}(\boldsymbol{p}) \leq \mathcal{I}(\boldsymbol{p})$ *for all* $\mathbf{p} \in \mathcal{P}$*, where* $\mathcal{P}$ *is the domain of* $\mathcal{I}$*. An interference function* $\overline{\mathcal{I}}(\boldsymbol{p})$ *is said to be a* majorant *if* $\overline{\mathcal{I}}(\boldsymbol{p}) \geq \mathcal{I}(\boldsymbol{p})$ *for all* $\mathbf{p} \in \mathcal{P}$*.*

Consider (3.10). By exchanging inf and sup, we obtain for all $\boldsymbol{p} > 0$

$$\mathcal{I}(\boldsymbol{p}) \geq \sup_{\boldsymbol{v}>0} \inf_{\boldsymbol{q}>0} G_{\mathcal{I}}(\boldsymbol{q},\boldsymbol{p},\boldsymbol{v}) = \underline{\mathcal{I}}(\boldsymbol{p}) \tag{3.12}$$

$$\mathcal{I}(\boldsymbol{p}) \leq \inf_{\boldsymbol{v}>0} \sup_{\boldsymbol{q}>0} G_{\mathcal{I}}(\boldsymbol{q},\boldsymbol{p},\boldsymbol{v}) = \overline{\mathcal{I}}(\boldsymbol{p}). \tag{3.13}$$

The resulting functions $\underline{\mathcal{I}}(\boldsymbol{p})$ and $\overline{\mathcal{I}}(\boldsymbol{p})$ are minorants and majorants, respectively. They will play an important role for the analysis of concave and convex interference functions in Subsections 3.3.4 and 3.4.4.

3.1.3 Representation Theorem based on Level Sets

In Subsection 3.1.1 we have shown that every general interference function has a sup-inf and inf-sup characterization based on functions $G_{\mathcal{I}}(\boldsymbol{q},\boldsymbol{p},\boldsymbol{v})$ and $F_{\mathcal{I}}(\boldsymbol{q},\boldsymbol{p},\boldsymbol{w})$. The optimization is unconstrained, i.e., the feasible region is the whole space $\mathbb{R}^K_{++}$.

Next, we show alternative min-max and max-min representations based on the following level sets.

$$\underline{L}(\mathcal{I}) = \{\hat{\boldsymbol{p}} > 0 : \mathcal{I}(\hat{\boldsymbol{p}}) \leq 1\} \tag{3.14}$$

$$\overline{L}(\mathcal{I}) = \{\hat{\boldsymbol{p}} > 0 : \mathcal{I}(\hat{\boldsymbol{p}}) \geq 1\} \tag{3.15}$$

$$B(\mathcal{I}) = \{\hat{\boldsymbol{p}} > 0 : \mathcal{I}(\hat{\boldsymbol{p}}) = 1\} . \tag{3.16}$$

With the continuity (Lemma 2.15 from Section 2.5), we know that the sets $\underline{L}(\mathcal{I})$, $B(\mathcal{I})$, and $\overline{L}(\mathcal{I})$ are relatively closed in $\mathbb{R}^K_{++}$.

Definition 3.4. *A set* $\mathcal{V} \subset \mathbb{R}^K_{++}$ *is said to be* relatively closed *in* $\mathbb{R}^K_{++}$ *if there exists a closed set* $\mathcal{A} \subset \mathbb{R}^K$ *such that* $\mathcal{V} = \mathcal{A} \bigcap \mathbb{R}^K_{++}$*. For the sake of simplicity we will refer to such sets as* closed *in the following.*

This leads to our first theorem, which will serve as a basis for some of the following results.

Theorem 3.5. *Let* $\mathcal{I}$ *be an arbitrary interference function. For any* $\mathbf{p} \in \mathbb{R}^K_{++}$*, we have*

$$\mathcal{I}(\boldsymbol{p}) = \min_{\hat{\boldsymbol{p}} \in \underline{L}(\mathcal{I})} \left(\max_{k \in \mathcal{K}} \frac{p_k}{\hat{p}_k} \right) \tag{3.17}$$

$$= \max_{\hat{\boldsymbol{p}} \in \overline{L}(\mathcal{I})} \left(\min_{k \in \mathcal{K}} \frac{p_k}{\hat{p}_k} \right) . \tag{3.18}$$

Proof. We first show (3.17). Consider an arbitrary fixed $\boldsymbol{p} > 0$ and $\hat{\boldsymbol{p}} \in \underline{L}(\mathcal{I})$. With Lemma 3.1 we have

$$\mathcal{I}(\boldsymbol{p}) \leq \left(\max_{k\in\mathcal{K}} \frac{p_k}{\hat{p}_k}\right) \cdot \mathcal{I}(\hat{\boldsymbol{p}}) \leq \max_{k\in\mathcal{K}} \frac{p_k}{\hat{p}_k} , \tag{3.19}$$

where the last inequality follows from the definition (3.14). This holds for arbitrary $\hat{\boldsymbol{p}} \in \underline{L}(\mathcal{I})$, thus

$$\mathcal{I}(\boldsymbol{p}) \leq \inf_{\hat{\boldsymbol{p}}\in\underline{L}(\mathcal{I})} \max_{k\in\mathcal{K}} \frac{p_k}{\hat{p}_k} . \tag{3.20}$$

Now, we choose $\hat{\boldsymbol{p}}'$ with $\hat{p}'_k = p_k/\mathcal{I}(\boldsymbol{p})$, $\forall k$. With A2 we have $\mathcal{I}(\hat{\boldsymbol{p}}') = 1$, so $\hat{\boldsymbol{p}}' \in \underline{L}(\mathcal{I})$. This particular choice fulfills $\max_{k\in\mathcal{K}}(p_k/\hat{p}'_k) = \mathcal{I}(\boldsymbol{p})$. Thus, $\hat{\boldsymbol{p}}'$ achieves the infimum (3.20) and (3.17) holds.

The second equality is shown in a similar way: With Lemma 3.1, we have

$$\mathcal{I}(\boldsymbol{p}) \geq \left(\min_{k\in\mathcal{K}} \frac{p_k}{\hat{p}_k}\right) \cdot \mathcal{I}(\hat{\boldsymbol{p}}) \geq \min_{k\in\mathcal{K}} \frac{p_k}{\hat{p}_k} \tag{3.21}$$

for all $\boldsymbol{p} > 0$ and $\hat{\boldsymbol{p}} \in \overline{L}(\mathcal{I})$. Similar to the first case, it can be observed that (3.21) is fulfilled with equality for $\hat{\boldsymbol{p}}' = \boldsymbol{p}/\mathcal{I}(\boldsymbol{p})$, with $\hat{\boldsymbol{p}}' \in \overline{L}(\mathcal{I})$. Thus, (3.18) is fulfilled. □

Theorem 3.5 states that every $\mathcal{I}(\boldsymbol{p})$ can be represented as an optimum over elementary building blocks

$$\overline{\mathcal{I}}(\boldsymbol{p}, \hat{\boldsymbol{p}}) = \max_{k\in\mathcal{K}} \frac{p_k}{\hat{p}_k} \tag{3.22}$$

$$\underline{\mathcal{I}}(\boldsymbol{p}, \hat{\boldsymbol{p}}) = \min_{k\in\mathcal{K}} \frac{p_k}{\hat{p}_k} . \tag{3.23}$$

Assume that $\hat{\boldsymbol{p}}$ is an arbitrary fixed parameter, then (3.22) and (3.23) are functions in $\boldsymbol{p}$. Both $\overline{\mathcal{I}}$ and $\underline{\mathcal{I}}$ fulfill the axioms A1-A3, thus they can be considered as *elementary interference functions*. Note, that the existence of an optimizer $\hat{\boldsymbol{p}}$ in (3.17) is ensured by A1. This rules out $\mathcal{I}(\boldsymbol{p}) = 0$, thus $\underline{L}(\mathcal{I}) = \mathbb{R}^K_{++}$ cannot occur.

Next, consider the set $B(\mathcal{I})$, as defined by (3.16). In the proof of Theorem 3.5 it was shown that $\hat{\boldsymbol{p}}' \in \overline{L}(\mathcal{I}) \cap \underline{L}(\mathcal{I}) = B(\mathcal{I})$. That is, we can restrict the optimization to the boundary $B(\mathcal{I})$.

Corollary 3.6. *Let $\mathcal{I}$ be an arbitrary interference function. For any $\boldsymbol{p} \in \mathbb{R}^K_{++}$, we have*

$$\mathcal{I}(\boldsymbol{p}) = \min_{\hat{\boldsymbol{p}}\in B(\mathcal{I})} \overline{\mathcal{I}}(\boldsymbol{p}, \hat{\boldsymbol{p}}) \tag{3.24}$$

$$= \max_{\hat{\boldsymbol{p}}\in B(\mathcal{I})} \underline{\mathcal{I}}(\boldsymbol{p}, \hat{\boldsymbol{p}}) . \tag{3.25}$$

Note, that the optimization domain $\overline{L}(\mathcal{I})$ in (3.18) cannot be replaced by $\underline{L}(\mathcal{I})$. Since $B(\mathcal{I}) \subseteq \underline{L}(\mathcal{I})$, relation (3.25) implies

$$\mathcal{I}(\boldsymbol{p}) \leq \sup_{\hat{\boldsymbol{p}} \in \underline{L}(\mathcal{I})} \underline{\mathcal{I}}(\boldsymbol{p}, \hat{\boldsymbol{p}}) = +\infty .$$

Likewise, $B(\mathcal{I}) \subseteq \overline{L}(\mathcal{I})$ and (3.24) implies

$$\mathcal{I}(\boldsymbol{p}) \geq \inf_{\hat{\boldsymbol{p}} \in \overline{L}(\mathcal{I})} \overline{\mathcal{I}}(\boldsymbol{p}, \hat{\boldsymbol{p}}) = 0 .$$

That is, by exchanging the respective optimization domain, we only obtain trivial bounds.

3.1.4 Elementary Sets and Interference Functions

In this section we will analyze the elementary interference functions $\overline{\mathcal{I}}(\boldsymbol{p}, \hat{\boldsymbol{p}})$ and $\underline{\mathcal{I}}(\boldsymbol{p}, \hat{\boldsymbol{p}})$ for an arbitrary and fixed parameter $\hat{\boldsymbol{p}} \in \mathbb{R}^K_{++}$. This approach helps to better understand the structure of interference functions and corresponding level sets.

We start by showing convexity.

Lemma 3.7. *Let $\hat{\boldsymbol{p}} > 0$ be arbitrary and fixed. The function $\overline{\mathcal{I}}(\boldsymbol{p}, \hat{\boldsymbol{p}})$ is convex on $\mathbb{R}^K_+$. The function $\underline{\mathcal{I}}(\boldsymbol{p}, \hat{\boldsymbol{p}})$ is concave on $\mathbb{R}^K_+$.*

Proof. The maximum of convex functions is convex. The minimum of concave functions is concave. □

As an immediate consequence of Theorem 3.5, every interference function $\mathcal{I}$ can be expressed as a minimum over elementary convex interference functions $\overline{\mathcal{I}}(\boldsymbol{p}, \hat{\boldsymbol{p}})$ with $\hat{\boldsymbol{p}} \in \underline{L}(\mathcal{I})$. Alternatively, $\mathcal{I}$ can be expressed as a maximum over concave interference functions. Note, that this behavior is due to the properties A1, A2, A3 and cannot be generalized to arbitrary functions.

Any sublevel set of a convex function is convex. Therefore, the following set is convex.

$$\underline{L}(\overline{\mathcal{I}}) = \{\boldsymbol{p} > 0 : \overline{\mathcal{I}}(\boldsymbol{p}, \hat{\boldsymbol{p}}) \leq 1\} \tag{3.26}$$

We have $\overline{\mathcal{I}}(\hat{\boldsymbol{p}}, \hat{\boldsymbol{p}}) = 1$, and $\overline{\mathcal{I}}(\boldsymbol{p}, \hat{\boldsymbol{p}}) = \max_{k \in \mathcal{K}} p_k / \hat{p}_k \leq 1$ for all $\boldsymbol{p} \in \underline{L}(\overline{\mathcal{I}})$. Thus,

$$p_k \leq \hat{p}_k , \quad \forall k \in \mathcal{K} . \tag{3.27}$$

The concave function $\underline{\mathcal{I}}(\boldsymbol{p}, \hat{\boldsymbol{p}})$ is associated with a convex superlevel set

$$\overline{L}(\underline{\mathcal{I}}) = \{\boldsymbol{p} > 0 : \underline{\mathcal{I}}(\boldsymbol{p}, \hat{\boldsymbol{p}}) \geq 1\} . \tag{3.28}$$

Every $\boldsymbol{p} \in \overline{L}(\underline{\mathcal{I}})$ fulfills

$$p_k \geq \hat{p}_k , \quad \forall k \in \mathcal{K} . \tag{3.29}$$

Both sets $\underline{L}(\overline{\mathcal{I}})$ and $\overline{L}(\underline{\mathcal{I}})$ are illustrated in Fig. 3.1.

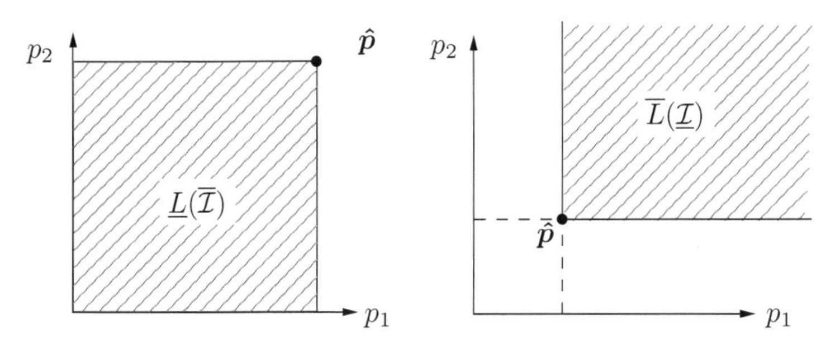

Fig. 3.1. Illustration of the convex comprehensive sets $\underline{L}(\overline{\mathcal{I}})$ and $\overline{L}(\underline{\mathcal{I}})$, as defined by (3.26) and (3.28), respectively.

Let us summarize the results. Starting from an interference function $\mathcal{I}$, we obtain the sublevel set $\underline{L}(\mathcal{I}) \subset \mathbb{R}^K_{++}$, as defined by (3.14). For any $\hat{\boldsymbol{p}} \in \underline{L}(\mathcal{I})$, there exists a sublevel set of the form (3.26), which is contained in $\underline{L}(\mathcal{I})$. Thus, the region $\underline{L}(\mathcal{I})$ is the union over convex downward-comprehensive sets. Therefore, $\underline{L}(\mathcal{I})$ is downward-comprehensive (this also follows from (3.14) with A3). However, $\underline{L}(\mathcal{I})$ is not necessarily convex. From Theorem 3.5 we know that we can use (3.17) to get back the original interference function $\mathcal{I}$.

The following corollary summarizes the properties of the elementary sets $\underline{L}(\mathcal{I})$ and $\overline{L}(\mathcal{I})$.

Corollary 3.8. *Let $\mathcal{I}$ be an arbitrary interference function. The sublevel set $\underline{L}(\mathcal{I})$, as defined by (3.14), is closed and downward-comprehensive. The superlevel set $\overline{L}(\mathcal{I})$, as defined by (3.15), is closed and upward-comprehensive.*

For any $\hat{\boldsymbol{p}} > 0$, there is a set of interference functions

$$I_{\hat{\boldsymbol{p}}} = \{\mathcal{I} : \mathcal{I}(\hat{\boldsymbol{p}}) = 1\} .$$

The following theorem shows the special role of the interference function $\underline{\mathcal{I}}(\boldsymbol{p}, \hat{\boldsymbol{p}}) \in I_{\hat{\boldsymbol{p}}}$.

Theorem 3.9. *Consider an arbitrary $\hat{\boldsymbol{p}} > 0$ and an interference function $\mathcal{I}$, with $\mathcal{I}(\hat{\boldsymbol{p}}) = 1$, such that*

$$\mathcal{I}(\boldsymbol{p}) \leq \underline{\mathcal{I}}(\boldsymbol{p}, \hat{\boldsymbol{p}}), \quad \forall \boldsymbol{p} > 0 , \tag{3.30}$$

then this can only be satisfied with equality.

Proof. Inequality (3.30) implies $\overline{L}(\mathcal{I}) \subseteq \overline{L}(\underline{\mathcal{I}})$, or in other words, every $\boldsymbol{p} \in \{\boldsymbol{p} : \mathcal{I}(\boldsymbol{p}) \geq 1\}$ fulfills $\underline{\mathcal{I}}(\boldsymbol{p}, \hat{\boldsymbol{p}}) \geq 1$. This can be written as $\min_k p_k/\hat{p}_k \geq 1$, or equivalently $\boldsymbol{p} \geq \hat{\boldsymbol{p}}$. With $\mathcal{I}(\hat{\boldsymbol{p}}) = 1$ and A3, it follows that $\mathcal{I}(\boldsymbol{p}) \geq \mathcal{I}(\hat{\boldsymbol{p}}) = 1$. Thus, the set $\overline{L}(\underline{\mathcal{I}}) = \{\boldsymbol{p} : \boldsymbol{p} \geq \hat{\boldsymbol{p}}\}$ also belongs to $\overline{L}(\mathcal{I})$. Consequently, $\overline{L}(\mathcal{I}) = \overline{L}(\underline{\mathcal{I}})$. With Theorem 3.5 we can conclude that $\mathcal{I}(\boldsymbol{p}) = \underline{\mathcal{I}}(\boldsymbol{p}, \hat{\boldsymbol{p}})$ for all $\boldsymbol{p} > 0$. □

Theorem 3.9 shows that $\underline{\mathcal{I}}(\boldsymbol{p},\hat{\boldsymbol{p}})$ is the *smallest* interference function from the set $I_{\hat{\boldsymbol{p}}}$. Here 'smallest' is used in the sense of a relation $\mathcal{I}_1 \leq \mathcal{I}_2$ meaning $\mathcal{I}_1(\boldsymbol{p}) \leq \mathcal{I}_2(\boldsymbol{p})$ for all $\boldsymbol{p} > 0$.

The following result is shown by similar arguments.

Theorem 3.10. *Consider* $\hat{\boldsymbol{p}} > 0$ *and an interference function* $\mathcal{I}$, *with* $\mathcal{I}(\hat{\boldsymbol{p}}) = 1$, *such that*

$$\mathcal{I}(\boldsymbol{p}) \geq \overline{\mathcal{I}}(\boldsymbol{p},\hat{\boldsymbol{p}}), \quad \forall \boldsymbol{p} > 0 \tag{3.31}$$

then this can only be satisfied with equality.

The interference function $\overline{\mathcal{I}}(\boldsymbol{p},\hat{\boldsymbol{p}})$ is the greatest interference function from the set $I_{\hat{\boldsymbol{p}}}$.

Theorems 3.9 and 3.10 show that only the interference functions $\overline{\mathcal{I}}(\boldsymbol{p},\hat{\boldsymbol{p}})$ and $\underline{\mathcal{I}}(\boldsymbol{p},\hat{\boldsymbol{p}})$ provide majorants and minorants for arbitrary interference functions (see Definition 3.3). This is a property by which general interference functions are characterized.

3.2 Synthesis of General Interference Functions

In the previous section we have analyzed the basic building blocks of an interference function $\mathcal{I}$, and its connection with level sets. Now, we study the converse approach, i.e., the synthesis of an interference function for a given set $\mathcal{V}$.

3.2.1 Interference Functions and Comprehensive Sets

We start by showing that for any closed downward-comprehensive set $\mathcal{V} \subset \mathbb{R}^K_{++}$, we can synthesize an interference function $\mathcal{I}_{\mathcal{V}}(\boldsymbol{p})$. By constructing the sublevel set $\underline{L}(\mathcal{I}_{\mathcal{V}})$ we get back the original set.

Theorem 3.11. *For any non-empty, closed, and downward-comprehensive set* $\mathcal{V} \subset \mathbb{R}^K_{++}$, $\mathcal{V} \neq \mathbb{R}^K_{++}$, *there exists an interference function*

$$\mathcal{I}_{\mathcal{V}}(\boldsymbol{p}) := \inf_{\hat{\boldsymbol{p}} \in \mathcal{V}} \max_{k \in \mathcal{K}} \frac{p_k}{\hat{p}_k} = \min_{\hat{\boldsymbol{p}} \in \mathcal{V}} \max_{k \in \mathcal{K}} \frac{p_k}{\hat{p}_k}, \tag{3.32}$$

and $\underline{L}(\mathcal{I}_{\mathcal{V}}) = \mathcal{V}$.

Proof. For any non-empty set $\mathcal{V} \subset \mathbb{R}^K_{++}$, the function $\mathcal{I}_{\mathcal{V}}$ fulfills properties A1, A2, A3. With the additional assumption $\mathcal{V} \neq \mathbb{R}^K_{++}$, we know that there exists a $\hat{\boldsymbol{p}} > 0$ such that $\mathcal{I}_{\mathcal{V}}(\hat{\boldsymbol{p}}) > 0$. Therefore, $\mathcal{I}_{\mathcal{V}}(\boldsymbol{p}) > 0$ for all $\boldsymbol{p} > 0$. We only need to show $\underline{L}(\mathcal{I}_{\mathcal{V}}) = \mathcal{V}$, then it follows from Theorem 3.5 that the infimum is attained, i.e., the right-hand equality in (3.32) holds.

Consider an arbitrary $\boldsymbol{p} \in \underline{L}(\mathcal{I}_{\mathcal{V}})$, i.e., $\mathcal{I}_{\mathcal{V}}(\boldsymbol{p}) \leq 1$. Defining $\boldsymbol{p}(\lambda) = \lambda\boldsymbol{p}$, with $0 < \lambda < 1$, we have $\mathcal{I}_{\mathcal{V}}\big(\boldsymbol{p}(\lambda)\big) = \lambda\mathcal{I}_{\mathcal{V}}(\boldsymbol{p}) < 1$. According to the definition (3.32), there exists a $\hat{\boldsymbol{p}} \in \mathcal{V}$ such that

$$\max_{k\in\mathcal{K}} \frac{p_k(\lambda)}{\hat{p}_k} < 1 \,. \tag{3.33}$$

Comprehensiveness implies $\boldsymbol{p}(\lambda) < \hat{\boldsymbol{p}}$ and therefore $\boldsymbol{p}(\lambda) \in \mathcal{V}$. Since $\mathcal{V}$ is closed, $\lim_{\lambda\to 1} \boldsymbol{p}(\lambda) = \boldsymbol{p}$ implies $\boldsymbol{p} \in \mathcal{V}$. thus,

$$\underline{L}(\mathcal{I}_\mathcal{V}) \subseteq \mathcal{V} \,. \tag{3.34}$$

Conversely, consider an arbitrary $\hat{\boldsymbol{p}} \in \mathcal{V}$, for which

$$\mathcal{I}_\mathcal{V}(\hat{\boldsymbol{p}}) = \inf_{\tilde{\boldsymbol{p}}\in\mathcal{V}} \max_{k\in\mathcal{K}} \frac{\hat{p}_k}{\tilde{p}_k} \le \max_{k\in\mathcal{K}} \frac{\hat{p}_k}{\hat{p}_k} = 1 \,. \tag{3.35}$$

This shows $\hat{\boldsymbol{p}} \in \underline{L}(\mathcal{I}_\mathcal{V})$ and therefore

$$\mathcal{V} \subseteq \underline{L}(\mathcal{I}_\mathcal{V}) \,. \tag{3.36}$$

Combining (3.34) and (3.36), we have $\underline{L}(\mathcal{I}_\mathcal{V}) = \mathcal{V}$. □

It can be observed that the restriction $\mathcal{V} \neq \mathbb{R}^K_{++}$ is closely linked with property A1. In particular, there exists a $\boldsymbol{p} > 0$ such that $\mathcal{I}_\mathcal{V}(\boldsymbol{p}) > 0$ if and only if the corresponding set $\mathcal{V}$ fulfills $\mathcal{V} \neq \mathbb{R}^K_{++}$.

Similar results exist for upward-comprehensive sets:

Theorem 3.12. *For any non-empty, closed, and upward-comprehensive set* $\mathcal{V} \subset \mathbb{R}^K_{++}$, $\mathcal{V} \neq \mathbb{R}^K_{++}$, *there exists an interference function*

$$\mathcal{I}_\mathcal{V}(\boldsymbol{p}) := \sup_{\hat{\boldsymbol{p}}\in\mathcal{V}} \min_{k\in\mathcal{K}} \frac{p_k}{\hat{p}_k} = \max_{\hat{\boldsymbol{p}}\in\mathcal{V}} \min_{k\in\mathcal{K}} \frac{p_k}{\hat{p}_k} \,, \tag{3.37}$$

and $\overline{L}(\mathcal{I}_\mathcal{V}) = \mathcal{V}$.

Proof. The proof is similar to the proof of Theorem 3.11. Every $\boldsymbol{p} \in \overline{L}(\mathcal{I}_\mathcal{V})$ is also contained in $\mathcal{V}$, thus implying $\overline{L}(\mathcal{I}_\mathcal{V}) \subseteq \mathcal{V}$. Conversely, it is shown that every $\hat{\boldsymbol{p}} \in \mathcal{V}$ is also contained in the set $\overline{L}(\mathcal{I}_\mathcal{V})$, thus $\mathcal{V} \subseteq \overline{L}(\mathcal{I}_\mathcal{V})$. □

The following corollary is an immediate consequence.

Corollary 3.13. *Let* $\mathcal{V}_1$, $\mathcal{V}_2$ *be two arbitrary closed comprehensive sets, as defined in the previous theorems. If* $\mathcal{I}_{\mathcal{V}_1} = \mathcal{I}_{\mathcal{V}_2}$, *then* $\mathcal{V}_1 = \mathcal{V}_2$.

Proof. If the sets are downward-comprehensive, then this is a direct consequence of Theorem 3.11, because $\mathcal{V}_1 = \underline{L}(\mathcal{I}_{\mathcal{V}_1}) = \underline{L}(\mathcal{I}_{\mathcal{V}_2}) = \mathcal{V}_2$. For upward-comprehensive sets, the result follows from Theorem 3.12. □

3.2.2 Comprehensive Hull

Next, assume that $\mathcal{V} \subset \mathbb{R}^K_{++}$, $\mathcal{V} \neq \mathbb{R}^K_{++}$, is an arbitrary non-empty closed set which is not necessarily comprehensive. In this case, (3.32) still yields an interference function. However, the properties stated by Theorem 3.11 and Theorem 3.12 need not be fulfilled. That is, $\underline{L}(\mathcal{I}_\mathcal{V}) \neq \mathcal{V}$ and $\overline{L}(\mathcal{I}_\mathcal{V}) \neq \mathcal{V}$ in general.

The next theorem shows that the level sets $\underline{L}(\mathcal{I}_\mathcal{V})$ and $\overline{L}(\mathcal{I}_\mathcal{V})$ provide comprehensive hulls of the original set $\mathcal{V}$.

Theorem 3.14. *Let $\mathcal{V}_0 \supseteq \mathcal{V}$ be the* downward-comprehensive hull *of $\mathcal{V}$, i.e., the smallest closed downward-comprehensive subset of $\mathbb{R}^K_{++}$ containing $\mathcal{V}$. Let $\mathcal{I}_\mathcal{V}(\boldsymbol{p})$ be defined by (3.32), then*

$$\underline{L}(\mathcal{I}_\mathcal{V}) = \mathcal{V}_0 \,. \tag{3.38}$$

Proof. From Corollary 3.8 we know that $\underline{L}(\mathcal{I}_\mathcal{V})$ is downward-comprehensive. By assumption, $\mathcal{V}_0$ is the smallest downward-comprehensive set containing $\mathcal{V}$, so together with (3.36) we have

$$\mathcal{V} \subseteq \mathcal{V}_0 \subseteq \underline{L}(\mathcal{I}_\mathcal{V}) \,. \tag{3.39}$$

We also have

$$\begin{aligned} \mathcal{V}_0 \supseteq \mathcal{V} &\implies \mathcal{I}_{\mathcal{V}_0}(\boldsymbol{p}) \leq \mathcal{I}_\mathcal{V}(\boldsymbol{p}), \quad \forall \boldsymbol{p} \in \mathbb{R}^K_{++} \\ &\implies \underline{L}(\mathcal{I}_{\mathcal{V}_0}) \supseteq \underline{L}(\mathcal{I}_\mathcal{V}) \,. \end{aligned} \tag{3.40}$$

From Theorem 3.11 we know that $\underline{L}(\mathcal{I}_{\mathcal{V}_0}) = \mathcal{V}_0$. Combining (3.39) and (3.40), the result (3.38) follows. □

To summarize, $\mathcal{V} \subseteq \underline{L}(\mathcal{I}_\mathcal{V})$ is fulfilled for any non-empty closed set $\mathcal{V} \subset \mathbb{R}^K_{++}$, $\mathcal{V} \neq \mathbb{R}^K_{++}$. The set $\underline{L}(\mathcal{I}_\mathcal{V})$ is the downward-comprehensive hull of $\mathcal{V}$. The set $\mathcal{V}$ is downward-comprehensive if and only if $\mathcal{V} = \underline{L}(\mathcal{I}_\mathcal{V})$. Examples are given in Fig. 3.2. Likewise, an upward-comprehensive hull can be constructed for any non-empty closed set $\mathcal{V} \subset \mathbb{R}^K_{++}$, $\mathcal{V} \neq \mathbb{R}^K_{++}$.

Theorem 3.15. *Let $\mathcal{V}_\infty \supseteq \mathcal{V}$ be the* upward-comprehensive hull *of $\mathcal{V}$, i.e., the smallest closed upward-comprehensive subset of $\mathbb{R}^K_{++}$ containing $\mathcal{V}$. Let $\mathcal{I}_\mathcal{V}(\boldsymbol{p})$ be defined by (3.37), then*

$$\overline{L}(\mathcal{I}_\mathcal{V}) = \mathcal{V}_\infty \,. \tag{3.41}$$

Proof. This is shown by arguments similar to those in Theorem 3.14. □

Next, we study interference functions with a special monotonicity property. To this end we need some definitions.

Definition 3.16. $\boldsymbol{p}^{(1)} \succ \boldsymbol{p}^{(2)}$ *means $p_k^{(1)} \geq p_k^{(2)}$, $\forall k \in \mathcal{K}$, and there exists at least one component k_0 such that $p_{k_0}^{(1)} > p_{k_0}^{(2)}$.*

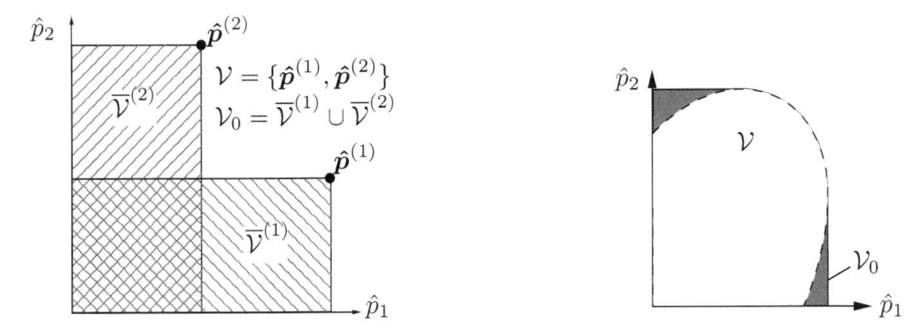

Fig. 3.2. Two examples illustrating Theorem 3.14: The set $\mathcal{V}_0 = \underline{L}(\mathcal{I}_{\mathcal{V}})$ is the comprehensive hull of an arbitrary non-comprehensive closed set $\mathcal{V} \subset \mathbb{R}^K_{++}$.

Definition 3.17. *An interference function* $\mathcal{I}(\boldsymbol{p})$ *is said to be* strictly monotone *if* $\boldsymbol{p}^{(1)} \succ \boldsymbol{p}^{(2)}$ *implies* $\mathcal{I}(\boldsymbol{p}^{(1)}) > \mathcal{I}(\boldsymbol{p}^{(2)})$.

The next theorem shows that strict monotonicity of $\mathcal{I}(\boldsymbol{p})$ corresponds to certain properties of the associated level sets $\underline{L}(\mathcal{I})$ and $\overline{L}(\mathcal{I})$, whose boundary is $B(\mathcal{I})$.

Theorem 3.18. *An interference function* $\mathcal{I}(\boldsymbol{p})$ *is strictly monotone if and only if no segment of the boundary* $B(\mathcal{I})$*, as defined by (3.16), is parallel to a coordinate axis.*

Proof. Assume that $\mathcal{I}(\boldsymbol{p})$ is strictly monotone. We will show by contradiction that there is no parallel segment. To this end, suppose that a segment of the boundary $B(\mathcal{I})$ is parallel to a coordinate axis. On this line, consider two arbitrary points $\boldsymbol{p}^{(1)}$, $\boldsymbol{p}^{(2)}$ with $\boldsymbol{p}^{(1)} \succ \boldsymbol{p}^{(2)}$. We have $1 = \mathcal{I}(\boldsymbol{p}^{(1)}) = \mathcal{I}(\boldsymbol{p}^{(2)})$, i.e., $\mathcal{I}$ is not strictly monotone, which is a contradiction.

Conversely, assume that there is no parallel segment. Consider a boundary point $\hat{\boldsymbol{p}}$ with $\mathcal{I}(\hat{\boldsymbol{p}}) = 1$. An arbitrary $\boldsymbol{p} \succ \hat{\boldsymbol{p}}$ does not belong to $B(\mathcal{I})$. That is, $\mathcal{I}(\boldsymbol{p}) > 1 = \mathcal{I}(\hat{\boldsymbol{p}})$, thus $\mathcal{I}$ is strictly monotone. □

This result is illustrated in Fig. 3.3.

We will now study under which condition the infimum (1.22) can be attained. This question is closely connected with the achievability of the boundary of the SIR region $\mathcal{S}$, which was already discussed. From Theorem 3.5 and [2, Thm. 2.14], we know that there exists a $\tilde{\boldsymbol{p}} \in \mathcal{S}$ such that the balanced level $C(\boldsymbol{\gamma})$ is achieved by all users, i.e.,

$$C(\boldsymbol{\gamma})\tilde{p}_k = \gamma_k \mathcal{I}_k(\tilde{\boldsymbol{p}}), \quad \forall k \in \mathcal{K}\,, \tag{3.42}$$

if and only if there exists a $\mu > 0$ and a $\tilde{\boldsymbol{p}} > 0$ such that

$$\mu \cdot \tilde{p}_k = \gamma_k \cdot \max_{l \in \mathcal{K}} \frac{\tilde{p}_l}{\hat{p}_l^{(k)}}\,, \quad \forall k \in \mathcal{K}\,, \tag{3.43}$$

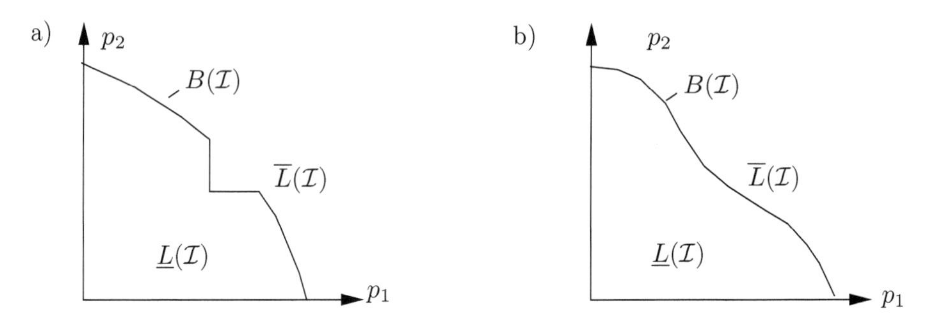

Fig. 3.3. Illustration of Theorem 3.18. Example a) leads to a non-strictly monotone interference function, whereas example b) is associated with a strictly monotone interference function, i.e., no segment of the boundary is parallel to the coordinate axes.

where $\hat{\boldsymbol{p}}^{(k)} = \arg\min_{\hat{\boldsymbol{p}} \in \underline{L}(\mathcal{I}_k)} \max_l \frac{\tilde{p}_l}{\hat{p}_l}$.

With Theorem 3.5 it is clear that (3.42) implies (3.43). Conversely, assume that (3.43) is fulfilled. By the uniqueness of the balanced optimum [2], $\mu = C(\boldsymbol{\gamma})$ can be concluded, so (3.42) is fulfilled.

For the special case of monotone interference functions, as studied in Section 2.3, we have the following result.

Theorem 3.19. *Let $\mathcal{I}_1, \ldots, \mathcal{I}_K$ be interference functions such that the boundaries of the the corresponding sets $\underline{L}(\mathcal{I}_k)$ do not contain segments parallel to the coordinate axes, and there is no self-interference, then for any $\boldsymbol{\gamma} > 0$ there exists a vector $\boldsymbol{p} > 0$ such that*

$$C(\boldsymbol{\gamma}) p_k = \gamma_k \mathcal{I}_k(\boldsymbol{p}), \quad k \in \mathcal{K}\,, \tag{3.44}$$

where $C(\boldsymbol{\gamma})$ is defined by (1.22).

Proof. This is a consequence of Theorem 3.18 and the result [2, Sec. 2.5]. □

One practical example for which the achievability of the boundary is important is the aforementioned problem of combined beamforming and power allocation. Some algorithms, like the ones proposed in [1, 33, 83], require that the chosen SINR target $\boldsymbol{\gamma}$ lies in the interior of the feasible SIR region $\mathcal{S}$. That is, $C(\boldsymbol{\gamma}) < 1$ must be fulfilled, otherwise the iteration diverges. This criterion can be checked by computing the min-max balancing problem (1.22). This requires the existence of a fixed point $\tilde{\boldsymbol{p}} > 0$ fulfilling (3.42).

3.3 Concave Interference Functions

In this section we analyze the structure of arbitrary *concave interference functions*, as defined in Subsection 1.3.1. Examples are the interference functions resulting from adaptive receive and transmit strategies (1.14), (1.15), and (1.17).

The results of Section 3.1 show that every interference function has fundamental max-min and min-max representations. Now, we consider interference functions that are concave in addition. In the following we will show that concavity leads to a particular structure. Namely, every concave interference function is represented by a minimum of linear interference functions, where the minimum is taken over an upward-comprehensive closed convex set $\mathcal{N}_0(\mathcal{I})$ depending on $\mathcal{I}$. That is, any concave $\mathcal{I}(\boldsymbol{p})$ has a representation

$$\mathcal{I}(\boldsymbol{p}) = \min_{\boldsymbol{w} \in \mathcal{N}_0(\mathcal{I})} \boldsymbol{w}^T \boldsymbol{p} , \quad \text{for all } \boldsymbol{p} > 0. \tag{3.45}$$

This stands in an interesting analogy with the adaptive receive strategies from Section 1.4.3. The coefficients $\boldsymbol{w}$ can be considered as "coupling coefficients" which model the cross-talk between the communication links, and the set $\mathcal{N}_0(\mathcal{I})$ depends on possible receive strategies. An example is the beamforming receiver, which leads to interference of the form (1.14).

This structure result, which will be derived and explained in the following, has some very interesting consequences. It shows that certain algorithms that were recently developed for the joint optimization of powers and receive (resp. transmit) strategies, are indeed applicable to arbitrary systems of *concave interference functions*. Algorithms based on the representation (3.45) will be studied in Chapters 5 and 6.

3.3.1 Representation of Concave Interference Functions

A useful concept for analyzing concave functions is the conjugate function (see e.g. [23, 84])

$$\underline{\mathcal{I}}^*(\boldsymbol{w}) = \inf_{\boldsymbol{p}>0} \Big(\sum_{l=1}^K w_l p_l - \mathcal{I}(\boldsymbol{p}) \Big) , \quad \boldsymbol{w} \in \mathbb{R}^K . \tag{3.46}$$

However, the function $\mathcal{I}$ is not just concave, we can also exploit that it fulfills the properties A1, A2, A3. This leads to the following observations.

Lemma 3.20. *For any given $\boldsymbol{w} \in \mathbb{R}^K$, the conjugate (3.46) is either minus infinity or zero, i.e.,*

$$\underline{\mathcal{I}}^*(\boldsymbol{w}) > -\infty \quad \Leftrightarrow \quad \underline{\mathcal{I}}^*(\boldsymbol{w}) = 0 . \tag{3.47}$$

Proof. The norm of $\boldsymbol{p}$ in (3.46) is not constrained, thus for all $\mu > 0$,

$$\begin{aligned} \underline{\mathcal{I}}^*(\boldsymbol{w}) &= \inf_{\boldsymbol{p}>0} \Big(\sum_{l=1}^K w_l \cdot \mu p_l - \mathcal{I}(\mu \boldsymbol{p}) \Big) , \\ &= \mu \cdot \inf_{\boldsymbol{p}>0} \Big(\sum_{l=1}^K w_l \cdot p_l - \mathcal{I}(\boldsymbol{p}) \Big) = \mu \cdot \underline{\mathcal{I}}^*(\boldsymbol{w}) . \end{aligned} \tag{3.48}$$

The second step follows from A2. Assume $\underline{\mathcal{I}}^*(\boldsymbol{w}) > -\infty$, then (3.48) can only hold for all $\mu > 0$ if $\underline{\mathcal{I}}^*(\boldsymbol{w}) = 0$. □

Lemma 3.21. *If $\boldsymbol{w}$ has a negative component then $\underline{\mathcal{I}}^*(\boldsymbol{w}) = -\infty$.*

Proof. Assume $w_r < 0$ for some arbitrary index r. Introducing a power vector $\boldsymbol{p}(\lambda)$ with $p_l(\lambda) = 1$, $l \neq r$ and $p_l(\lambda) = \lambda$, $l = r$, where $\lambda \in \mathbb{R}_{++}$, we have

$$\begin{aligned}\underline{\mathcal{I}}^*(\boldsymbol{w}) &\leq \lambda \cdot w_r + \sum_{l \neq r} w_l - \mathcal{I}(\boldsymbol{p}(\lambda)) ,\\ &\leq \lambda \cdot w_r + \sum_{l \neq r} w_l = -\lambda \cdot |w_r| + \sum_{l \neq r} w_l .\end{aligned}$$

The first inequality follows from $\underline{\mathcal{I}}^*(\boldsymbol{w})$ being the infimum over all power vectors. The second inequality follows from axiom A1. Letting $\lambda \to \infty$, the right-hand side of the inequality tends to $-\infty$. □

From Lemmas 3.20 and 3.21 it can be concluded that the set of vectors $\boldsymbol{w}$ leading to a finite conjugate $\underline{\mathcal{I}}^*(\boldsymbol{w}) > -\infty$ is

$$\mathcal{N}_0(\mathcal{I}) = \{\boldsymbol{w} \in \mathbb{R}^K_+ : \underline{\mathcal{I}}^*(\boldsymbol{w}) = 0\} . \tag{3.49}$$

Next, it is shown that every $\boldsymbol{w} \in \mathcal{N}_0(\mathcal{I})$ is associated with a hyperplane upper-bounding the interference function.

Lemma 3.22. *For any $\boldsymbol{w} \in \mathcal{N}_0(\mathcal{I})$, we have*

$$\mathcal{I}(\boldsymbol{p}) \leq \sum_{l \in \mathcal{K}} w_l p_l , \quad \forall \boldsymbol{p} > 0 . \tag{3.50}$$

Proof. With definition (3.49) we have

$$0 = \underline{\mathcal{I}}^*(\boldsymbol{w}) = \inf_{\hat{\boldsymbol{p}}>0} \Big(\sum_{l=1}^{K} w_l \cdot \hat{p}_l - \mathcal{I}(\hat{\boldsymbol{p}}) \Big) \leq \sum_{k \in \mathcal{K}} w_l \cdot p_l - \mathcal{I}(\boldsymbol{p})$$

for all $\boldsymbol{p} > 0$, thus (3.50) holds. □

This leads to our first main result, which shows that every concave interference function is characterized as a minimum over a sum of weighted powers.

Theorem 3.23. *Let $\mathcal{I}$ be an arbitrary concave interference function, then*

$$\mathcal{I}(\boldsymbol{p}) = \min_{\boldsymbol{w} \in \mathcal{N}_0(\mathcal{I})} \sum_{k \in \mathcal{K}} w_k p_k , \quad \textit{for all } \boldsymbol{p} > 0. \tag{3.51}$$

Proof. Consider an arbitrary fixed $\boldsymbol{p} > 0$. Since $\mathcal{I}(\boldsymbol{p})$ is concave, we know that (see e.g. [23, 84]), there exists a vector $\tilde{\boldsymbol{w}} \in \mathbb{R}^K$ such that

$$\tilde{\boldsymbol{w}}^T \hat{\boldsymbol{p}} - \mathcal{I}(\hat{\boldsymbol{p}}) \geq \tilde{\boldsymbol{w}}^T \boldsymbol{p} - \mathcal{I}(\boldsymbol{p}) \quad \text{for all } \hat{\boldsymbol{p}} > 0 \, . \tag{3.52}$$

The vector $\tilde{\boldsymbol{w}}$ must be non-negative, otherwise (3.52) cannot be fulfilled for all $\hat{\boldsymbol{p}} > 0$. This can be shown by contradiction. Suppose that $\tilde{w}_r < 0$ for some index r, and we choose $\hat{\boldsymbol{p}}_\epsilon$ such that $[\hat{\boldsymbol{p}}_\epsilon]_l = p_l$, $l \neq r$, and $[\hat{\boldsymbol{p}}_\epsilon]_r = p_r + \epsilon$, with $\epsilon > 0$. With A3 (monotonicity) we know that $\hat{\boldsymbol{p}}_\epsilon \geq \boldsymbol{p}$ implies $\mathcal{I}(\hat{\boldsymbol{p}}_\epsilon) \geq \mathcal{I}(\boldsymbol{p})$. Thus, (3.52) leads to $0 \leq \tilde{\boldsymbol{w}}^T(\hat{\boldsymbol{p}}_\epsilon - \boldsymbol{p}) = \epsilon \cdot \tilde{w}_r$. This contradicts the assumption $\tilde{w}_r < 0$.

It was shown in [2] that the function $\mathcal{I}(\boldsymbol{p})$ is continuous on $\mathbb{R}^K_{++}$, thus $\boldsymbol{p} < +\infty$ implies $\mathcal{I}(\boldsymbol{p}) < +\infty$. Therefore,

$$\tilde{\boldsymbol{w}}^T \boldsymbol{p} - \mathcal{I}(\boldsymbol{p}) > -\infty \, . \tag{3.53}$$

Inequality (3.52) holds for all $\hat{\boldsymbol{p}} > 0$. Taking the infimum and using (3.53), we have

$$\inf_{\hat{\boldsymbol{p}} > 0} \Big(\sum_{l \in \mathcal{K}} \tilde{w}_l \cdot \hat{p}_l - \mathcal{I}(\hat{\boldsymbol{p}}) \Big) \geq \tilde{\boldsymbol{w}}^T \boldsymbol{p} - \mathcal{I}(\boldsymbol{p}) > -\infty \, . \tag{3.54}$$

Comparison with (3.46) shows that $\underline{\mathcal{I}}^*(\tilde{\boldsymbol{w}}) > -\infty$ and therefore $\tilde{\boldsymbol{w}} \in \mathcal{N}_0(\mathcal{I})$. Lemma 3.22 implies

$$\mathcal{I}(\boldsymbol{p}) \leq \sum_{l \in \mathcal{K}} \tilde{w}_l p_l \quad \text{for all } \boldsymbol{p} > 0 \, . \tag{3.55}$$

Now, (3.52) holds for all $\hat{\boldsymbol{p}} > 0$, so it holds as well for $\lambda \hat{\boldsymbol{p}}$, with $\lambda > 0$. Because of property A2, we have $\mathcal{I}(\lambda \hat{\boldsymbol{p}}) = \lambda \mathcal{I}(\hat{\boldsymbol{p}})$, and thus

$$0 = \lim_{\lambda \to 0} \Big(\lambda \tilde{\boldsymbol{w}}^T \hat{\boldsymbol{p}} - \lambda \mathcal{I}(\hat{\boldsymbol{p}}) \Big) \geq \tilde{\boldsymbol{w}}^T \boldsymbol{p} - \mathcal{I}(\boldsymbol{p}) \, . \tag{3.56}$$

Thus, $\mathcal{I}(\boldsymbol{p}) \geq \tilde{\boldsymbol{w}}^T \boldsymbol{p}$. Comparison with (3.55) shows that this inequality can only be fulfilled with equality. It can be concluded that for any $\boldsymbol{p} > 0$, there exists a $\tilde{\boldsymbol{w}} \in \mathcal{N}_0(\mathcal{I})$ which minimizes $\boldsymbol{w}^T \boldsymbol{p}$, such that the lower bound $\mathcal{I}(\boldsymbol{p})$ is achieved. Hence, (3.51) holds. □

The proof shows that every $\tilde{\boldsymbol{w}}$ fulfilling (3.52) for a given point $\boldsymbol{p}$, is a minimizer of (3.51). Conversely, any $\tilde{\boldsymbol{w}} \in \mathcal{N}_0(\mathcal{I})$ which fulfills

$$\mathcal{I}(\boldsymbol{p}) = \min_{\boldsymbol{w} \in \mathcal{N}_0(\mathcal{I})} \sum_{l \in \mathcal{K}} w_l p_l = \sum_{l \in \mathcal{K}} \tilde{w}_l p_l \tag{3.57}$$

also fulfills the inequality (3.52). This is a consequence of Lemma 3.22, which leads to

$$\mathcal{I}(\hat{\boldsymbol{p}}) - \mathcal{I}(\boldsymbol{p}) = \mathcal{I}(\hat{\boldsymbol{p}}) - \sum_{l \in \mathcal{K}} \tilde{w}_l p_l \leq \sum_{l \in \mathcal{K}} \tilde{w}_l (\hat{p} - p_l) \quad \text{for all } \hat{\boldsymbol{p}} > 0 \, .$$

Thus, for any given $\boldsymbol{p} > 0$, the set of optimal coefficients $\tilde{\boldsymbol{w}}$ achieving the minimum (3.57), is identical to the set of $\tilde{\boldsymbol{w}} \in \mathcal{N}_0(\mathcal{I})$ for which (3.52) is fulfilled.

Theorem 3.23 opens up new perspectives for a more general understanding of interference functions. For example, $\mathcal{I}(\boldsymbol{p})$ in (3.51) can be the optimum of a weighted cost minimization problem from some strategey set $\mathcal{N}_0(\mathcal{I})$, with weighting factors p_k.

3.3.2 Properties of the Set $\mathcal{N}_0(\mathcal{I})$

Theorem 3.23 shows that an arbitrary concave interference function $\mathcal{I}$ can be characterized as the minimum of a weighted sum of powers, optimized over the set $\mathcal{N}_0(\mathcal{I})$. In this section we will further analyze the relationship between $\mathcal{I}$ and $\mathcal{N}_0(\mathcal{I})$. The results will be needed later, e.g., in Section 3.8.2 where convex approximations will be studied.

Lemma 3.24. *Let $\mathcal{I}$ be a concave interference function, then $\mathcal{N}_0(\mathcal{I}) \subseteq \mathbb{R}^K_+$, as defined by (3.49), is a non-empty* upward-comprehensive closed convex *set.*

Proof. From the proof of Theorem 3.23 it is clear that $\mathcal{N}_0(\mathcal{I})$ is non-empty. This is a consequence of the concavity of $\mathcal{I}$.

Now, we show convexity of $\mathcal{N}_0(\mathcal{I})$. Let $\hat{\boldsymbol{w}}, \check{\boldsymbol{w}} \in \mathcal{N}_0(\mathcal{I})$ and $\boldsymbol{w}(\lambda) = (1 - \lambda)\hat{\boldsymbol{w}} + \lambda\check{\boldsymbol{w}}$. Using $\mathcal{I}(\boldsymbol{p}) = (1-\lambda)\mathcal{I}(\boldsymbol{p}) + \lambda\mathcal{I}(\boldsymbol{p})$, we have

$$\begin{aligned} \underline{\mathcal{I}}^*\big(\boldsymbol{w}(\lambda)\big) &= \inf_{\boldsymbol{p}>0}\Big((1-\lambda)\sum_{l\in\mathcal{K}}\hat{w}_l p_l + \lambda\sum_{l\in\mathcal{K}}\check{w}_l p_l - \mathcal{I}(\boldsymbol{p})\Big) \\ &\geq (1-\lambda)\inf_{\boldsymbol{p}>0}\big(\sum_{l\in\mathcal{K}}\hat{w}_l p_l - \mathcal{I}(\boldsymbol{p})\big) + \qquad (3.58) \\ &\quad + \lambda\inf_{\boldsymbol{p}>0}\big(\sum_{l\in\mathcal{K}}\check{w}_l p_l - \mathcal{I}(\boldsymbol{p})\big) \\ &= (1-\lambda)\underline{\mathcal{I}}^*(\hat{\boldsymbol{w}}) + \lambda\underline{\mathcal{I}}^*(\check{\boldsymbol{w}}) > -\infty\,. \qquad (3.59) \end{aligned}$$

Thus, $\boldsymbol{w}(\lambda) \in \mathcal{N}_0(\mathcal{I})$, which proves convexity.

Now, we show that $\mathcal{N}_0(\mathcal{I})$ is closed. Let $\boldsymbol{w}^{(n)}$ be an arbitrary convergent Cauchy sequence in $\mathcal{N}_0(\mathcal{I})$, i.e., there exists a $\boldsymbol{w}^*$ such that $\lim_{n\to\infty} w_k^{(n)} = w_k^*$ for all components $k \in \mathcal{K}$. We need to show that the limit $\boldsymbol{w}^*$ is also contained in $\mathcal{N}_0(\mathcal{I})$.

Since $\boldsymbol{w}^{(n)} \in \mathbb{R}^K_+$, also $\boldsymbol{w}^* \in \mathbb{R}^K_+$. For an arbitrary fixed $\boldsymbol{p} > 0$, we have

$$\begin{aligned} \sum_{k\in\mathcal{K}} w_k^* p_k - \mathcal{I}(\boldsymbol{p}) &= \lim_{n\to\infty}\big(\sum_{k\in\mathcal{K}} w_k^{(n)} p_k - \mathcal{I}(\boldsymbol{p})\big) \\ &\geq \liminf_{n\to\infty}\Big(\inf_{\tilde{\boldsymbol{p}}>0}\big(\sum_{k\in\mathcal{K}} w_k^{(n)}\tilde{p}_k - \mathcal{I}(\tilde{\boldsymbol{p}})\big)\Big) \\ &= \liminf_{n\to\infty}\big(\underline{\mathcal{I}}^*(\boldsymbol{w}^{(n)})\big) = 0\,. \qquad (3.60) \end{aligned}$$

The last step follows from $\boldsymbol{w}^{(n)} \in \mathcal{N}_0(\mathcal{I})$, which implies $\underline{\mathcal{I}}^*(\boldsymbol{w}^{(n)}) = 0$ for all n. Since inequality (3.60) holds for all $\boldsymbol{p} > 0$, we have

$$\underline{\mathcal{I}}^*(\boldsymbol{w}^*) = \inf_{\boldsymbol{p}>0} \Big(\sum_{l \in \mathcal{K}} w_l^* p_l - \mathcal{I}(\boldsymbol{p}) \Big) \geq 0 > -\infty \,. \tag{3.61}$$

Thus, $\boldsymbol{w}^* \in \mathcal{N}_0(\mathcal{I})$, which proves that $\mathcal{N}_0(\mathcal{I})$ is closed.

It remains to show upward-comprehensiveness. Consider an arbitrary $\hat{\boldsymbol{w}} \in \mathcal{N}_0(\mathcal{I})$. If $\boldsymbol{w} \geq \hat{\boldsymbol{w}}$ then

$$\sum_{l \in \mathcal{K}} p_l w_l - \mathcal{I}(\boldsymbol{p}) \geq \sum_{l \in \mathcal{K}} p_l \hat{w}_l - \mathcal{I}(\boldsymbol{p}) \geq \underline{\mathcal{I}}^*(\hat{\boldsymbol{w}}) > -\infty$$

for all $\boldsymbol{p} > 0$. Thus, $\boldsymbol{w} \in \mathcal{N}_0(\mathcal{I})$. □

Remark 3.25. The proof of Lemma 3.24 does not rely on concavity, except for the comment on non-emptiness. Thus, $\mathcal{N}_0(\mathcal{I})$ is a upward-comprehensive closed convex set for any interference function fulfilling A1, A2, A3.

Thus far, we have analyzed the elementary building blocks of concave interference functions. Lemma 3.24 shows that any concave interference function $\mathcal{I}$ is associated with a upward-comprehensive closed convex coefficient set $\mathcal{N}_0(\mathcal{I})$, as illustrated in Fig. 3.4.

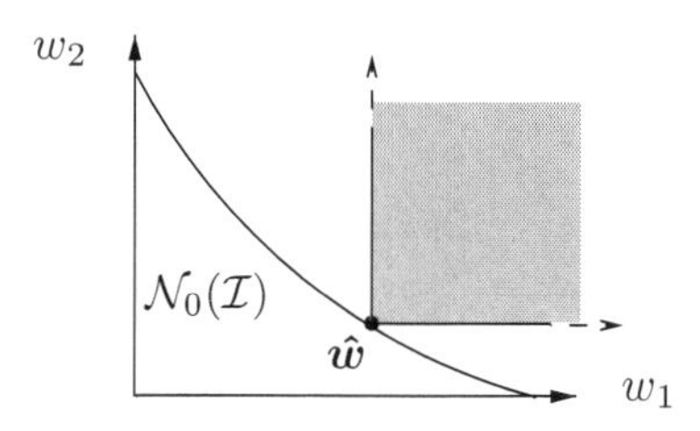

Fig. 3.4. Illustration of Lemma 3.24: the coefficient set $\mathcal{N}_0(\mathcal{I})$ is upward-comprehensive closed convex. For any $\hat{\boldsymbol{w}} \in \mathcal{N}_0(\mathcal{I})$, all points $\boldsymbol{w} \geq \hat{\boldsymbol{w}}$ (the shaded box) are also contained in $\mathcal{N}_0(\mathcal{I})$.

3.3.3 Synthesis of Concave Interference Functions

Representation (3.51) shows a fundamental structure of interference functions. Such a search for elementary building blocks is sometimes referred to as *analysis*.

Next, we study the converse approach, namely the *synthesis* of a concave interference function. Starting from an arbitrary non-empty upward-comprehensive closed convex set $\mathcal{V} \subseteq \mathbb{R}^K_+$, we can construct a function

$$\mathcal{I}_{\mathcal{V}}(\boldsymbol{p}) = \min_{\boldsymbol{w} \in \mathcal{V}} \sum_{l \in \mathcal{K}} w_l p_l \,. \tag{3.62}$$

It is easily verified that $\mathcal{I}_{\mathcal{V}}$ is concave and fulfills the properties A1, A2, A3. Thus, every upward-comprehensive closed convex set $\mathcal{V}$ is associated with a concave interference function $\mathcal{I}_{\mathcal{V}}$.

The next theorem shows that the operations *analysis* and *synthesis* are reversible. From $\mathcal{V}$ we obtain an interference function $\mathcal{I}_{\mathcal{V}}$, then $\mathcal{N}_0(\mathcal{I}_{\mathcal{V}})$ yields back the original set $\mathcal{V}$.

Theorem 3.26. *For any non-empty* upward-comprehensive closed convex *set* $\mathcal{V} \subseteq \mathbb{R}^K_+$ *we have*

$$\mathcal{V} = \mathcal{N}_0(\mathcal{I}_{\mathcal{V}}) \,. \tag{3.63}$$

Proof. Consider an arbitrary $\boldsymbol{v} \in \mathcal{V}$. With (3.62), we have

$$\begin{aligned} \underline{\mathcal{I}}^*(\boldsymbol{v}) &= \inf_{\boldsymbol{p}>0} \Big(\sum_{l \in \mathcal{K}} v_l p_l - \mathcal{I}_{\mathcal{V}}(\boldsymbol{p}) \Big) \\ &\geq \inf_{\boldsymbol{p}>0} \Big(\sum_{l \in \mathcal{K}} v_l p_l - \sum_{l \in \mathcal{K}} v_l p_l \Big) = 0 \,. \end{aligned} \tag{3.64}$$

Thus, $\boldsymbol{v} \in \mathcal{N}_0(\mathcal{I}_{\mathcal{V}})$, and consequently $\mathcal{V} \subseteq \mathcal{N}_0(\mathcal{I}_{\mathcal{V}})$. Next, equality is shown by contradiction. Suppose $\mathcal{V} \neq \mathcal{N}_0(\mathcal{I}_{\mathcal{V}})$. This implies the existence of a $\hat{\boldsymbol{w}} > 0$ with $\hat{\boldsymbol{w}} \notin \mathcal{V}$ and $\hat{\boldsymbol{w}} \in \mathcal{N}_0(\mathcal{I}_{\mathcal{V}})$. Note, that $\hat{\boldsymbol{w}}$ can be assumed to be strictly positive since $\mathbb{R}^K_{++} \cap \mathcal{V} \neq \mathbb{R}^K_{++} \cap \mathcal{N}_0(\mathcal{I}_{\mathcal{V}})$, otherwise we would have the contradiction

$$\mathcal{V} = \overline{\mathbb{R}^K_{++} \cap \mathcal{V}} = \overline{\mathbb{R}^K_{++} \cap \mathcal{N}_0(\mathcal{I}_{\mathcal{V}})} = \mathcal{N}_0(\mathcal{I}_{\mathcal{V}}) \,.$$

Next, we can exploit that the set $\mathcal{V}$ is convex and its intersection with $\mathbb{R}^K_{++}$ is non-empty (this follows from comprehensiveness). From the *separating hyperplanes theorem* (see e.g. [23] or [84, Thm. 4.1.1, p. 51]), we know that there is a $\hat{\boldsymbol{p}} > 0$ such that

$$\begin{aligned} \mathcal{I}_{\mathcal{V}}(\hat{\boldsymbol{p}}) = \min_{\boldsymbol{v} \in \mathcal{V}} \sum_{l \in \mathcal{K}} v_l \hat{p}_l &> \sum_{l \in \mathcal{K}} \hat{w}_l \hat{p}_l \\ &\geq \min_{\boldsymbol{w} \in \mathcal{N}_0(\mathcal{I}_{\mathcal{V}})} \sum_{l \in \mathcal{K}} w_l \hat{p}_l = \mathcal{I}_{\mathcal{V}}(\hat{\boldsymbol{p}}) \,, \end{aligned} \tag{3.65}$$

where the last equality follows from Theorem 3.23. This is a contradiction, thus $\mathcal{V} = \mathcal{N}_0(\mathcal{I}_{\mathcal{V}})$. □

The next corollary shows that different sets $\mathcal{V}^{(1)}$ and $\mathcal{V}^{(2)}$ always lead to different interference functions $\mathcal{I}_{\mathcal{V}^{(1)}}(\boldsymbol{p})$ and $\mathcal{I}_{\mathcal{V}^{(2)}}(\boldsymbol{p})$, respectively.

Corollary 3.27. *Let* $\mathcal{V}^{(1)}$ *and* $\mathcal{V}^{(2)}$ *be two arbitrary* upward-comprehensive closed convex *sets from* $\mathbb{R}^K_+$. *If* $\mathcal{I}_{\mathcal{V}^{(1)}}(\boldsymbol{p}) = \mathcal{I}_{\mathcal{V}^{(2)}}(\boldsymbol{p})$ *for all* $\boldsymbol{p} > 0$, *then* $\mathcal{V}^{(1)} = \mathcal{V}^{(2)}$.

Proof. The assumption implies $\mathcal{N}_0(\mathcal{I}_{\mathcal{V}^{(1)}}) = \mathcal{N}_0(\mathcal{I}_{\mathcal{V}^{(2)}})$. The result follows with Theorem 3.26, which shows $\mathcal{V} = \mathcal{N}_0(\mathcal{I}_{\mathcal{V}})$. □

These results show a one-to-one correspondence between concave interference functions and upward-comprehensive closed convex sets. Every concave interference function $\mathcal{I}$ is uniquely associated with an upward-comprehensive closed convex set $\mathcal{N}_0(\mathcal{I})$. Conversely, every upward-comprehensive closed convex set $\mathcal{V}$ is uniquely associated with an interference function $\mathcal{I}_{\mathcal{V}}$. We have $\mathcal{I} = \mathcal{I}_{\mathcal{N}_0(\mathcal{I})}$ and $\mathcal{V} = \mathcal{N}_0(\mathcal{I}_{\mathcal{V}})$.

The representation (3.51) has an interesting interpretation in the context of network resource allocation. Suppose that w_k stands for some QoS measure, like bit error rate, or delay. The variables p_l are weighting fectors that account for individual user priorities. Then, $\mathcal{I}(\boldsymbol{p})$ can be interpreted as the minimum network cost obtained by optimizing over the boundary of the convex cost region $\mathcal{N}_0(\mathcal{I})$, as illustrated in Fig. 3.5. This shows a connection between the axiomatic framework of interference functions and resource allocation problems.

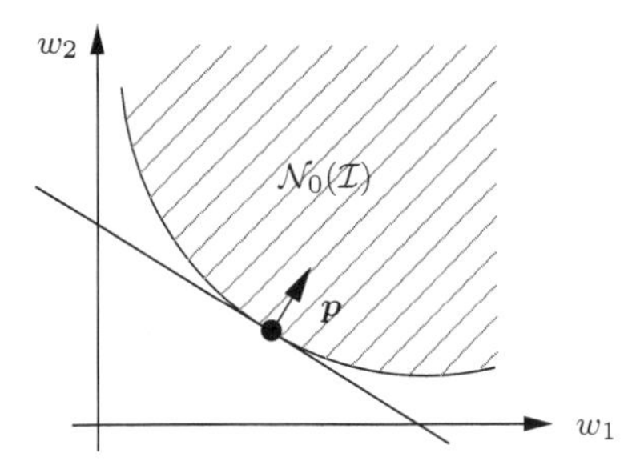

Fig. 3.5. The concave interference function $\mathcal{I}(\boldsymbol{p})$ can be interpreted as the minimum of a weighted sum-cost function optimized over the convex set $\mathcal{N}_0$. The "weighting vector" $\boldsymbol{p}$ controls the tradeoff between the utilities w_k.

3.3.4 Least Concave Majorant and Alternative Representation

Theorem 3.23 shows that any concave interference function can be expressed as the minimum over linear functions, where the optimization is over the upward-comprehensive closed convex set $\mathcal{N}_0(\mathcal{I})$.

In this subsection we will explore an alternative approach, based on the function

$$\begin{aligned} \overline{g}_{\mathcal{I}}(\boldsymbol{v}) &= \sup_{\boldsymbol{q}>0} \left(\frac{\mathcal{I}(\boldsymbol{q})}{\sum_{l\in\mathcal{K}} v_l q_l} \right), \quad \boldsymbol{v} \gneq \boldsymbol{0} \\ &= \sup_{\substack{\boldsymbol{q}>0 \\ \|\boldsymbol{q}\|_1=1}} \left(\frac{\mathcal{I}(\boldsymbol{q})}{\sum_{l\in\mathcal{K}} v_l q_l} \right). \end{aligned} \tag{3.66}$$

With $\overline{g}_{\mathcal{I}}(\boldsymbol{v})$ we can rewrite the majorant (3.13) as

$$\begin{aligned}\overline{\mathcal{I}}(\boldsymbol{p}) &= \inf_{\boldsymbol{v}>0} \sup_{\boldsymbol{q}>0} \left(\frac{\mathcal{I}(\boldsymbol{q})}{\sum_{l\in\mathcal{K}} v_l q_l} \right) \sum_{k\in\mathcal{K}} v_k p_k \\ &= \inf_{\boldsymbol{v}>0} \overline{g}_{\mathcal{I}}(\boldsymbol{v}) \sum_{k\in\mathcal{K}} v_k p_k \,. \end{aligned} \tag{3.67}$$

The point-wise infimum of linear functions is concave, thus $\overline{\mathcal{I}}(\boldsymbol{p})$ is a concave interference function. In the remainder of this section we will discuss properties of $\overline{g}_{\mathcal{I}}(\boldsymbol{v})$ and the majorant $\overline{\mathcal{I}}(\boldsymbol{p})$. In particular, it will be shown that $\overline{\mathcal{I}}(\boldsymbol{p})$ is a best-possible concave majorant for any interference function, and it provides an equivalent representation for any concave interference function.

We begin by showing that the supremum (3.66) is always attained.

Lemma 3.28. *For any $\boldsymbol{v} > 0$, there is a $\hat{\boldsymbol{q}} := \hat{\boldsymbol{q}}(\boldsymbol{v}) \geq 0$, with $\|\hat{\boldsymbol{q}}\|_1 = 1$, such that*

$$\overline{g}_{\mathcal{I}}(\boldsymbol{v}) = \frac{\mathcal{I}(\hat{\boldsymbol{q}})}{\sum_{l\in\mathcal{K}} v_l \hat{q}_l} = \max_{\substack{\boldsymbol{q}\geq 0 \\ \|\boldsymbol{q}\|_1=1}} \left(\frac{\mathcal{I}(\boldsymbol{q})}{\sum_{l\in\mathcal{K}} v_l q_l} \right) . \tag{3.68}$$

Proof. Since $\boldsymbol{v} > 0$, we we have $\sum_{l\in\mathcal{K}} v_l q_l > 0$ for all $\boldsymbol{q} \geq 0$. Thus we can take the supremum (3.67) over the compact domain $\{\boldsymbol{q} \geq 0 : \|\boldsymbol{q}\|_1 = 1\}$. We have $\sum_{l\in\mathcal{K}} v_l q_l > 0$. The inverse of a continous positive function is continuous. Also, it was shown in Section 2.5 that $\mathcal{I}(\boldsymbol{p})$ is continuous on $\mathbb{R}^K_{++}$. Theorem 2.17 shows that $\mathcal{I}(\boldsymbol{p})$ has a unique continuation on the boundary, thus continuity extends to $\mathbb{R}^K_+$. Any continuous real-valued function attains a maximum over a compact set, thus (3.68) holds. □

The following theorem and corollary show that the concave majorant $\overline{\mathcal{I}}(\boldsymbol{p})$ is best possible, and any concave interference function has a saddle-point characterization. We use $G_{\mathcal{I}}(\boldsymbol{q}, \boldsymbol{p}, \boldsymbol{v})$, as defined by (3.8).

Theorem 3.29. *$\mathcal{I}$ is a concave interference function if and only if $\overline{\mathcal{I}}(\boldsymbol{p}) = \mathcal{I}(\boldsymbol{p})$ for all $\boldsymbol{p} > 0$. The following identity holds.*

$$\mathcal{I}(\boldsymbol{p}) = \sup_{\boldsymbol{q}>0} \inf_{\boldsymbol{v}>0} G_{\mathcal{I}}(\boldsymbol{q}, \boldsymbol{p}, \boldsymbol{v}) = \inf_{\boldsymbol{v}>0} \sup_{\boldsymbol{q}>0} G_{\mathcal{I}}(\boldsymbol{q}, \boldsymbol{p}, \boldsymbol{v}) \,. \tag{3.69}$$

Proof. With (3.13) we have $\mathcal{I}(\boldsymbol{p}) \leq \overline{\mathcal{I}}(\boldsymbol{p})$ for all $\boldsymbol{p} > 0$. Assume that $\mathcal{I}(\boldsymbol{p})$ is concave, and recall Lemma A.12 from Appendix A.7. From Theorem 3.23 we know that there exists a $\underline{\mathcal{V}}$ such that

$$
\begin{aligned}
\overline{\mathcal{I}}(\boldsymbol{p}) &= \inf_{\boldsymbol{v}>0}\left(\sup_{\boldsymbol{q}>0}\left(\frac{\min_{\underline{\boldsymbol{v}}\in\underline{\mathcal{V}}}\sum_{l\in\mathcal{K}}\underline{v}_l q_l}{\sum_{l\in\mathcal{K}} v_l q_l}\right)\sum_{k\in\mathcal{K}} v_k p_k\right) \\
&\le \inf_{\boldsymbol{v}>0}\left(\min_{\underline{\boldsymbol{v}}\in\underline{\mathcal{V}}}\left(\sup_{\boldsymbol{q}>0}\frac{\sum_{l\in\mathcal{K}}\underline{v}_l q_l}{\sum_{l\in\mathcal{K}} v_l q_l}\right)\sum_{k\in\mathcal{K}} v_k p_k\right) \\
&= \inf_{\boldsymbol{v}>0}\left(\min_{\underline{\boldsymbol{v}}\in\underline{\mathcal{V}}}\left(\max_{l\in\mathcal{K}}\frac{\underline{v}_l}{v_l}\right)\sum_{k\in\mathcal{K}} v_k p_k\right) \\
&= \inf_{\underline{\boldsymbol{v}}\in\underline{\mathcal{V}}}\left(\min_{\boldsymbol{v}>0}\left(\max_{l\in\mathcal{K}}\frac{\underline{v}_l}{v_l}\right)\sum_{k\in\mathcal{K}} v_k p_k\right) \\
&\le \inf_{\underline{\boldsymbol{v}}\in\underline{\mathcal{V}}}\left(\left(\max_{l\in\mathcal{K}}\frac{\underline{v}_l}{\underline{v}_l}\right)\sum_{k\in\mathcal{K}} \underline{v}_k p_k\right) \\
&= \inf_{\underline{\boldsymbol{v}}\in\underline{\mathcal{V}}}\left(\sum_{k\in\mathcal{K}} \underline{v}_k p_k\right) = \mathcal{I}(\boldsymbol{p}) \,.
\end{aligned}
$$

Thus, $\mathcal{I}(\boldsymbol{p}) = \overline{\mathcal{I}}(\boldsymbol{p})$ for all $\boldsymbol{p} > 0$.

Conversely, assume that $\mathcal{I}(\boldsymbol{p}) = \overline{\mathcal{I}}(\boldsymbol{p})$ holds. Because $\overline{\mathcal{I}}(\boldsymbol{p})$ is concave, also $\mathcal{I}(\boldsymbol{p})$ is concave. □

Recall the definition of the majorant from Subsection 3.1.2. Among all concave majorants, the *least concave majorant* provides the minimum upper limit of the original interference function $\mathcal{I}$. It can be regarded as the "best" concave approximation of $\mathcal{I}$.

Corollary 3.30. *Let $\mathcal{I}$ be an arbitrary interference function, then $\overline{\mathcal{I}}(\boldsymbol{p})$ is the least concave majorant of $\mathcal{I}(\boldsymbol{p})$.*

Proof. Let $\mathcal{I}'$ be the least concave majorant of $\mathcal{I}$. Then for all $\boldsymbol{p} > 0$, we have $\overline{\mathcal{I}}(\boldsymbol{p}) \ge \mathcal{I}'(\boldsymbol{p}) \ge \mathcal{I}(\boldsymbol{p})$. With $\overline{g}_{\mathcal{I}}(\boldsymbol{v})$ defined by (3.66), we have

$$
\overline{g}_{\overline{\mathcal{I}}}(\boldsymbol{v}) \ge \overline{g}_{\mathcal{I}'}(\boldsymbol{v}) \ge \overline{g}_{\mathcal{I}}(\boldsymbol{v}) \quad \text{for all } \boldsymbol{v} > 0 \,. \tag{3.70}
$$

Consequently,

$$
\begin{aligned}
\overline{\mathcal{I}}(\boldsymbol{p}) &= \inf_{\substack{\boldsymbol{v}>0\\ \|\boldsymbol{v}\|_1=1}} \overline{g}_{\mathcal{I}}(\boldsymbol{v})\sum_{k\in\mathcal{K}} v_k p_k \\
&\le \inf_{\substack{\boldsymbol{v}>0\\ \|\boldsymbol{v}\|_1=1}} \overline{g}_{\mathcal{I}'}(\boldsymbol{v})\sum_{k\in\mathcal{K}} v_k p_k = \mathcal{I}'(\boldsymbol{p}) \,.
\end{aligned} \tag{3.71}
$$

Thus, $\overline{\mathcal{I}}(\boldsymbol{p}) = \mathcal{I}'(\boldsymbol{p})$ for all $\boldsymbol{p} > 0$. □

Next, consider the inverse function

$$
\mathcal{I}_2(\boldsymbol{v}) = \frac{1}{\overline{g}_{\mathcal{I}}(\boldsymbol{v})} = \inf_{\substack{\boldsymbol{q}>0\\ \|\boldsymbol{q}\|_1=1}} \frac{\sum_{l\in\mathcal{K}} v_l q_l}{\mathcal{I}(\boldsymbol{q})} \,. \tag{3.72}
$$

We show that $\mathcal{I}_2(\boldsymbol{v})$ is a concave interference function. The function $\mathcal{I}_2(\boldsymbol{v})$ is always defined because $\mathcal{I}$ is strictly positive (axiom A1). For arbitrary $\boldsymbol{v} > 0$ and $\lambda > 0$ we have

$$\overline{g}_{\mathcal{I}}(\lambda\boldsymbol{v}) = \frac{1}{\lambda} \cdot \overline{g}_{\mathcal{I}}(\boldsymbol{v}) \,. \tag{3.73}$$

If $\boldsymbol{v}^{(1)} \geq \boldsymbol{v}^{(2)}$, then $\overline{g}_{\mathcal{I}}(\boldsymbol{v}^{(1)}) \leq \overline{g}_{\mathcal{I}}(\boldsymbol{v}^{(2)})$, thus $\mathcal{I}_2(\boldsymbol{v})$ fulfills axioms A2, A3. Next, positivity (A1) is shown. With Lemma 3.28, the following identity is fulfilled for any $\boldsymbol{v} > 0$, with $\|\boldsymbol{v}\|_1 = 1$.

$$\overline{g}_{\mathcal{I}}(\boldsymbol{v}) = \max_{\substack{\boldsymbol{q}\geq 0 \\ \|\boldsymbol{q}\|_1=1}} \left(\frac{\mathcal{I}(\boldsymbol{q})}{\sum_{l\in\mathcal{K}} v_l q_l} \right) > 0 \,.$$

This inequality holds because $\mathcal{I}(\boldsymbol{q})$ is strictly positive by definition (axiom A1). Thus, $\mathcal{I}_2(\boldsymbol{v}) > 0$ for every $\boldsymbol{v} > 0$. With Theorem 2.16 (continuation) we can extend $\mathcal{I}_2$ to non-negative vectors $\boldsymbol{v} \geq 0$, with $\|\boldsymbol{v}\|_1 = 1$. Then, $\overline{g}_{\mathcal{I}}(\boldsymbol{v}) = 1/\mathcal{I}_2(\boldsymbol{v})$ is continuous on $\{\boldsymbol{v} \geq 0 : \|\boldsymbol{v}\|_1 = 1\}$. It can be observed from (3.72) that $\mathcal{I}_2(\boldsymbol{v})$ is concave as the pointwise infimum of linear functions. Hence, $\mathcal{I}_2(\boldsymbol{v})$ is a concave interference function. This enables us to prove the following result.

Theorem 3.31. *Let $\mathcal{I}$ be an arbitrary interference function, then $\overline{g}_{\mathcal{I}}(\boldsymbol{v})$ is continuous on $\mathbb{R}^K_+$, and there exists a non-empty upward-comprehensive closed convex set $\overline{\mathcal{W}} \subset \mathbb{R}^K_+$ such that*

$$\overline{g}_{\mathcal{I}}(\boldsymbol{v}) = \frac{1}{\min_{\boldsymbol{w}\in\overline{\mathcal{W}}} \sum_{k\in\mathcal{K}} w_k v_k} = \max_{\boldsymbol{w}\in\overline{\mathcal{W}}} \frac{1}{\sum_{k\in\mathcal{K}} w_k v_k} \,. \tag{3.74}$$

Proof. Since $\mathcal{I}_2(\boldsymbol{v})$ is a concave interference function, we know from Theorem 3.23 that (3.74) holds for any $\boldsymbol{v} \in \mathbb{R}^K_{++}$. The denominator in (3.74) is strictly positive, thus $g_{\mathcal{I}}$ is continuous as the pointwise minimum of continuous functions. From Theorem 2.17 we know that continuity extends to $\mathbb{R}^K_+$. □

With the continuity shown by Theorem 3.31 and (3.67) we know that the majorant $\overline{\mathcal{I}}$ can be rewritten as

$$\overline{\mathcal{I}}(\boldsymbol{p}) = \min_{\substack{\boldsymbol{v}\geq 0 \\ \|\boldsymbol{v}\|_1=1}} \overline{g}_{\mathcal{I}}(\boldsymbol{v}) \cdot \sum_{k\in\mathcal{K}} v_k p_k \,, \quad \boldsymbol{p} > 0 \,. \tag{3.75}$$

With (3.75) and Theorem 3.29 we obtain the following result.

Theorem 3.32. *$\mathcal{I}$ is a concave interference function if and only if*

$$\mathcal{I}(\boldsymbol{p}) = \min_{\substack{\boldsymbol{v}\geq 0 \\ \|\boldsymbol{v}\|_1=1}} \overline{g}_{\mathcal{I}}(\boldsymbol{v}) \cdot \sum_{k\in\mathcal{K}} v_k p_k \quad \textit{for all } \boldsymbol{p} > 0 \,. \tag{3.76}$$

Comparing (3.76) with (3.51) in Theorem 3.23, we observe two different ways of expressing a concave interference function as the minimum of linear functions. In (3.51), the coefficient set $\mathcal{N}_0(\mathcal{I})$ is used to incorporate the properties of $\mathcal{I}$, while (3.76) uses the function $\overline{g}_{\mathcal{I}}$. An alternative way of obtaining the least concave majorant will be discussed in Section 3.8.

3.4 Convex Interference Functions

In this section we analyze the structure of *convex* interference functions, as defined in Section 1.3.1. The results are similar to the concave case studied in Section 3.3.

A main result of this section is Theorem 3.35, which states that every convex interference function can be expressed as a maximum of linear functions. An example is the "worst-case" interference discussed in Subsection 1.4.7.

3.4.1 Representation of Convex Interference Functions

We begin by introducing the convex conjugate function [23].

$$\bar{\mathcal{I}}^*(\boldsymbol{w}) = \sup_{\boldsymbol{p}>0}\Big(\sum_{l\in\mathcal{K}} w_l p_l - \mathcal{I}(\boldsymbol{p})\Big) . \tag{3.77}$$

Exploiting the special properties A1, A2, A3, we obtain the following result.

Lemma 3.33. *The conjugate function (3.77) is either infinity or zero, i.e.,*

$$\bar{\mathcal{I}}^*(\boldsymbol{w}) < +\infty \quad \Leftrightarrow \quad \bar{\mathcal{I}}^*(\boldsymbol{w}) = 0 . \tag{3.78}$$

Proof. This is shown in a similar way to the proof of Lemma 3.20. □

Due to the monotonicity axiom A3, the coefficients $\boldsymbol{w}$ of interest are non-negative. This will become clear later, from the proof of Theorem 3.35. Therefore, the coefficient set of interest is

$$\mathcal{W}_0(\mathcal{I}) = \{\boldsymbol{w} \in \mathbb{R}^K_+ : \bar{\mathcal{I}}^*(\boldsymbol{w}) = 0\} . \tag{3.79}$$

Every $\boldsymbol{w} \in \mathcal{W}_0(\mathcal{I})$ is associated with a hyperplane which lower bounds the interference function.

Lemma 3.34. *For any $\boldsymbol{w} \in \mathcal{W}_0(\mathcal{I})$,*

$$\sum_{l\in\mathcal{K}} w_l p_l \le \mathcal{I}(\boldsymbol{p}) , \quad \forall \boldsymbol{p} > 0 . \tag{3.80}$$

Proof. For all $\boldsymbol{p} > 0$, we have

$$0 = \bar{\mathcal{I}}^*(\boldsymbol{w}) = \sup_{\hat{\boldsymbol{p}}>0}\Big(\sum_{l\in\mathcal{K}} w_l \cdot \hat{p}_l - \mathcal{I}(\hat{\boldsymbol{p}})\Big) \ge \sum_{l\in\mathcal{K}} w_l \cdot p_l - \mathcal{I}(\boldsymbol{p}) .$$

Thus, (3.80) holds. □

Based on this lemma, we will now show that every convex interference function can always be characterized as a maximum sum of weighted powers.

Theorem 3.35. *Let $\mathcal{I}$ be an arbitrary convex interference function, then*

$$\mathcal{I}(\boldsymbol{p}) = \max_{\boldsymbol{w} \in \mathcal{W}_0(\mathcal{I})} \sum_{k \in \mathcal{K}} w_k \cdot p_k \,, \quad \text{for all } \boldsymbol{p} > 0. \tag{3.81}$$

Proof. Consider an arbitrary fixed $\boldsymbol{p} > 0$. Since $\mathcal{I}(\boldsymbol{p})$ is convex, there exists a vector $\tilde{\boldsymbol{w}} \in \mathbb{R}^K$ such that [84, Thm. 1.2.1, p. 77]

$$\tilde{\boldsymbol{w}}^T \hat{\boldsymbol{p}} - \mathcal{I}(\hat{\boldsymbol{p}}) \leq \tilde{\boldsymbol{w}}^T \boldsymbol{p} - \mathcal{I}(\boldsymbol{p}) \quad \text{for all } \hat{\boldsymbol{p}} > 0 \,. \tag{3.82}$$

The vector $\tilde{\boldsymbol{w}}$ must be non-negative, otherwise (3.82) cannot be fulfilled for all $\hat{\boldsymbol{p}} > 0$. This can be shown by contradiction. Suppose that $\tilde{w}_r < 0$ for some index r, and we choose $\hat{\boldsymbol{p}}_\epsilon > 0$ such that $[\hat{\boldsymbol{p}}_\epsilon]_l = p_l$, $l \neq r$, and $[\hat{\boldsymbol{p}}_\epsilon]_r = p_r - \epsilon$, with $0 < \epsilon < p_r$. With A3 (monotonicity) we know that $\hat{\boldsymbol{p}}_\epsilon \leq \boldsymbol{p}$ implies $\mathcal{I}(\hat{\boldsymbol{p}}_\epsilon) \leq \mathcal{I}(\boldsymbol{p})$. Thus, (3.82) leads to $0 \geq \tilde{\boldsymbol{w}}^T(\hat{\boldsymbol{p}}_\epsilon - \boldsymbol{p}) = -\epsilon \cdot \tilde{w}_r$. This contradicts the assumption $\tilde{w}_r < 0$.

Because of the non-negativity of $\mathcal{I}(\boldsymbol{p})$, we have

$$\tilde{\boldsymbol{w}}^T \boldsymbol{p} - \mathcal{I}(\boldsymbol{p}) < +\infty \,. \tag{3.83}$$

Inequality (3.82) holds for all $\hat{\boldsymbol{p}} > 0$. Taking the supremum and using (3.83), we have

$$\sup_{\hat{\boldsymbol{p}} > 0} \Big(\sum_{l \in \mathcal{K}} \tilde{w}_l \cdot \hat{p}_l - \mathcal{I}(\hat{\boldsymbol{p}}) \Big) \leq \tilde{\boldsymbol{w}}^T \boldsymbol{p} - \mathcal{I}(\boldsymbol{p}) < +\infty \,. \tag{3.84}$$

Comparison with the conjugate (3.77) shows that $\bar{\mathcal{I}}^*(\tilde{\boldsymbol{w}}) < +\infty$ and therefore $\tilde{\boldsymbol{w}} \in \mathcal{W}_0(\mathcal{I})$. Lemma 3.34 implies

$$\tilde{\boldsymbol{w}}^T \boldsymbol{p} \leq \mathcal{I}(\boldsymbol{p}) \,, \quad \forall \boldsymbol{p} > 0 \,. \tag{3.85}$$

Inequality (3.82) holds for all $\hat{\boldsymbol{p}}$, so it holds as well for $\lambda \hat{\boldsymbol{p}}$, with an arbitrary $\lambda > 0$. With A2, we have

$$\tilde{\boldsymbol{w}}^T \boldsymbol{p} - \mathcal{I}(\boldsymbol{p}) \geq \lim_{\lambda \to 0} \Big(\tilde{\boldsymbol{w}}^T \lambda \hat{\boldsymbol{p}} - \lambda \mathcal{I}(\hat{\boldsymbol{p}}) \Big) = 0 \,. \tag{3.86}$$

By combining (3.85) and (3.86), it can be concluded that $\mathcal{I}(\boldsymbol{p}) = \tilde{\boldsymbol{w}}^T \boldsymbol{p}$. Thus, $\tilde{\boldsymbol{w}}$ is the maximizer of (3.81). □

From the proof of Theorem 3.35 it becomes clear that the maximizer of (3.81) is always non-negative. Also, the set $\mathcal{W}_0(\mathcal{I})$ is non-empty.

Example 3.36. In order to illustrate Theorem 3.35, consider the simple interference function $\mathcal{I}(\boldsymbol{p}) = \max_{k \in \mathcal{K}} p_k$, which can be written as

$$\mathcal{I}(\boldsymbol{p}) = \max_{k \in \mathcal{K}} p_k = \max_{\boldsymbol{w} \in \mathbb{R}_+^K : \|\boldsymbol{w}\|_1 = 1} \boldsymbol{w}^T \boldsymbol{p} \,.$$

In this case, $\mathcal{W}_0(\mathcal{I}) = \{\boldsymbol{w} \in \mathbb{R}_+^K : \|\boldsymbol{w}\|_1 = 1\}$.

Theorem 3.35 also provides a direct means for characterizing interference coupling. Recall the definition of the asymptotic coupling matrix $\boldsymbol{A}_{\mathcal{I}}$ introduced in Subsection 2.2.1. Since we are dealing with convex interference functions, the asymptotic coupling matrix is equivalent to the dependency matrix $\boldsymbol{D}_{\mathcal{I}}$, introduced in Subsection 2.2.2.

From Theorem 3.35 we know that every convex interference function can be expressed as

$$\mathcal{I}(\boldsymbol{p}) = \max_{\boldsymbol{w} \in \mathcal{W}(\mathcal{I})} \boldsymbol{w}^T \boldsymbol{p} \,. \tag{3.87}$$

For every choice of $\boldsymbol{p}$ there exists a coupling matrix matrix $\boldsymbol{W} = [\boldsymbol{w}_1, \dots, \boldsymbol{w}_K]^T$, with $\boldsymbol{w}_k \in \mathcal{W}(\mathcal{I}_k)$, such that $\mathcal{I}_k(\boldsymbol{p}) = \boldsymbol{w}_k^T \boldsymbol{p}$ for all k. In order for two interference functions $\mathcal{I}_k$ and $\mathcal{I}_l$ to be coupled, it suffices that there exists one $\boldsymbol{W}$ such that $[\boldsymbol{W}]_{kl} > 0$. This already implies $[\boldsymbol{A}_{\mathcal{I}}]_{kl} > 0$.

3.4.2 Properties of the set $\mathcal{W}_0(\mathcal{I})$

Consider an arbitrary convex interference function $\mathcal{I}$. The resulting coefficient set $\mathcal{W}_0(\mathcal{I})$ has a particular structure. Recall the definition of a downward-comprehensive set introduced in Subsection 2.6.2.

Lemma 3.37. *Let $\mathcal{I}$ be a convex interference function, then the set $\mathcal{W}_0(\mathcal{I})$, as defined by (3.79), is non-empty, bounded, and* downward-comprehensive closed convex.

Proof. From the proof of Theorem 3.23 it is clear that $\mathcal{N}_0(\mathcal{I})$ is non-empty. This is a consequence of the concavity of $\mathcal{I}$.

First, convexity is shown. Let $\hat{\boldsymbol{w}}, \check{\boldsymbol{w}} \in \mathcal{W}_0(\mathcal{I})$ and $\boldsymbol{w}(\lambda) = (1-\lambda)\hat{\boldsymbol{w}} + \lambda\check{\boldsymbol{w}}$. Similar to (3.59) we can show

$$\bar{\mathcal{I}}^*\big(\boldsymbol{w}(\lambda)\big) \le (1-\lambda)\bar{\mathcal{I}}^*(\hat{\boldsymbol{w}}) + \lambda\bar{\mathcal{I}}^*(\check{\boldsymbol{w}}) < +\infty \,. \tag{3.88}$$

Thus, $\boldsymbol{w}(\lambda) \in \mathcal{W}_0(\mathcal{I})$.

Next, we show that the set is upper-bounded. Consider an arbitrary $\boldsymbol{w} \in \mathcal{W}_0(\mathcal{I})$. With (3.81), we have

$$\sum_{l \in \mathcal{K}} w_l \le \max_{\boldsymbol{w} \in \mathcal{W}_0(\mathcal{I})} \sum_{k \in \mathcal{K}} w_k = \mathcal{I}(\boldsymbol{1}) \,. \tag{3.89}$$

Here, $\boldsymbol{1} = [1, \dots, 1]^T$ is the all-ones vector. The function $\mathcal{I}$ is continuous monotone, thus $\mathcal{I}(\boldsymbol{1}) < +\infty$ and $\mathcal{W}_0(\mathcal{I})$ is bounded.

Now, we show that $\mathcal{W}_0(\mathcal{I})$ is closed. Let $\boldsymbol{w}^{(n)}$ be an arbitrary convergent Cauchy sequence in $\mathcal{W}_0(\mathcal{I})$, i.e., there exists a $\boldsymbol{w}^*$ such that $\lim_{n\to\infty} w_k^{(n)} = w_k^*$ for all components $k \in \mathcal{K}$. We need to show that the limit $\boldsymbol{w}^*$ is also contained in $\mathcal{W}_0(\mathcal{I})$.

Since $\boldsymbol{w}^{(n)} \in \mathbb{R}_+^K$, also $\boldsymbol{w}^* \in \mathbb{R}_+^K$. For an arbitrary fixed $\boldsymbol{p} > 0$, we have

$$\begin{aligned}\sum_{k\in\mathcal{K}} w_k^* p_k - \mathcal{I}(\boldsymbol{p}) &= \lim_{n\to\infty}\big(\sum_{k\in\mathcal{K}} w_k^{(n)} p_k - \mathcal{I}(\boldsymbol{p})\big)\\ &\le \limsup_{n\to\infty}\Big(\sup_{\tilde{\boldsymbol{p}}>0}\big(\sum_{k\in\mathcal{K}} w_k^{(n)}\tilde{p}_k - \mathcal{I}(\tilde{\boldsymbol{p}})\big)\Big)\\ &= \limsup_{n\to\infty}\big(\bar{\mathcal{I}}^*(\boldsymbol{w}^{(n)})\big) = 0\,. \end{aligned} \tag{3.90}$$

The last step follows from $\boldsymbol{w}^{(n)} \in \mathcal{W}_0(\mathcal{I})$, which implies $\bar{\mathcal{I}}^*(\boldsymbol{w}^{(n)}) = 0$. Since inequality (3.90) holds for all $\boldsymbol{p} > 0$, we have

$$\bar{\mathcal{I}}^*(\boldsymbol{w}^*) = \sup_{\boldsymbol{p}>0}\Big(\sum_{l\in\mathcal{K}} w_l^* p_l - \mathcal{I}(\boldsymbol{p})\Big) \le 0 < +\infty\,. \tag{3.91}$$

Thus, $\boldsymbol{w}^* \in \mathcal{W}_0(\mathcal{I})$, which proves that $\mathcal{W}_0(\mathcal{I})$ is closed.

In order to show downward-comprehensiveness, consider an arbitrary $\hat{\boldsymbol{w}} \in \mathcal{W}_0(\mathcal{I})$. For any $\boldsymbol{w} \in \mathbb{R}_+^K$ with $\boldsymbol{w} \le \hat{\boldsymbol{w}}$, we have

$$\sum_{l\in\mathcal{K}} p_l w_l - \mathcal{I}(\boldsymbol{p}) \le \sum_{l\in\mathcal{K}} p_l \hat{w}_l - \mathcal{I}(\boldsymbol{p}) \le \bar{\mathcal{I}}^*(\hat{\boldsymbol{w}}) < +\infty$$

for all $\boldsymbol{p} > 0$, thus $\boldsymbol{w} \in \mathcal{W}_0(\mathcal{I})$. □

The proof of Lemma 3.37 does not rely on convexity, except for showing non-emptiness and boundedness. Thus, $\mathcal{W}_0(\mathcal{I})$ is a downward-comprehensive closed convex set for any non-trivial interference function fulfilling A1, A2, A3. The result is illustrated in Fig. 3.6.

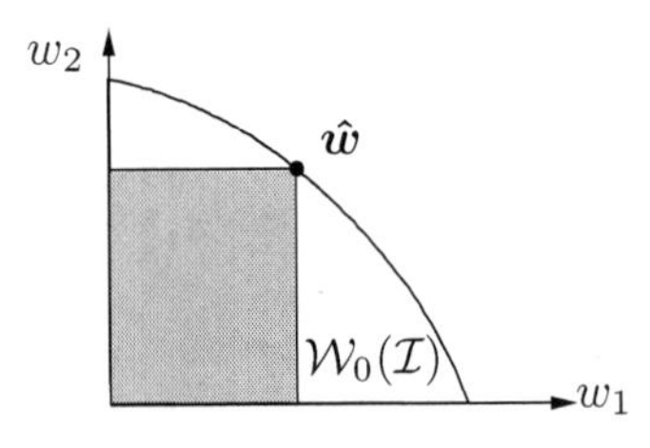

Fig. 3.6. Illustration of Lemma 3.37: the coefficient set $\mathcal{W}_0(\mathcal{I})$ is downward-comprehensive closed convex. For any $\hat{\boldsymbol{w}} \in \mathcal{W}_0(\mathcal{I})$, all points $\boldsymbol{w} \le \hat{\boldsymbol{w}}$ (shaded box) are also contained in $\mathcal{W}_0(\mathcal{I})$.

3.4.3 Synthesis of Convex Interference Functions

Next, consider the converse approach, i.e., the *synthesis* of a convex interference function from a bounded downward-comprehensive closed convex set $\mathcal{V}$.

The maximum of linear functions is convex, thus

$$\mathcal{I}_{\mathcal{V}}(\boldsymbol{p}) = \max_{\boldsymbol{w}\in\mathcal{V}} \sum_{l\in\mathcal{K}} p_l w_l \tag{3.92}$$

is a convex interference function which fulfills A1, A2, A3.

Similar to the results of Section 3.3, the operations *analysis* and *synthesis* are shown to be reversible:

Theorem 3.38. *For any non-empty* downward-comprehensive closed convex *set $\mathcal{V} \subseteq \mathbb{R}^K_+$ we have*

$$\mathcal{V} = \mathcal{W}_0(\mathcal{I}_{\mathcal{V}}) . \tag{3.93}$$

Proof. Consider an arbitrary $\boldsymbol{v} \in \mathcal{V}$. Lemma 3.34 implies

$$\begin{aligned} \bar{\mathcal{I}}^*(\boldsymbol{v}) &= \sup_{\boldsymbol{p}>0}\Big(\sum_{l\in\mathcal{K}} v_l p_l - \mathcal{I}_{\mathcal{V}}(\boldsymbol{p})\Big) \\ &\le \sup_{\boldsymbol{p}>0}\Big(\sum_{l\in\mathcal{K}} v_l p_l - \sum_{l\in\mathcal{K}} v_l p_l\Big) = 0 . \end{aligned} \tag{3.94}$$

With Lemma 3.33 we have $\boldsymbol{v} \in \mathcal{W}_0(\mathcal{I}_{\mathcal{V}})$, and consequently $\mathcal{V} \subseteq \mathcal{W}_0(\mathcal{I}_{\mathcal{V}})$. Similar to the proof of Theorem 3.26, we can show by contradiction that this can only be fulfilled with equality. Suppose that $\mathcal{V} \neq \mathcal{W}_0(\mathcal{I}_{\mathcal{V}})$, then this implies the existence of a $\hat{\boldsymbol{w}} \in \mathcal{W}_0(\mathcal{I}_{\mathcal{V}})$ with $\hat{\boldsymbol{w}} \notin \mathcal{V}$ and $\hat{\boldsymbol{w}} > 0$. Applying the theorem of separating hyperplanes, we know that there is a $\hat{\boldsymbol{p}} > 0$ such that

$$\begin{aligned} \mathcal{I}_{\mathcal{V}}(\hat{\boldsymbol{p}}) = \max_{\boldsymbol{w}\in\mathcal{V}} \sum_{l\in\mathcal{K}} w_l \hat{p}_l &< \sum_{l\in\mathcal{K}} \hat{w}_l \hat{p}_l \\ &= \max_{\boldsymbol{w}\in\mathcal{W}_0(\mathcal{I}_{\mathcal{V}})} \sum_{l\in\mathcal{K}} w_l \hat{p}_l = \mathcal{I}_{\mathcal{V}}(\hat{\boldsymbol{p}}) \end{aligned} \tag{3.95}$$

where the last equality follows from (3.35). This is a contradiction, thus $\mathcal{V} = \mathcal{W}_0(\mathcal{I}_{\mathcal{V}})$. □

The next corollary shows that there is a direct correspondence between any convex interference function $\mathcal{I}$ and the respective downward-comprehensive closed convex set $\mathcal{W}_0(\mathcal{I})$.

Corollary 3.39. *Let $\mathcal{W}_1$ and $\mathcal{W}_2$, be two arbitrary* downward-comprehensive closed convex *sets from $\mathbb{R}^K_+$. If $\mathcal{I}_{\mathcal{W}_1}(\boldsymbol{p}) = \mathcal{I}_{\mathcal{W}_2}(\boldsymbol{p})$ for all $\boldsymbol{p} > 0$, then $\mathcal{W}_1 = \mathcal{W}_2$.*

Proof. The proof follows from Theorem 3.38. □

The results show that any convex interference function $\mathcal{I}(\boldsymbol{p})$ can be interpreted as the maximum of the linear function $\sum_l p_l w_l$ over a bounded downward-comprehensive closed convex set $\mathcal{W}_0(\mathcal{I})$. This can be interpreted as the maximum of a weighted sum utility, where $\mathcal{W}_0(\mathcal{I})$ is the utility set and $\boldsymbol{p}$ is a vector of weighting factors that account for individual user priorities, as illustrated in Figure 3.7.

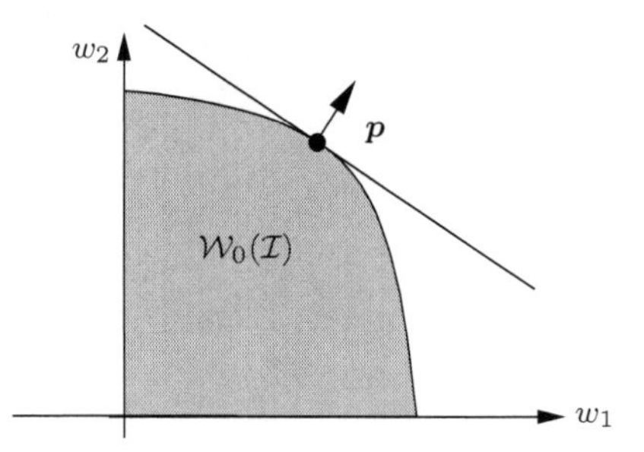

Fig. 3.7. Every convex interference function $\mathcal{I}(\boldsymbol{p})$ can be interpreted as the maximum of a weighted sum-utility function optimized over the convex set $\mathcal{W}_0(\mathcal{I})$. The "weighting vector" $\boldsymbol{p}$ controls the tradeoff between the utilities w_k.

3.4.4 Greatest Convex Minorant and Alternative Representation

Theorem 3.35 shows that every convex interference function $\mathcal{I}$ can be expressed as the maximum of linear functions, where the optimization is over the downward-comprehensive compact convex set $\mathcal{W}_0(\mathcal{I})$. The properties of $\mathcal{I}$ depend on the structure of $\mathcal{W}_0(\mathcal{I})$.

In this subsection we will discuss an alternative approach. Again, $\mathcal{I}$ is expressed as the maximum over linear functions. However, instead of optimizing over a constrained set, the properties of $\mathcal{I}$ are captured by a function $\underline{g}_{\mathcal{I}}$ defined as follows.

$$\begin{aligned}\underline{g}_{\mathcal{I}}(\boldsymbol{v}) &= \inf_{\boldsymbol{q}>0}\left(\frac{\mathcal{I}(\boldsymbol{q})}{\sum_{l\in\mathcal{K}} v_l q_l}\right), \quad \boldsymbol{v} \gneq \mathbf{0} \\ &= \inf_{\substack{\boldsymbol{q}>0 \\ \|\boldsymbol{q}\|_1=1}}\left(\frac{\mathcal{I}(\boldsymbol{q})}{\sum_{l\in\mathcal{K}} v_l q_l}\right). \end{aligned} \tag{3.96}$$

With (3.96), we can rewrite the minorant (3.12) as follows.

$$\begin{aligned}\underline{\mathcal{I}}(\boldsymbol{p}) &= \sup_{\boldsymbol{v}>0}\left(\inf_{\boldsymbol{q}>0}\left(\frac{\mathcal{I}(\boldsymbol{q})}{\sum_{l\in\mathcal{K}} v_l q_l}\right)\sum_{k\in\mathcal{K}} v_k p_k\right) \\ &= \sup_{\boldsymbol{v}>0}\left(\underline{g}_{\mathcal{I}}(\boldsymbol{v})\sum_{k\in\mathcal{K}} v_k p_k\right). \end{aligned} \tag{3.97}$$

The pointwise supremum of linear functions is convex, thus $\underline{\mathcal{I}}(\boldsymbol{p})$ is a convex interference function. In the remainder of this section we will discuss properties of $\underline{g}_{\mathcal{I}}(\boldsymbol{v})$ and the majorant $\underline{\mathcal{I}}(\boldsymbol{p})$. In particular, it will be shown that $\underline{\mathcal{I}}(\boldsymbol{p})$ is a best-possible convex minorant for any interference function, and it provides an equivalent representation for any convex interference function.

We begin by showing that the infimum (3.96) is attained.

Lemma 3.40. *For any $\boldsymbol{v} > 0$, there is a $\hat{\boldsymbol{q}} = \hat{\boldsymbol{q}}(\boldsymbol{v}) \geq 0$, with $\|\hat{\boldsymbol{q}}\|_1 = 1$, such that*

$$\underline{g}_{\mathcal{I}}(\boldsymbol{v}) = \frac{\mathcal{I}(\hat{\boldsymbol{q}})}{\sum_{l\in\mathcal{K}} v_l \hat{q}_l} = \min_{\substack{\boldsymbol{q}\geq 0\\ \|\boldsymbol{q}\|_1=1}} \left(\frac{\mathcal{I}(\boldsymbol{q})}{\sum_{l\in\mathcal{K}} v_l q_l} \right) . \tag{3.98}$$

Proof. Since $\boldsymbol{v} > 0$, we we have $\sum_{l\in\mathcal{K}} v_l q_l > 0$ for all $\boldsymbol{q} \geq 0$. Thus, we can take the infimum (3.96) over the compact domain $\{\boldsymbol{q} \geq 0 : \|\boldsymbol{q}\|_1 = 1\}$. We have $\sum_{l\in\mathcal{K}} v_l q_l > 0$. The inverse of a continous positive function is continuous. Also, $\mathcal{I}(\boldsymbol{p})$ is continuous on $\mathbb{R}^K_{++}$ [2]. Theorem 2.17 shows that $\mathcal{I}(\boldsymbol{p})$ has a unique continuation on the boundary, thus continuity extends to $\mathbb{R}^K_{+}$. Any continuous real-valued function attains a minimum over a compact set, thus (3.98) holds. □

The next theorem and the following corollary show that the convex minorant $\underline{\mathcal{I}}(\boldsymbol{p})$ is best possible, and any convex interference function has a saddle-point characterization. We use $G_{\mathcal{I}}(\boldsymbol{q}, \boldsymbol{p}, \boldsymbol{v})$, as defined by (3.8).

Theorem 3.41. *$\mathcal{I}$ is a convex interference function if and only if $\underline{\mathcal{I}}(\mathbf{p}) = \mathcal{I}(\mathbf{p})$ for all $\mathbf{p} > 0$, i.e.,*

$$\mathcal{I}(\boldsymbol{p}) = \inf_{\boldsymbol{q}>0} \sup_{\boldsymbol{v}>0} G_{\mathcal{I}}(\boldsymbol{q}, \boldsymbol{p}, \boldsymbol{v}) = \sup_{\boldsymbol{v}>0} \inf_{\boldsymbol{q}>0} G_{\mathcal{I}}(\boldsymbol{q}, \boldsymbol{p}, \boldsymbol{v}) . \tag{3.99}$$

Proof. With (3.12) we have $\mathcal{I}(\boldsymbol{p}) \geq \underline{\mathcal{I}}(\boldsymbol{p})$ for all $\boldsymbol{p} > 0$. Assume that $\mathcal{I}(\boldsymbol{p})$ is convex. From Lemma 3.35 we know that there exists a $\overline{\mathcal{V}}$ such that

$$\begin{aligned}
\underline{\mathcal{I}}(\boldsymbol{p}) &= \sup_{\boldsymbol{v}>0} \left(\inf_{\boldsymbol{q}>0} \left(\frac{\max_{\overline{\boldsymbol{v}}\in\overline{\mathcal{V}}} \sum_{l\in\mathcal{K}} \overline{v}_l q_l}{\sum_{l\in\mathcal{K}} v_l q_l} \right) \sum_{k\in\mathcal{K}} v_k p_k \right) \\
&\geq \sup_{\boldsymbol{v}>0} \left(\max_{\overline{\boldsymbol{v}}\in\overline{\mathcal{V}}} \left(\inf_{\boldsymbol{q}>0} \frac{\sum_{l\in\mathcal{K}} \overline{v}_l q_l}{\sum_{l\in\mathcal{K}} v_l q_l} \right) \sum_{k\in\mathcal{K}} v_k p_k \right) \\
&= \sup_{\boldsymbol{v}>0} \left(\max_{\overline{\boldsymbol{v}}\in\overline{\mathcal{V}}} \left(\min_{l\in\mathcal{K}} \frac{\overline{v}_l}{v_l} \right) \sum_{k\in\mathcal{K}} v_k p_k \right) \\
&= \sup_{\overline{\boldsymbol{v}}\in\overline{\mathcal{V}}} \left(\max_{\boldsymbol{v}>0} \left(\min_{l\in\mathcal{K}} \frac{\overline{v}_l}{v_l} \right) \sum_{k\in\mathcal{K}} v_k p_k \right) \\
&\geq \sup_{\overline{\boldsymbol{v}}\in\overline{\mathcal{V}}} \left(\left(\min_{l\in\mathcal{K}} \frac{\overline{v}_l}{\overline{v}_l} \right) \sum_{k\in\mathcal{K}} \overline{v}_k p_k \right) \\
&= \sup_{\overline{\boldsymbol{v}}\in\overline{\mathcal{V}}} \left(\sum_{k\in\mathcal{K}} \overline{v}_k p_k \right) = \mathcal{I}(\boldsymbol{p}) .
\end{aligned}$$

Thus, $\mathcal{I}(\boldsymbol{p}) = \underline{\mathcal{I}}(\boldsymbol{p})$ for all $\boldsymbol{p} > 0$.

Conversely, assume that $\mathcal{I}(\boldsymbol{p}) = \underline{\mathcal{I}}(\boldsymbol{p})$ holds. Because $\underline{\mathcal{I}}(\boldsymbol{p})$ is convex, also $\mathcal{I}(\boldsymbol{p})$ is convex. □

Recall the definition of the minorant from Subsection 3.1.2. Among all convex minorants, the *greatest convex minorant* provides the maximum lower limit of the original interference function $\mathcal{I}$. It can be regarded as the “best” convex approximation of $\mathcal{I}$.

Corollary 3.42. *Let $\mathcal{I}$ be an arbitrary interference function, then $\underline{\mathcal{I}}(\boldsymbol{p})$ is the greatest convex minorant of $\mathcal{I}(\boldsymbol{p})$.*

Proof. Let $\mathcal{I}'$ be the greatest convex minorant of $\mathcal{I}$. For all $\boldsymbol{p} > 0$, we have $\underline{\mathcal{I}}(\boldsymbol{p}) \leq \mathcal{I}'(\boldsymbol{p}) \leq \mathcal{I}(\boldsymbol{p})$, and thus

$$g_{\underline{\mathcal{I}}}(\boldsymbol{v}) \leq g_{\mathcal{I}'}(\boldsymbol{v}) \leq g_{\mathcal{I}}(\boldsymbol{v}) \quad \text{for all } \boldsymbol{v} > 0 \,. \tag{3.100}$$

Consequently,

$$\begin{aligned} \underline{\mathcal{I}}(\boldsymbol{p}) &= \sup_{\substack{\boldsymbol{v}>0 \\ \|\boldsymbol{v}\|_1=1}} g_{\mathcal{I}}(\boldsymbol{v}) \sum_{k\in\mathcal{K}} v_k p_k \\ &\geq \sup_{\substack{\boldsymbol{v}>0 \\ \|\boldsymbol{v}\|_1=1}} g_{\mathcal{I}'}(\boldsymbol{v}) \sum_{k\in\mathcal{K}} v_k p_k = \mathcal{I}'(\boldsymbol{p}) \,. \end{aligned} \tag{3.101}$$

Thus, $\underline{\mathcal{I}}(\boldsymbol{p}) = \mathcal{I}'(\boldsymbol{p})$ for all $\boldsymbol{p} > 0$. □

Next, consider the inverse function

$$\mathcal{I}_1(\boldsymbol{v}) = \frac{1}{g_{\mathcal{I}}(\boldsymbol{v})} = \sup_{\substack{\boldsymbol{q}>0 \\ \|\boldsymbol{q}\|_1=1}} \frac{\sum_{l\in\mathcal{K}} v_l q_l}{\mathcal{I}(\boldsymbol{q})} \,. \tag{3.102}$$

We show that $\mathcal{I}_1(\boldsymbol{v})$ is a convex interference function. The function $\mathcal{I}_1(\boldsymbol{v})$ is always defined because $\mathcal{I}$ is strictly positive (axiom A1). For arbitrary $\boldsymbol{v} > 0$ and $\lambda > 0$ we have

$$g_{\mathcal{I}}(\lambda \boldsymbol{v}) = \frac{1}{\lambda} \cdot g_{\mathcal{I}}(\boldsymbol{v}) \,. \tag{3.103}$$

If $\boldsymbol{v}^{(1)} \geq \boldsymbol{v}^{(2)}$, then $g_{\mathcal{I}}(\boldsymbol{v}^{(1)}) \leq g_{\mathcal{I}}(\boldsymbol{v}^{(2)})$, thus $\mathcal{I}_1(\boldsymbol{v})$ fulfills axioms A2, A3. Next, positivity (A1) is shown. With Lemma 3.40 the following identity is fulfiled for any $\boldsymbol{v} > 0$, with $\|\boldsymbol{v}\|_1 = 1$.

$$g_{\mathcal{I}}(\boldsymbol{v}) = \min_{\substack{\boldsymbol{q}\geq 0 \\ \|\boldsymbol{q}\|_1=1}} \left(\frac{\mathcal{I}(\boldsymbol{q})}{\sum_{l\in\mathcal{K}} v_l q_l} \right) \leq \frac{\frac{1}{K}\mathcal{I}(\boldsymbol{1})}{\frac{1}{K}\sum_{l\in\mathcal{K}} v_l} = \mathcal{I}(\boldsymbol{1}) \,.$$

That is, $g_{\mathcal{I}}(\boldsymbol{v})$ is bounded from above by some constant $\mathcal{I}(\boldsymbol{1})$, and we have

$$\mathcal{I}_1(\boldsymbol{v}) = \frac{1}{g_{\mathcal{I}}(\boldsymbol{v})} \geq \frac{1}{\mathcal{I}(\boldsymbol{1})} > 0 \,.$$

Thus, $\mathcal{I}_1$ is lower bounded for any $\boldsymbol{v}$ and A1 is fulfilled. With Theorem 2.16 (continuation) we can extend $\mathcal{I}_2$ to non-negative vectors $\boldsymbol{v} \geq 0$, with $\|\boldsymbol{v}\|_1 = 1$. Then, $g_{\mathcal{I}}(\boldsymbol{v}) = 1/\mathcal{I}_1(\boldsymbol{v})$ is continuous on $\{\boldsymbol{v} \geq 0 : \|\boldsymbol{v}\|_1 = 1\}$. It can be observed from (3.102) that $\mathcal{I}_1(\boldsymbol{v})$ is convex as the pointwise supremum of linear functions. Hence, $\mathcal{I}_1(\boldsymbol{v})$ is a convex interference function.

Theorem 3.43. *Let $\mathcal{I}$ be an arbitrary interference function, then $\underline{g}_{\mathcal{I}}(\boldsymbol{v})$ is continuous on $\mathbb{R}^K_+$, and there exists a non-empty bounded downward-comprehensive closed convex set $\underline{\mathcal{W}} \subset \mathbb{R}^K_+$ such that*

$$\underline{g}_{\mathcal{I}}(\boldsymbol{v}) = \frac{1}{\max_{\boldsymbol{w} \in \underline{\mathcal{W}}} \sum_{k \in \mathcal{K}} w_k v_k} = \min_{\boldsymbol{w} \in \underline{\mathcal{W}}} \frac{1}{\sum_{k \in \mathcal{K}} w_k v_k} . \tag{3.104}$$

Proof. Since $\mathcal{I}_1(\boldsymbol{v})$ is a convex interference function, we know from Lemma 3.35 that (3.104) holds for any $\boldsymbol{v} \in \mathbb{R}^K_{++}$. The denominator in (3.104) is strictly positive, so $\underline{g}_{\mathcal{I}}$ is continuous as the pointwise minimum of continuous functions. From Theorem 2.17 we know that continuity extends to $\mathbb{R}^K_+$. □

With the continuity shown by Theorem 3.43 and property (3.97), we know that the minorant $\underline{\mathcal{I}}$ can be rewritten as

$$\underline{\mathcal{I}}(\boldsymbol{p}) = \max_{\substack{\boldsymbol{v} \geq 0 \\ \|\boldsymbol{v}\|_1 = 1}} \underline{g}_{\mathcal{I}}(\boldsymbol{v}) \cdot \sum_{k \in \mathcal{K}} v_k p_k , \quad \boldsymbol{p} > 0 . \tag{3.105}$$

That is, the supremum can be replaced by a maximum over a compact set, and $\underline{g}_{\mathcal{I}}$ is defined as in (3.96). With Theorem 3.41 and (3.105), the following result is shown.

Theorem 3.44. *$\mathcal{I}$ is a convex interference function if and only if*

$$\mathcal{I}(\boldsymbol{p}) = \max_{\substack{\boldsymbol{v} \geq 0 \\ \|\boldsymbol{v}\|_1 = 1}} \underline{g}_{\mathcal{I}}(\boldsymbol{v}) \cdot \sum_{k \in \mathcal{K}} v_k p_k \quad \text{for all } \boldsymbol{p} > 0 . \tag{3.106}$$

Comparing (3.106) with (3.81) in Theorem 3.35, we observe two different ways of expressing a convex interference function as the maximum of linear functions. In (3.81), the coefficient set $\mathcal{W}_0(\mathcal{I})$ is used to incorporate the properties of $\mathcal{I}$, while (3.106) uses the function $\underline{g}_{\mathcal{I}}$.

An alternative way of obtaining the greatest convex minorant will be discussed in Section 3.8.

3.5 Expressing Utility Sets as Sub-/Superlevel Sets of Convex/Concave Interference Functions

In the previous sections we have discussed the relationship between convex comprehensive sets and convex or concave interference functions. In this section we discuss an alternative approach. It will be shown that any convex compact downward-comprehensive set from $\mathbb{R}^K_{++}$ can be expressed as a sublevel set of a convex interference function, and any closed upward-comprehensive convex set can be expressed as a superlevel set of a concave interference function. Later, in Section 4.1.6, it will be shown how this result can be applied to cooperative game theory.

Consider a convex interference function $\mathcal{I}(\boldsymbol{p})$ generated from a non-empty convex compact downward-comprehensive set $\mathcal{V} \subset \mathbb{R}^K_{++}$, $\mathcal{V} \neq \mathbb{R}^K_{++}$ as follows.

$$\mathcal{I}(\boldsymbol{p}) = \max_{\boldsymbol{w} \in \mathcal{V}} \sum_{k \in \mathcal{K}} w_k \cdot p_k \ , \quad \text{for all } \boldsymbol{p} > 0. \tag{3.107}$$

Recall the following definition from Subsection 3.1.3.

$$\underline{L}(\mathcal{I}) = \{\boldsymbol{p} > 0 : \mathcal{I}(\boldsymbol{p}) \leq 1\} \ .$$

From Corollary 3.8, we know that $\underline{L}(\mathcal{I})$ is closed and downward-comprehensive. Exploiting the convexity of $\mathcal{I}$, it can be shown that $\underline{L}(\mathcal{I})$ is upper-bounded. The set $\underline{L}(\mathcal{I})$ is also convex, since it is a sublevel set of a convex function.

However, $\underline{L}(\mathcal{I}) \neq \mathcal{V}$ in general. That is, $\mathcal{V}$ is *not* always a sublevel set of the convex interference function $\mathcal{I}$. While Theorem 3.11 shows that every closed downward-comprehensive set can be expressed as a sublevel set of an interference function, there is no corresponding result for convex sets and functions of the form (3.107).

In order to express $\mathcal{V}$ as a sublevel set of a convex interference function, we need to introduce another interference function

$$\mathcal{I}_1(\boldsymbol{p}) = \max_{\boldsymbol{v} \in \underline{L}(\mathcal{I})} \sum_{k \in \mathcal{K}} v_k p_k \ . \tag{3.108}$$

Unlike $\mathcal{I}$, the new function $\mathcal{I}_1$ is constructed from the level set $\underline{L}(\mathcal{I})$, thus it depends on the original set $\mathcal{V}$ only indirectly. The maximum (3.108) is guaranteed to exist since $\underline{L}(\mathcal{I})$ is a compact set (relatively in $\mathbb{R}^K_{++}$). The function $\mathcal{I}_1$ is also a convex interference function. The next theorem shows that the sublevel set $\underline{L}(\mathcal{I}_1)$ equals the original set $\mathcal{V}$.

Theorem 3.45. *Consider an arbitrary non-empty compact downward-comprehensive convex set $\mathcal{V} \subset \mathbb{R}^K_{++}$, $\mathcal{V} \neq \mathbb{R}^K_{++}$, from which we synthesize a convex interference function $\mathcal{I}$, as defined by (3.87). Let $\mathcal{I}_1$ be defined by (3.108), then*

$$\mathcal{V} = \underline{L}(\mathcal{I}_1) \ . \tag{3.109}$$

Proof. Let $\boldsymbol{v} \in \mathcal{V}$, then it can be observed from (3.87) that $\sum_k v_k p_k \leq 1$ for all $\boldsymbol{p} \in \underline{L}(\mathcal{I})$. Thus,

$$1 \geq \max_{\boldsymbol{p} \in \underline{L}(\mathcal{I})} \sum_{k \in \mathcal{K}} v_k p_k = \mathcal{I}_1(\boldsymbol{v}) \ .$$

That is, $\boldsymbol{v} \in \mathcal{V}$ is also contained in the sublevel set of $\mathcal{I}_1$, i.e., $\boldsymbol{v} \in \underline{L}(\mathcal{I}_1)$, thus implying $\mathcal{V} \subseteq \underline{L}(\mathcal{I}_1)$.

It remains to show the converse, i.e., $\mathcal{V} \supseteq \underline{L}(\mathcal{I}_1)$. Consider an arbitrary $\boldsymbol{v} \in \underline{L}(\mathcal{I}_1)$. It can be observed from (3.108) that $\sum_k v_k p_k \leq 1$ for all $\boldsymbol{p} \in \mathbb{R}^K_{++}$ such that $\mathcal{I}(\boldsymbol{p}) \leq 1$. Now we choose $\boldsymbol{p} > 0$ such that $\mathcal{I}(\boldsymbol{p}) = 1$. This implies

$$\sum_{k \in \mathcal{K}} v_k p_k - \mathcal{I}(\boldsymbol{p}) \leq 1 - 1 = 0 \ .$$

Thus,

$$\sup_{p>0\,:\,\mathcal{I}(p)=1} \Big(\sum_k v_k p_k - \mathcal{I}(p)\Big) \le 0 \tag{3.110}$$

Let $\hat{p} > 0$ be arbitrary. Because of the properties of the set $\mathcal{V}$, we have $\mathcal{I}(\hat{p}) > 0$ and $\hat{\lambda} := 1/\mathcal{I}(\hat{p}) < +\infty$. Defining $\tilde{p} = \hat{\lambda}\hat{p}$ and exploiting A2, we have

$$\begin{aligned}\sum_{k\in\mathcal{K}} v_k \hat{p}_k - \mathcal{I}(\hat{p}) &= \frac{1}{\hat{\lambda}} \cdot \hat{\lambda}\Big(\sum_{k\in\mathcal{K}} v_k \hat{p}_k - \mathcal{I}(\hat{p})\Big) \\ &= \frac{1}{\hat{\lambda}}\Big(\sum_{k\in\mathcal{K}} v_k \tilde{p}_k - \mathcal{I}(\tilde{p})\Big) \le 0 \end{aligned} \tag{3.111}$$

The last inequality follows from $\mathcal{I}(\tilde{p}) = 1$ and (3.110). Consequently,

$$\bar{\mathcal{I}}^*(v) := \sup_{\hat{p}>0}\Big(\sum_{k\in\mathcal{K}} v_k \hat{p}_k - \mathcal{I}(\hat{p})\Big) \le 0 \,.$$

The function $\bar{\mathcal{I}}^*(v)$ is the *conjugate* of $\mathcal{I}$. It was shown in [4] that $\bar{\mathcal{I}}^*(v) < +\infty$ implies $v \in \mathcal{V}$. That is, every $v \in \underline{L}(\mathcal{I}_1)$ is also contained in $\mathcal{V}$, which concludes the proof. □

Theorem 3.45 shows that *any* convex compact downward-comprehensive set from $\mathbb{R}^K_{++}$ can be expressed as a sublevel set of a convex interference function. Conversely, it is clear from the results of Section 3.1 that any sublevel set of a convex interference function is compact downward-comprehensive convex.

Similar results can be derived for *concave* interference functions. Consider a non-empty convex closed upward-comprehensive set $\mathcal{V} \subset \mathbb{R}^K_{++}$, $\mathcal{V} \neq \mathbb{R}^K_{++}$. This set is associated with a concave interference function

$$\mathcal{I}(p) = \min_{v\in\mathcal{V}} \sum_{k\in\mathcal{K}} v_k p_k \,. \tag{3.112}$$

The superlevel set $\overline{L}(\mathcal{I})$ is upward-comprehensive closed convex. However, $\overline{L}(\mathcal{I}) \neq \mathcal{V}$ in general. In order to express $\mathcal{V}$ as a superlevel set, we need to introduce an additional interference function

$$\mathcal{I}_2(p) = \min_{v\in\overline{L}(\mathcal{I})} \sum_{k\in\mathcal{K}} v_k p_k \,. \tag{3.113}$$

We have the following result.

Theorem 3.46. *Consider an arbitrary non-empty upward-comprehensive closed convex set $\mathcal{V} \subset \mathbb{R}^K_{++}$, $\mathcal{V} \neq \mathbb{R}^K_{++}$, from which we synthesize a concave interference function $\mathcal{I}$, as defined by (3.112). Let $\mathcal{I}_2$ be defined by (3.113), then*

$$\mathcal{V} = \overline{L}(\mathcal{I}_2) \,. \tag{3.114}$$

Proof. The proof is similar to the proof of Theorem 3.45. □

Theorem 3.46 shows that *every* upward-comprehensive closed convex set from $\mathbb{R}^K_{++}$ can be expressed as a superlevel set of a concave interference function. Conversely, every superlevel set of a concave interference function is closed downward-comprehensive convex.

Theorems 3.45 and 3.46 have an interesting interpretation in terms of resource allocation problems: Every convex interference function has a representation (3.87). This can be interpreted as the maximum weighted total *network utility* from a utility set $\mathcal{V} = \{\boldsymbol{v} > 0 : \mathcal{I}_1(\boldsymbol{v}) \leq 1\}$. Here, the convex interference function $\mathcal{I}_1(\boldsymbol{v})$ can be seen as an indicator function measuring the feasibility of the utilities $\boldsymbol{v}$. Likewise, every concave interference function has a representation (3.112). This can be interpreted as the minimum weighted total *network cost* from a feasible set $\mathcal{V} = \{\boldsymbol{v} > 0 : \mathcal{I}_2(\boldsymbol{v}) \geq 1\}$. The concave interference function $\mathcal{I}_2(\boldsymbol{v})$ can be seen as an indicator function providing a single measure for the feasibility of a given cost vector $\boldsymbol{v}$.

3.6 Log-Convex Interference Functions

In this section we will study elementary building blocks of log-convex interference functions. To this end, we use the variable substitution $\boldsymbol{p} = \mathrm{e}^{\boldsymbol{s}}$ (see Subsection 1.3.2). One main result is Theorem 3.53 [5], which shows that any log-convex interference function $\mathcal{I}(\boldsymbol{p})$, on $\mathbb{R}^K_{++}$, can be represented as

$$\mathcal{I}(\boldsymbol{p}) = \max_{\boldsymbol{w} \in \mathcal{L}(\mathcal{I})} \Big(\underline{f}_{\mathcal{I}}(\boldsymbol{w}) \cdot \prod_{l \in \mathcal{K}} (p_l)^{w_l} \Big) \tag{3.115}$$

where

$$\mathcal{L}(\mathcal{I}) = \big\{ \boldsymbol{w} \in \mathbb{R}^K_+ : \underline{f}_{\mathcal{I}}(\boldsymbol{w}) > 0 \big\} , \tag{3.116}$$

and $\underline{f}_{\mathcal{I}}(\boldsymbol{w})$ is defined as follows.

$$\underline{f}_{\mathcal{I}}(\boldsymbol{w}) = \inf_{\boldsymbol{p} > 0} \frac{\mathcal{I}(\boldsymbol{p})}{\prod_{l \in \mathcal{K}} (p_l)^{w_l}} , \qquad \boldsymbol{w} \in \mathbb{R}^K_+ . \tag{3.117}$$

Note, that $\boldsymbol{w}$ is required to be non-negative, for reasons that will become clear later. Since $\boldsymbol{p} > 0$, we have $\underline{f}_{\mathcal{I}}(\boldsymbol{w}) \geq 0$.

Conversely, log-convex interference functions can be synthesized from certain utility sets. The results allow for some interesting interpretations. For example, connections with the Kullback-Leibler distance (Subsection 3.6.4) and cooperative game theory (Subsection 4.1.6) will be shown. Some of the properties will be used later in Chapter 4, where the boundary of the QoS region will be analyzed.

3.6.1 Basic Building Blocks of Log-Convex Interference Functions

The following function $\xi(\boldsymbol{p})$ is a basic building block of any log-convex interference function.

$$\xi(\boldsymbol{p}) = \prod_{l\in\mathcal{K}} (p_l)^{w_l} \tag{3.118}$$

where $\boldsymbol{w} = [w_1, \ldots, w_K]^T \in \mathbb{R}^K_+$, are some given coefficients with $\|\boldsymbol{w}\|_1 = 1$. Using the substitution $\boldsymbol{p} = \mathrm{e}^{\boldsymbol{s}}$, it can be verified that $\xi(\mathrm{e}^{\boldsymbol{s}})$ is log-convex on $\mathbb{R}^K$. In addition, $\xi(\mathrm{e}^{\boldsymbol{s}})$ fulfills property A1 (positivity) because $\boldsymbol{p} = \mathrm{e}^{\boldsymbol{s}} > 0$. Property A2 (scale-invariance) follows from the assumption $\|\boldsymbol{w}\|_1 = \sum_l w_l = 1$, which leads to

$$\begin{aligned}\xi(\alpha\boldsymbol{p}) &= \prod_{l\in\mathcal{K}} (\alpha p_l)^{w_l} = \alpha^{(\sum_l w_l)} \cdot \prod_{l\in\mathcal{K}} (p_l)^{w_l} \\ &= \alpha \prod_{l\in\mathcal{K}} (p_l)^{w_l} = \alpha \cdot \xi(\boldsymbol{p}) \,.\end{aligned} \tag{3.119}$$

Property A3 (monotonicity) follows from $\boldsymbol{w} \geq 0$. The assumption $\boldsymbol{w} \geq 0$ is necessary since otherwise A3 would be violated. Furthermore, $\|\boldsymbol{w}\|_1 = 1$ is necessary for A2 to hold, as can be seen from (3.119). Thus, $\xi(\boldsymbol{p})$ is a log-convex interference function if and only if $\|\boldsymbol{w}\|_1 = 1$ and $\boldsymbol{w} \geq 0$.

Next, consider the function $\underline{f}_{\mathcal{I}}(\boldsymbol{w})$, defined by (3.117). The following lemma shows that $\underline{f}_{\mathcal{I}}(\boldsymbol{w})$ has an interpretation in the context of convex analysis.

Lemma 3.47. *The function* $\log \underline{f}_{\mathcal{I}}(\boldsymbol{w})$ *is the conjugate of the convex function* $\log \mathcal{I}(\mathrm{e}^{\boldsymbol{s}})$.

Proof. By monotonicity of the log function, we have

$$\log \underline{f}_{\mathcal{I}}(\boldsymbol{w}) = \inf_{\boldsymbol{s}\in\mathbb{R}^K} \Big(\log \mathcal{I}(\mathrm{e}^{\boldsymbol{s}}) - \sum_{l\in\mathcal{K}} w_l s_l\Big) \,, \tag{3.120}$$

which is the definition of the conjugate [23, 84]. □

In (3.117), the function $\underline{f}_{\mathcal{I}}(\boldsymbol{w})$ was defined on $\mathbb{R}^K_+$. This is justified by the following lemma, which shows that negative components lead to the trivial case $\underline{f}_{\mathcal{I}}(\boldsymbol{w}) = 0$. Recall that $I(\boldsymbol{p})$ is defined on $\mathbb{R}^K_{++}$, thus $I(\boldsymbol{p}) > 0$ is always fulfilled as a consequence of A1. In order to achieve the maximum (3.115), we are only interested in those $\boldsymbol{w}$ for which $\underline{f}_{\mathcal{I}}(\boldsymbol{w}) > 0$. This is the set $\mathcal{L}(\mathcal{I})$ defined in (3.116).

Lemma 3.48. *Let* $\mathcal{I}$ *be an arbitrary interference function, and let* $\boldsymbol{w}$ *be some vector with a negative component then* $\underline{f}_{\mathcal{I}}(\boldsymbol{w}) = 0$.

Proof. Consider an arbitrary $\boldsymbol{w} \in \mathbb{R}^K$, with a negative component $w_r < 0$ for some index r. Defining a power vector $\boldsymbol{p}(\lambda)$ with $p_l(\lambda) = 1$, $l \neq r$ and $p_r(\lambda) = \lambda$, with $\lambda > 0$, we have

$$\underline{f}_{\mathcal{I}}(\boldsymbol{w}) \leq \frac{\mathcal{I}(\boldsymbol{p}(\lambda))}{\prod_{l\in\mathcal{K}} \big(p_l(\lambda)\big)^{w_l}} = (\lambda)^{|w_r|} \cdot \mathcal{I}(\boldsymbol{p}(\lambda)) \,.$$

Because $\mathcal{I}(\boldsymbol{p}(\lambda)) \leq \mathcal{I}(\mathbf{1})$ for all $\lambda \in (0,1]$, we have

$$\underline{f}_{\mathcal{I}}(\boldsymbol{w}) \leq \lim_{\lambda \to 0} (\lambda)^{|w_r|} \cdot \mathcal{I}(\mathbf{1}) = 0 \,.$$

This can only be fulfilled with equality. □

The next lemma shows an additional property of the vectors $\boldsymbol{w} \in \mathcal{L}(\mathcal{I})$.

Lemma 3.49. *Let $\mathcal{I}$ be an interference function, and $\boldsymbol{w} \in \mathbb{R}^K_+$. If $\underline{f}_{\mathcal{I}}(\boldsymbol{w}) > 0$ then $\|\boldsymbol{w}\|_1 = 1$.*

Proof. The proof is by contradiction. Suppose that $\underline{f}_{\mathcal{I}}(\boldsymbol{w}) > 0$ and $\|\boldsymbol{w}\|_1 \neq 1$. From (3.117) we know that for an arbitrary constant $\hat{\boldsymbol{p}} > 0$ and a scalar $\lambda > 0$ we have

$$\underline{f}_{\mathcal{I}}(\boldsymbol{w}) \leq \frac{\mathcal{I}(\lambda \hat{\boldsymbol{p}})}{\prod_{l \in \mathcal{K}} (\lambda \hat{p}_l)^{w_l}} = \frac{1}{\lambda^{(\|\boldsymbol{w}\|_1 - 1)}} \cdot C_1 \,, \tag{3.121}$$

with a constant $C_1 = \mathcal{I}(\hat{\boldsymbol{p}}) / \prod_l (\hat{p}_l)^{w_l}$. Inequality (3.121) holds for all $\lambda > 0$, thus

$$\|\boldsymbol{w}\|_1 > 1 \quad \Rightarrow \quad 0 = \lim_{\lambda \to \infty} \frac{1}{\lambda^{(\|\boldsymbol{w}\|_1 - 1)}} \cdot C_1 \geq \underline{f}_{\mathcal{I}}(\boldsymbol{w}) \geq 0$$

$$\|\boldsymbol{w}\|_1 < 1 \quad \Rightarrow \quad 0 = \lim_{\lambda \to 0} \frac{1}{\lambda^{(\|\boldsymbol{w}\|_1 - 1)}} \cdot C_1 \geq \underline{f}_{\mathcal{I}}(\boldsymbol{w}) \geq 0 \,.$$

This leads to the contradiction $\underline{f}_{\mathcal{I}}(\boldsymbol{w}) = 0$, thus implying $\|\boldsymbol{w}\|_1 = 1$. □

From Lemmas 3.48 and 3.49 we know that the coefficients of interest are contained in the set $\mathcal{L}(\mathcal{I})$. We know from Lemma 3.49 that every $\boldsymbol{w} \in \mathcal{L}(\mathcal{I})$ fulfills $\|\boldsymbol{w}\|_1 = 1$. The structure of $\mathcal{L}(\mathcal{I})$ is further characterized by the following lemmas.

Lemma 3.50. *The function $\underline{f}_{\mathcal{I}}(\boldsymbol{w})$, as defined by (3.117), is log-concave on $\mathbb{R}^K_+$.*

Proof. The function $\prod_{l \in \mathcal{K}} (p_l)^{w_l}$ is log-convex and log-concave in $\boldsymbol{w}$, and so is its inverse. Point-wise minimization preserves log-concavity, thus $\underline{f}_{\mathcal{I}}(\boldsymbol{w})$ is log-concave. □

Notice that $\underline{f}_{\mathcal{I}}(\boldsymbol{w})$ is not an interference function. With Lemma 3.50 we can prove the following result.

Lemma 3.51. *The set $\mathcal{L}(\mathcal{I})$, as defined by (3.116), is convex.*

Proof. Consider two points $\hat{\boldsymbol{w}}, \check{\boldsymbol{w}} \in \mathcal{L}(\mathcal{I})$, and the line

$$\boldsymbol{w}(\lambda) = (1-\lambda)\hat{\boldsymbol{w}}_k + \lambda \check{\boldsymbol{w}}_k, \quad \lambda \in [0,1] \,.$$

We have $\boldsymbol{w}(\lambda) \in \mathbb{R}^K_+$. The function $\underline{f}_{\mathcal{I}}(\boldsymbol{w})$ is log-concave on $\mathbb{R}^K_+$ (Lemma 3.50), thus

$$\underline{f}_{\mathcal{I}}(\boldsymbol{w}(\lambda)) \geq \underline{f}_{\mathcal{I}}(\hat{\boldsymbol{w}})^{1-\lambda} \cdot \underline{f}_{\mathcal{I}}(\check{\boldsymbol{w}})^{\lambda} . \tag{3.122}$$

Because $\underline{f}_{\mathcal{I}_k}(\hat{\boldsymbol{w}}_k) > 0$ and $\underline{f}_{\mathcal{I}_k}(\check{\boldsymbol{w}}_k) > 0$, we have $\underline{f}_{\mathcal{I}}(\boldsymbol{w}(\lambda)) > 0$, thus $\underline{f}_{\mathcal{I}}(\boldsymbol{w}(\lambda)) \in \mathcal{L}(\mathcal{I})$. □

Another property will be needed later:

Lemma 3.52. *The function $\underline{f}_{\mathcal{I}}(\boldsymbol{w})$ is upper semi-continuous. That is, for every sequence $\boldsymbol{w}^{(n)} \geq 0$, with $\|\boldsymbol{w}^{(n)}\|_1 = 1$ and $\lim_{n\to\infty} \boldsymbol{w}^{(n)} = \boldsymbol{w}^*$, we have*

$$\underline{f}_{\mathcal{I}}(\boldsymbol{w}^*) \geq \limsup_{n\to\infty} \underline{f}_{\mathcal{I}}(\boldsymbol{w}^{(n)}) . \tag{3.123}$$

Proof. By definition (3.117), we have

$$\frac{\mathcal{I}(\boldsymbol{p})}{\prod_l (p_l)^{w_l^{(n)}}} \geq \underline{f}_{\mathcal{I}}(\boldsymbol{w}^{(n)}) , \quad \forall \boldsymbol{p} > 0 , \forall n \in \mathbb{N} . \tag{3.124}$$

The denominator in (3.124) is a continuous function of $\boldsymbol{w}$, thus

$$\frac{\mathcal{I}(\boldsymbol{p})}{\prod_l (p_l)^{w_l^*}} = \lim_{n\to\infty} \frac{\mathcal{I}(\boldsymbol{p})}{\prod_l (p_l)^{w_l^{(n)}}} \geq \limsup_{n\to\infty} \underline{f}_{\mathcal{I}}(\boldsymbol{w}^{(n)}) . \tag{3.125}$$

This holds for all $\boldsymbol{p} > 0$. The right side of this inequality is independent of $\boldsymbol{p}$, thus

$$\inf_{\boldsymbol{p}>0} \frac{\mathcal{I}(\boldsymbol{p})}{\prod_l (p_l)^{w_l^*}} = \underline{f}_{\mathcal{I}}(\boldsymbol{w}^*) \geq \limsup_{n\to\infty} \underline{f}_{\mathcal{I}}(\boldsymbol{w}^{(n)}) .$$

□

To summarize, any strictly positive log-convex interference function $\mathcal{I}(\boldsymbol{p})$ is associated with a function $\underline{f}_{\mathcal{I}}(\boldsymbol{w}) > 0$, with the following properties:

- $\underline{f}_{\mathcal{I}}(\boldsymbol{w})$ is log-concave and upper semi-continuous. The resulting superlevel set $\mathcal{L}(\mathcal{I})$ is convex.
- $\underline{f}_{\mathcal{I}}(\boldsymbol{w}) > 0$ implies $\|\boldsymbol{w}\|_1 = 1$, thus all elements of $\mathcal{L}(\mathcal{I})$ have this property.

Additional properties and interpretations of the function $\underline{f}_{\mathcal{I}}(\boldsymbol{w})$ will be discussed later.

3.6.2 Analysis of Log-Convex Interference Functions

With the results of the previous section, we are now in a position to prove the main representation theorem.

Theorem 3.53. *Every log-convex interference function $\mathcal{I}(\boldsymbol{p})$, on $\mathbb{R}^K_{++}$, can be represented as*

$$\mathcal{I}(\boldsymbol{p}) = \max_{\boldsymbol{w}\in\mathcal{L}(\mathcal{I})} \Big(\underline{f}_{\mathcal{I}}(\boldsymbol{w}) \cdot \prod_{l\in\mathcal{K}} (p_l)^{w_l} \Big) . \tag{3.126}$$

Proof. According to (3.117), we have for all $\boldsymbol{p} > 0$ and $\boldsymbol{w} \in \mathcal{L}(\mathcal{I})$,

$$\mathcal{I}(\boldsymbol{p}) \geq \underline{f}_{\mathcal{I}}(\boldsymbol{w}) \cdot \prod_{l \in \mathcal{K}} (p_l)^{w_l} \tag{3.127}$$

Thus,

$$\sup_{\boldsymbol{w} \in \mathcal{L}(\mathcal{I})} \Big(\underline{f}_{\mathcal{I}}(\boldsymbol{w}) \cdot \prod_{l \in \mathcal{K}} (p_l)^{w_l} \Big) \leq \mathcal{I}(\boldsymbol{p}) \,. \tag{3.128}$$

It will turn out later that the supremum (3.128) is actually attained.

The function $\log \mathcal{I}(\mathrm{e}^{\boldsymbol{s}})$ is convex, so for any $\hat{\boldsymbol{s}} \in \mathbb{R}^K$, there is a finite $\hat{\boldsymbol{w}} \in \mathbb{R}^K$ such that (see e.g. [84, Thm. 1.2.1, p. 77])

$$\log \mathcal{I}(\mathrm{e}^{\boldsymbol{s}}) - \log \mathcal{I}(\mathrm{e}^{\hat{\boldsymbol{s}}}) \geq \sum_{l \in \mathcal{K}} \hat{w}_l (s_l - \hat{s}_l) \,, \quad \text{for all } \boldsymbol{s} \in \mathbb{R}^K.$$

Using $\boldsymbol{p} = \mathrm{e}^{\boldsymbol{s}}$, this can be rewritten as

$$\frac{\mathcal{I}(\boldsymbol{p})}{\prod_{l \in \mathcal{K}} (p_l)^{\hat{w}_l}} \geq \frac{\mathcal{I}(\hat{\boldsymbol{p}})}{\prod_{l \in \mathcal{K}} (\hat{p}_l)^{\hat{w}_l}} = \hat{C}_1 \,, \quad \forall \boldsymbol{p} > 0 \,. \tag{3.129}$$

with a constant $\hat{C}_1 \in \mathbb{R}_{++}$. With (3.117) we have $\underline{f}_{\mathcal{I}}(\boldsymbol{w}) \geq \mathrm{e}^{\hat{C}_1} > 0$, thus $\hat{\boldsymbol{w}} \in \mathcal{L}(\mathcal{I})$. We can rewrite (3.129) as

$$\mathcal{I}(\hat{\boldsymbol{p}}) \leq \frac{\mathcal{I}(\boldsymbol{p})}{\prod_{l \in \mathcal{K}} (p_l)^{\hat{w}_l}} \prod_{l \in \mathcal{K}} (\hat{p}_l)^{\hat{w}_l} \,, \quad \forall \boldsymbol{p} > 0 \,. \tag{3.130}$$

Inequality (3.130) holds for all $\boldsymbol{p} > 0$, thus

$$\mathcal{I}(\hat{\boldsymbol{p}}) \leq \underline{f}_{\mathcal{I}}(\hat{\boldsymbol{w}}) \cdot \prod_{l \in \mathcal{K}} (\hat{p}_l)^{\hat{w}_l} \,, \tag{3.131}$$

which shows that inequality (3.128) must be fulfilled with equality, thus

$$\mathcal{I}(\boldsymbol{p}) = \sup_{\boldsymbol{w} \in \mathcal{L}(\mathcal{I})} \Big(\underline{f}_{\mathcal{I}}(\boldsymbol{w}) \cdot \prod_{l \in \mathcal{K}} (p_l)^{w_l} \Big) \,. \tag{3.132}$$

It remains to show that this supremum is attained. Consider an arbitrary $\boldsymbol{p} > 0$. From (3.132) we know that there is a sequence $\boldsymbol{w}^{(n)} \in \mathcal{L}(\mathcal{I})$, $n \in \mathbb{N}$, such that

$$\mathcal{I}(\boldsymbol{p}) - \frac{1}{n} \leq \underline{f}_{\mathcal{I}}(\boldsymbol{w}^{(n)}) \cdot \prod_{l \in \mathcal{K}} (p_l)^{w_l^{(n)}} \,, \quad \forall n \in \mathbb{N} \,. \tag{3.133}$$

There is a sub-sequence $\boldsymbol{w}^{(n_m)}$, $m \in \mathbb{N}$, which converges to a limit $\boldsymbol{w}^* = \lim_{m \to \infty} \boldsymbol{w}^{(n_m)}$. Now, we show that $\boldsymbol{w}^*$ is also contained in $\mathcal{L}(\mathcal{I})$. With $p_l \leq \|\boldsymbol{p}\|_\infty$ we can bound (3.133)

$$\mathcal{I}(\boldsymbol{p}) - \frac{1}{n} \leq \underline{f}_{\mathcal{I}}(\boldsymbol{w}^{(n)}) \cdot (\|\boldsymbol{p}\|_\infty)^{\sum_l w_l^{(n)}} \,. \tag{3.134}$$

Exploiting $\|\boldsymbol{w}^{(n)}\|_1 = 1$, we have

$$\underline{f}_{\mathcal{I}}(\boldsymbol{w}^{(n)}) \geq \frac{\mathcal{I}(\boldsymbol{p}) - \frac{1}{n}}{\|\boldsymbol{p}\|_\infty}, \quad \text{for all } n \in \mathbb{N}. \tag{3.135}$$

The function $\mathcal{I}$ is positive because of A1, thus

$$\liminf_{m\to\infty} \underline{f}_{\mathcal{I}}(\boldsymbol{w}^{(n_m)}) \geq \frac{\mathcal{I}(\boldsymbol{p})}{\|\boldsymbol{p}\|_\infty} > 0\,. \tag{3.136}$$

By combining Lemma 3.52 and (3.136) we obtain $\underline{f}_{\mathcal{I}}(\boldsymbol{w}^*) > 0$, thus $\boldsymbol{w}^* \in \mathcal{L}(\mathcal{I})$. With (3.132) we have

$$\begin{aligned}
\mathcal{I}(\boldsymbol{p}) &\geq \underline{f}_{\mathcal{I}}(\boldsymbol{w}^*) \cdot \prod_{l\in\mathcal{K}} (p_l)^{w_l^*} \\
&\geq \liminf_{m\to\infty} \Big(\underline{f}_{\mathcal{I}}(\boldsymbol{w}^{(n_m)}) \cdot \prod_{l\in\mathcal{K}} (p_l)^{w_l^{(n_m)}} \Big) \\
&\geq \liminf_{m\to\infty} \Big(\mathcal{I}(\boldsymbol{p}) - \frac{1}{n_m} \Big) = \mathcal{I}(\boldsymbol{p})\,,
\end{aligned} \tag{3.137}$$

where the last inequality follows from (3.133). Hence,

$$\mathcal{I}(\boldsymbol{p}) = \underline{f}_{\mathcal{I}}(\boldsymbol{w}^*) \cdot \prod_{l\in\mathcal{K}} (p_l)^{w_l^*} = \max_{\boldsymbol{w}\in\mathcal{L}(\mathcal{I})} \Big(\underline{f}_{\mathcal{I}}(\boldsymbol{w}) \cdot \prod_{l\in\mathcal{K}} (p_l)^{w_l} \Big)\,.$$

□

Theorem 3.53 shows that every log-convex interference function can be represented as (3.126). From Lemma 3.50 we know that $\underline{f}_{\mathcal{I}}(\boldsymbol{w})$ is log-concave. The product of log-concave functions is log-concave, thus $\underline{f}_{\mathcal{I}}(\boldsymbol{w}) \cdot \prod_{l\in\mathcal{K}} (p_l)^{w_l}$ is log-concave in $\boldsymbol{w}$. Consequently, problem (3.126) consists of maximizing a log-concave function over a convex set $\mathcal{L}(\mathcal{I})$.

3.6.3 Synthesis of Log-Convex Interference Functions

In the previous section we have analyzed log-convex interference functions. Any log-convex interference function can be broken down into elementary building blocks. Next, we will study the reverse approach: the *synthesis* of a log-convex interference function. To this end, consider the coefficient set

$$\mathcal{M} = \{\boldsymbol{w} \in \mathbb{R}_+^K : \|\boldsymbol{w}\|_1 = 1\}\,, \tag{3.138}$$

and an arbitrary non-negative bounded function $g(\boldsymbol{w}) : \mathcal{M} \mapsto \mathbb{R}_+$. We can synthesize a function

$$\mathcal{I}_g(\mathrm{e}^{\boldsymbol{s}}) = \sup_{\boldsymbol{w}\in\mathcal{M}:g(\boldsymbol{w})>0} \Big(g(\boldsymbol{w}) \prod_{l\in\mathcal{K}} (\mathrm{e}^{s_l})^{w_l} \Big)\,. \tag{3.139}$$

Notice, that $g(\boldsymbol{w}) \prod_{l\in\mathcal{K}} (\mathrm{e}^{s_l})^{w_l}$ is log-convex in $\boldsymbol{s}$ for any choice of $\boldsymbol{w}$. Maximization preserves log-convexity, thus $\mathcal{I}_g(\boldsymbol{p})$ is a log-convex interference function in the sense of Definition 1.4.

Lemma 3.54. *The convex function* $\log \mathcal{I}_g(\mathrm{e}^{s})$ *is the conjugate of the function* $\log(1/g(w))$.

Proof. Because of the monotonicity of the logarithm, we can exchange the order of sup and log, thus

$$\begin{aligned}\log \mathcal{I}_g(\mathrm{e}^{s}) &= \sup_{w \in \mathcal{M}: g(w)>0} \Big(\log g(w) + \sum_{l \in \mathcal{K}} w_l s_l \Big) \\ &= \sup_{w \in \mathcal{M}: g(w)>0} \Big(\sum_{l \in \mathcal{K}} w_l s_l - \log \frac{1}{g(w)} \Big) , \end{aligned} \quad (3.140)$$

which is the definition of the conjugate function [84]. □

Next, consider the analysis of the function $\mathcal{I}_g(\mathrm{e}^{s})$, for which there exists a function $\underline{f}_{\mathcal{I}_g}(w)$, as defined by (3.117). An interesting question is: when does $g = \underline{f}_{\mathcal{I}_g}$ hold? In other words: are analysis and synthesis reverse operations?

Theorem 3.55. $g = \underline{f}_{\mathcal{I}_g}$ *if and only if* $g(w)$ *is log-concave on* $\mathcal{M}$ *and upper semi-continuous.*

Proof. The function $\mathcal{I}_g$ is a log-convex interference function, thus $\underline{f}_{\mathcal{I}_g}$ is log-concave and upper semi-continuous. The result follows from Corollary 1.3.6 in [84, p. 219]. □

In the remainder of this section we will show application examples and additional interpretations of $\underline{f}_{\mathcal{I}}(w)$.

3.6.4 Connection with the Kullback-Leibler Distance

In Section 1.4 we have discussed the example of the linear interference function $\mathcal{I}(p) = v^T p$. For this special log-convex interference function, we will now show that the function $\underline{f}_{\mathcal{I}}(w)$ has an interesting interpretation. With the definition (3.117) we have

$$\underline{f}_{\mathcal{I}}(w) = \inf_{p>0} \frac{\sum_{l \in \mathcal{K}} v_l p_l}{\prod_{l \in \mathcal{K}} (p_l)^{w_l}} . \quad (3.141)$$

If two or more components of v are non-zero, then the optimization (3.141) is strictly convex after the substitution $p = \mathrm{e}^{s}$, as shown in [85]. Thus, there exists a unique optimizer p^*, which is found by computing the partial derivatives and setting the result to zero. A necessary and sufficient condition for optimality is

$$p_r^* = \frac{w_r}{v_r} \cdot \sum_{l \in \mathcal{K}} v_l p_l^* , \quad \forall r \in \mathcal{K} . \quad (3.142)$$

With (3.142), the minimum (3.141) can be written as

$$\underline{f}_{\mathcal{I}}(\boldsymbol{w}) = \frac{\sum_l v_l p_l^*}{\prod_r \left(\frac{w_r}{v_r} \cdot \sum_l v_l p_l^*\right)^{w_r}}$$
$$= \frac{\sum_l v_l p_l^*}{\prod_r \left(\frac{w_r}{v_r}\right)^{w_r} \cdot \left(\sum_l v_l p_l^*\right)^{\sum_r w_r}} . \tag{3.143}$$

Exploiting $\sum_r w_r = 1$, we have

$$\log \underline{f}_{\mathcal{I}}(\boldsymbol{w}) = \log \prod_{l \in \mathcal{K}} \left(\frac{w_l}{v_l}\right)^{-w_l} = - \sum_{l \in \mathcal{K}} w_l \log \frac{w_l}{v_l} . \tag{3.144}$$

It can be observed that $-\log \underline{f}_{\mathcal{I}}(\boldsymbol{w})$ is the Kullback-Leibler distance between the vectors $\boldsymbol{v}$ and $\boldsymbol{w}$. This connects the function $\underline{f}_{\mathcal{I}}(\boldsymbol{w})$ with a known measure. For related results on the connection between the Kullback-Leibler distance and the Perron root of non-negative matrices, see [86].

Next, consider K users with coupling coefficients $\boldsymbol{V} = [\boldsymbol{v}_1, \ldots, \boldsymbol{v}_K]^T$, and a spectral radius $\rho_{\boldsymbol{V}}(\boldsymbol{\gamma})$. The SIR region $\mathcal{S}$ is defined in (1.21). Since $\rho_{\boldsymbol{V}}(\boldsymbol{\gamma})$ is a log-convex interference function (see Subsection 1.4.4), all properties derived so far can be applied. The following corollary follows directly from the structure result Theorem 3.53.

Corollary 3.56. *Consider an arbitrary square irreducible matrix $\boldsymbol{V} \geq 0$ with interference functions $\mathcal{I}_k^{(\boldsymbol{V})}$, as defined by (1.10). Then there exists a log-concave function $f_{\boldsymbol{V}}(\boldsymbol{w})$, defined on $\mathbb{R}_+$, with $\|\boldsymbol{w}\|_1 = 1$, such that*

$$\rho_{\boldsymbol{V}}(\boldsymbol{\gamma}) = \max_{\boldsymbol{w} \in \mathcal{L}(\mathcal{I}^{(\boldsymbol{V})})} \left(f_{\boldsymbol{V}}(\boldsymbol{w}) \prod_{l \in \mathcal{K}} (\gamma_l)^{w_l} \right) . \tag{3.145}$$

As an example, consider the 2-user case, with

$$\rho_{\boldsymbol{V}}(\boldsymbol{\gamma}) = \rho\left(\begin{bmatrix} 0 & \gamma_1 V_{12} \\ \gamma_1 V_{21} & 0 \end{bmatrix} \right) = \sqrt{\gamma_1 \gamma_2 V_{12} V_{21}} . \tag{3.146}$$

The spectral radius of an irreducible non-negative matrix is given by its maximal eigenvalue. For $K = 2$, we obtain the function (3.146), which is log-convex after a substitution $\gamma_k = \exp q_k$ [41]. Here, we assume that there is no self interference, so the main diagonal is set to zero. Comparing (3.145) with (3.146) we have

$$f_{\boldsymbol{V}}(\boldsymbol{w}) = \begin{cases} \sqrt{V_{12} V_{21}}, & w_1 = w_2 = 1/2 \\ 0, & \text{otherwise.} \end{cases} \tag{3.147}$$

This shows how (3.146) can be understood as a special case of the more general representation (3.145).

3.6.5 Every Convex Interference Function is a Log-Convex Interference Function

It was shown in [2] that every convex interference function is a log-convex interference function. At first glance, this might seem contradictory since any log-convex function is convex, but not the other way round [23]. This apparent contradiction is explained by the special definition of a *log-convex interference function* (Definition 1.4) involving the change of variable $\boldsymbol{p} = \exp\{\boldsymbol{s}\}$.

The same result can be shown in a simpler and more direct way by exploiting the structure result of Theorem 3.35.

Theorem 3.57. *Every convex interference function is a log-convex interference function in the sense of Definition 1.4.*

Proof. Theorem 3.35 shows that every convex function $\mathcal{I}(\boldsymbol{p})$ can be expressed as $\max_{\boldsymbol{w} \in \mathcal{W}_0(\mathcal{I})} \sum_k w_k p_k$. The function $g(\mathrm{e}^{\boldsymbol{s}}) = \sum_k w_k \mathrm{e}^{s_k}$ is log-convex, i.e., $\log g(\mathrm{e}^{\boldsymbol{s}})$ is convex. Maximization preserves convexity, thus $\max_{\boldsymbol{w} \in \mathcal{W}_0(\mathcal{I})} \log g(\mathrm{e}^{\boldsymbol{s}})$ is convex as well. The result follows from interchanging log and max. □

Theorem 3.57 shows that the class of log-convex interference functions contains convex interference functions as a special case.

3.6.6 Greatest Log-Convex Minorant

It was shown in Subsection 3.1.1 that every interference function has sup-inf and inf-sup characterizations (3.10) and (3.11), involving functions $G_{\mathcal{I}}(\boldsymbol{q}, \boldsymbol{p}, \boldsymbol{v})$ and $F_{\mathcal{I}}(\boldsymbol{q}, \boldsymbol{p}, \boldsymbol{w})$, respectively. The function $G_{\mathcal{I}}(\boldsymbol{q}, \boldsymbol{p}, \boldsymbol{v})$ allowed us to derive best-possible concave majorants and convex minorants.

Next, similar results are shown for log-convex interference functions on the basis of $F_{\mathcal{I}}(\boldsymbol{q}, \boldsymbol{p}, \boldsymbol{w})$ and the conjugate

$$\underline{f}_{\mathcal{I}}(\boldsymbol{w}) = \inf_{\substack{\boldsymbol{q}>0 \\ \|\boldsymbol{q}\|_1=1}} \frac{\mathcal{I}(\boldsymbol{q})}{\prod_{l \in \mathcal{K}} (q_l)^{w_l}}, \quad \boldsymbol{w} \gneq \mathbf{0} .$$

By exchanging the order of inf and sup in (3.11) we obtain for all $\boldsymbol{p} > 0$

$$\begin{aligned} \mathcal{I}(\boldsymbol{p}) &\geq \sup_{\substack{\boldsymbol{w}>0 \\ \|\boldsymbol{w}\|_1=1}} \inf_{\boldsymbol{q}>0} F_{\mathcal{I}}(\boldsymbol{q}, \boldsymbol{p}, \boldsymbol{w}) \\ &= \sup_{\substack{\boldsymbol{w}>0 \\ \|\boldsymbol{w}\|_1=1}} \underline{f}_{\mathcal{I}}(\boldsymbol{w}) \prod_{k \in \mathcal{K}} (p_k)^{w_k} =: \underline{\mathcal{I}}^{(lcnvx)}(\boldsymbol{p}) . \end{aligned} \tag{3.148}$$

The function $\underline{\mathcal{I}}^{(lcnvx)}$ is a log-convex minorant of $\mathcal{I}$. Since $\mathcal{I}$ is a log-convex interference function by assumption, we know from Theorem 3.53 that the supremum (3.148) is attained, i.e.,

$$\underline{\mathcal{I}}^{(lcnvx)}(\boldsymbol{p}) = \max_{\substack{\boldsymbol{w}>0 \\ \|\boldsymbol{w}\|_1=1}} \underline{f}_{\mathcal{I}}(\boldsymbol{w}) \prod_{k \in \mathcal{K}} (p_k)^{w_k} . \tag{3.149}$$

Note, that the maximizer of (3.149) is contained in $\mathcal{L}(\mathcal{I})$. This follows from the positivity if $\mathcal{I}$ (Axiom A1). Thus, $\underline{\mathcal{I}}^{(lcnvx)}(\boldsymbol{p}) = \mathcal{I}(\boldsymbol{p})$. This leads to the following result.

Theorem 3.58. *$\mathcal{I}$ is a log-convex interference function if and only if $\underline{\mathcal{I}}^{(lcnvx)}(\boldsymbol{p}) = \mathcal{I}(\boldsymbol{p})$, i.e.,*

$$\mathcal{I}(\boldsymbol{p}) = \inf_{\boldsymbol{q}>0} \sup_{\substack{\boldsymbol{w}>0 \\ \|\boldsymbol{w}\|_1=1}} F_{\mathcal{I}}(\boldsymbol{q},\boldsymbol{p},\boldsymbol{w}) = \sup_{\substack{\boldsymbol{w}>0 \\ \|\boldsymbol{w}\|_1=1}} \inf_{\boldsymbol{q}>0} F_{\mathcal{I}}(\boldsymbol{q},\boldsymbol{p},\boldsymbol{w}) .$$

Proof. The proof follows from the previous discussion. It can also be obtained in a similar way as that of Theorem 3.41, with the max-min characterization (3.11). □

For arbitrary interference functions, the minorant $\underline{\mathcal{I}}^{(lcnvx)}(\boldsymbol{p})$ is best-possible. That is, it is not possible to find a tighter log-convex minorant.

Theorem 3.59. *Let $\mathcal{I}$ be an arbitrary interference function, then (3.149) is its* greatest log-convex minorant. *Precisely, let $\tilde{\mathcal{I}}$ be a log-convex interference function which fulfills*

$$0 < \underline{\mathcal{I}}^{(lcnvx)}(\boldsymbol{p}) \leq \tilde{\mathcal{I}}(\boldsymbol{p}) \leq \mathcal{I}(\boldsymbol{p}) , \quad \forall \boldsymbol{p} > 0 , \tag{3.150}$$

then $\underline{\mathcal{I}}^{(lcnvx)}(\boldsymbol{p}) = \tilde{\mathcal{I}}(\boldsymbol{p})$.

Proof. The functions $f_{\underline{\mathcal{I}}^{(lcnvx)}}(\boldsymbol{w})$, $f_{\tilde{\mathcal{I}}}(\boldsymbol{w})$, and $\underline{f}_{\mathcal{I}}(\boldsymbol{w})$ are defined as in (3.117). Because of (3.150) we have

$$f_{\underline{\mathcal{I}}^{(lcnvx)}}(\boldsymbol{w}) \leq f_{\tilde{\mathcal{I}}}(\boldsymbol{w}) \leq \underline{f}_{\mathcal{I}}(\boldsymbol{w}) , \quad \text{for all } \boldsymbol{w} \geq 0, \ \|\boldsymbol{w}\|_1 = 1.$$

This implies

$$\begin{aligned} \underline{\mathcal{I}}^{(lcnvx)}(\boldsymbol{p}) &= \sup_{\boldsymbol{w}\in\mathcal{L}(\underline{\mathcal{I}}^{(lcnvx)})} \Big(f_{\underline{\mathcal{I}}^{(lcnvx)}}(\boldsymbol{w}) \cdot \prod_{l\in\mathcal{K}} (p_l)^{w_l} \Big) \\ &\leq \sup_{\boldsymbol{w}\in\mathcal{L}(\tilde{\mathcal{I}})} \Big(f_{\tilde{\mathcal{I}}}(\boldsymbol{w}) \cdot \prod_{l\in\mathcal{K}} (p_l)^{w_l} \Big) \\ &\leq \sup_{\boldsymbol{w}\in\mathcal{L}(\mathcal{I})} \Big(\underline{f}_{\mathcal{I}}(\boldsymbol{w}) \cdot \prod_{l\in\mathcal{K}} (p_l)^{w_l} \Big) = \underline{\mathcal{I}}^{(lcnvx)}(\boldsymbol{p}) , \end{aligned} \tag{3.151}$$

from which we can conclude $\underline{\mathcal{I}}^{(lcnvx)}(\boldsymbol{p}) = \tilde{\mathcal{I}}(\boldsymbol{p})$. □

To conclude, every interference function can be approximated by best-possible convex and log-convex minorants. These minorants are also interference functions. It was already discussed in Subsection 3.6.5 that every convex interference function is a log-convex interference function, but the converse is false. That is, the class of log-convex interference functions is broader than the class of convex interference functions. Therefore, log-convex approximations can be generally tighter. This will be discussed in more detail later in Subsection 3.8.3.

3.6.7 Least Log-Concave Majorant

Log-concave interference functions were formally introduced in Subsection 1.3.2, but all the results presented thus far are on log-convex interference functions. This is because there is a certain asymmetry between log-convex and log-concave interference functions. Not all results of the previous sections are directly transferrable to the log-concave case. For example, every linear interference function is convex and log-convex. However, linear interference functions are concave but not log-concave, at least not in the sense used here.

Next, we complement the results of the previous subsection by showing that every interference function has a *least log-concave majorant.* To this end we introduce the following log-concave function.

$$\overline{f}_{\mathcal{I}}(\boldsymbol{w}) = \sup_{\boldsymbol{q}>0} \frac{\mathcal{I}(\boldsymbol{q})}{\prod_{l\in\mathcal{K}}(q_l)^{w_l}} = \sup_{\substack{\boldsymbol{q}>0\\ \|\boldsymbol{q}\|_1=1}} \frac{\mathcal{I}(\boldsymbol{q})}{\prod_{l\in\mathcal{K}}(q_l)^{w_l}}\,, \quad \boldsymbol{w} \gneq \boldsymbol{0}\,. \tag{3.152}$$

By exchanging inf and sup in (3.11) we obtain

$$\begin{aligned} \mathcal{I}(\boldsymbol{p}) &\leq \inf_{\substack{\boldsymbol{w}>0\\ \|\boldsymbol{w}\|_1=1}} \sup_{\boldsymbol{q}>0} F_{\mathcal{I}}(\boldsymbol{q},\boldsymbol{p},\boldsymbol{w}) \\ &= \inf_{\substack{\boldsymbol{w}>0\\ \|\boldsymbol{w}\|_1=1}} \overline{f}_{\mathcal{I}}(\boldsymbol{w}) \prod_{l\in\mathcal{K}} (q_l)^{w_l} =: \overline{\mathcal{I}}_2(\boldsymbol{w})\,. \end{aligned} \tag{3.153}$$

It is observed that $\overline{\mathcal{I}}_2(\boldsymbol{w})$ is an interference function and a log-concave majorant.

Theorem 3.60. *$\mathcal{I}$ is a log-concave interference function if and only if $\overline{\mathcal{I}}_2(\boldsymbol{p}) = \mathcal{I}(\boldsymbol{p})$, i.e.,*

$$\mathcal{I}(\boldsymbol{p}) = \sup_{\boldsymbol{q}>0} \inf_{\substack{\boldsymbol{w}>0\\ \|\boldsymbol{w}\|_1=1}} F_{\mathcal{I}}(\boldsymbol{q},\boldsymbol{p},\boldsymbol{w}) = \inf_{\substack{\boldsymbol{w}>0\\ \|\boldsymbol{w}\|_1=1}} \sup_{\boldsymbol{q}>0} F_{\mathcal{I}}(\boldsymbol{q},\boldsymbol{p},\boldsymbol{w}) \tag{3.154}$$

Proof. The proof is similar to that of Theorem 3.29. □

Corollary 3.61. *Let $\mathcal{I}$ be an arbitrary interference function, then $\overline{\mathcal{I}}_2$ is the least log-concave majorant.*

Proof. The proof is similar to that of Corollary 3.30. □

3.7 Application to Standard Interference Functions

Next, we discuss how the structure results from the previous sections can be applied to standard interference functions introduced in Section 2.4.1. The results appeared in [11]. In order to keep the discussion simple, we confine ourselves to $\underline{\boldsymbol{p}} \in \mathbb{R}^{K_u+1}_{++}$. This is a technical restriction compared to the set $\underline{\mathcal{P}}$ defined in (2.23), which includes zeros. However, the results can be extended by using the continuation discussed in Section 2.5.

3.7.1 Convex Weakly Standard Interference Functions

The connection between general interference functions and weakly standard interference functions has been studied in Subsection 2.4.2. The main result is Theorem 2.14.

In order to analyze the structure of convex weakly standard interference functions, we use the results from Section 3.1, with the conjugate function $\underline{g}_{\mathcal{I}}$ defined by (3.96). Assume that $\mathcal{J}$ is a weakly standard interference function. With (2.24) we obtain an interference function $\mathcal{I}_{\mathcal{J}}$ with dimension $K = K_u + 1$, where the last component models the effect of possible noise. We have

$$\begin{aligned} \underline{g}_{\mathcal{I}_{\mathcal{J}}}(\boldsymbol{v}) &= \inf_{\boldsymbol{q}\in\mathbb{R}^{K_u+1}_{++}} \frac{q_{K_u+1}\cdot\mathcal{J}\Big(\frac{q_1}{q_{K_u+1}},\ldots,\frac{q_{K_u}}{q_{K_u+1}}\Big)}{\sum_{l=1}^{K_u+1} v_l q_l} \\ &= \inf_{\boldsymbol{q}\in\mathbb{R}^{K_u+1}_{++}} \frac{\mathcal{J}\Big(\frac{q_1}{q_{K_u+1}},\ldots,\frac{q_{K_u}}{q_{K_u+1}}\Big)}{\sum_{l=1}^{K_u} v_l \frac{q_l}{q_{K_u+1}} + v_{K_u+1}} \\ &= \inf_{\tilde{\boldsymbol{q}}\in\mathbb{R}^{K_u}_{++}} \frac{\mathcal{J}\Big(\tilde{q}_1,\ldots,\tilde{q}_{K_u}\Big)}{\sum_{l=1}^{K_u} v_l \tilde{q}_{K_u+1} + v_{K_u+1}} =: \underline{g}_{\mathcal{J}}(\boldsymbol{v})\,. \end{aligned} \tag{3.155}$$

Using the structure results from Section 3.1, we can provide necessary and sufficient conditions for the convexity of standard interference functions:

Theorem 3.62. *A weakly standard interference function $\mathcal{J}$ is convex on $\mathbb{R}^{K_u}_+$ if and only if one of the following equivalent statements hold.*

- *The interference function $\mathcal{I}_{\mathcal{J}}$, as defined by (2.24), is convex.*
- *There exists a non-empty convex compact downward-comprehensive set $\mathcal{V} \subset \mathbb{R}^{K_u+1}_+$ such that for all $\boldsymbol{p} > 0$*

$$\mathcal{J}(\boldsymbol{p}) = \max_{\boldsymbol{v}\in\mathcal{V}} \Big(\sum_{k\in\mathcal{K}_u} v_k p_k + v_{K_u+1} \Big)\,. \tag{3.156}$$

  Note, that v_{K_u+1} does not need to be positive because $\mathcal{J}$ is only weakly standard.
- *There is a function $\underline{g}_{\mathcal{J}}(\boldsymbol{v})$, as defined by (3.155), such that*

$$\mathcal{J}(\boldsymbol{p}) = \max_{\substack{\boldsymbol{v}>0 \\ \|\boldsymbol{v}\|_1=1}} \underline{g}_{\mathcal{J}}(\boldsymbol{v}) \Big(\sum_{k\in\mathcal{K}_u} v_k p_k + v_{K_u+1} \Big)\,. \tag{3.157}$$

Proof. We have $\mathcal{J}(\boldsymbol{p}) = \mathcal{I}_{\mathcal{J}}(\boldsymbol{p}, 1)$, thus the second statement follows directly from Theorem 3.35, and the last statement follows from Theorem 3.44.

It remains to prove the first statement. If $\mathcal{I}_{\mathcal{J}}$ is convex then $\mathcal{J}(\boldsymbol{p}) = \mathcal{I}_{\mathcal{J}}(\boldsymbol{p}, 1)$ is convex on $\mathbb{R}^{K_u}_+$, since one coordinate is constant. Conversely, we need to show that any convex weakly standard interference function leads to a convex interference function $\mathcal{I}_{\mathcal{J}}$, as defined by (2.24). To this end, we introduce the conjugate function

$$\overline{\mathcal{J}}^*(\boldsymbol{v}) = \sup_{\boldsymbol{p}\in\mathbb{R}^{K_u}_{++}} \Big(\sum_{l\in\mathcal{K}_u} v_l p_l - \mathcal{J}(\boldsymbol{p}) \Big) \tag{3.158}$$

Corollary 2.18 states that $\mathcal{J}$ is continuous on $\mathbb{R}^{K_u}_{++}$ and the results of Section 2.5 show that it has a unique continuation on the boundary. The conjugate $\overline{\mathcal{J}}^*(\boldsymbol{v})$ is lower semi-continuous as the pointwise supremum of continuous functions. It is also convex on the domain

$$\mathcal{V}_0 = \{\boldsymbol{v} \in \mathbb{R}^{K_u} : \overline{\mathcal{J}}^*(\boldsymbol{v}) < +\infty\} . \tag{3.159}$$

For arbitrary $\boldsymbol{v} \in \mathcal{V}_0$ and $\lambda > 1$ we have

$$\begin{aligned} \overline{\mathcal{J}}^*(\boldsymbol{v}) &= \sup_{\boldsymbol{p}\in\mathbb{R}^{K_u}_{++}} \Big(\lambda \sum_{l\in\mathcal{K}_u} v_l p_l - \mathcal{J}(\lambda\boldsymbol{p}) \Big) \\ &\geq \sup_{\boldsymbol{p}\in\mathbb{R}^{K_u}_{++}} \Big(\lambda \sum_{l\in\mathcal{K}_u} v_l p_l - \lambda\mathcal{J}(\boldsymbol{p}) \Big) = \lambda \overline{\mathcal{J}}^*(\boldsymbol{v}) . \end{aligned} \tag{3.160}$$

Thus $(\lambda-1)\overline{\mathcal{J}}^*(\boldsymbol{v}) \leq 0$. This can only be fulfilled if $\overline{\mathcal{J}}^*(\boldsymbol{v}) \leq 0$. Thus, $\mathcal{V}_0$ can be expressed as the set of all $\boldsymbol{v} \in \mathbb{R}^{K_u}$ such that $\overline{\mathcal{J}}^*(\boldsymbol{v}) \leq 0$. Using similar arguments as in [4], it can also be shown that $\boldsymbol{v} \in \mathcal{V}_0$ implies $\boldsymbol{v} \geq 0$.

Since $\mathcal{J}$ is convex and continuous, the conjugate of the conjugate (the biconjugate) of $\mathcal{J}$ is $\mathcal{J}$ again, i.e.,

$$\begin{aligned} \mathcal{J}(\boldsymbol{p}) &= \sup_{\boldsymbol{v}\in\mathcal{V}_0} \Big(\sum_{l=1}^{K_u} v_l p_l - \overline{\mathcal{J}}^*(\boldsymbol{v}) \Big) \\ &= \sup_{\boldsymbol{v}>0:\overline{\mathcal{J}}^*(\boldsymbol{v})\leq 0} \Big(\sum_{l=1}^{K_u} v_l p_l - \overline{\mathcal{J}}^*(\boldsymbol{v}) \Big) . \end{aligned} \tag{3.161}$$

With (3.161) and (2.24) we have

$$\begin{aligned} \mathcal{I}_{\mathcal{J}}(\underline{\boldsymbol{p}}) &= \underline{p}_{K_u+1} \sup_{\boldsymbol{v}>0:\overline{\mathcal{J}}^*(\boldsymbol{v})\leq 0} \Big(\sum_{l=1}^{K_u} v_l \frac{p_l}{\underline{p}_{K_u+1}} - \overline{\mathcal{J}}^*(\boldsymbol{v}) \Big) \\ &= \sup_{\boldsymbol{v}>0:\overline{\mathcal{J}}^*(\boldsymbol{v})\leq 0} \Big(\sum_{l=1}^{K_u} v_l p_l - \underline{p}_{K_u+1} \overline{\mathcal{J}}^*(\boldsymbol{v}) \Big) . \end{aligned} \tag{3.162}$$

The supremum of linear functions is convex, thus $\mathcal{I}$ is a convex interference function.

□

The property $\overline{\mathcal{J}}^*(\boldsymbol{v}) \leq 0$ is important since otherwise monotonicity would not be fulfilled and $\mathcal{I}$ would not be an interference function. Showing this property is actually not required for the proof because Theorem 2.14 already states that $\mathcal{I}$ is an interference function. However, the proof shows this result directly. It thereby provides a better understanding of the fundamental structure of interference functions.

3.7.2 Concave Weakly Standard Interference Functions

For concave interference functions similar results can be derived. Consider an arbitrary weakly standard interference function $\mathcal{J}$. With the conjugate function $\overline{g}_{\mathcal{I}_\mathcal{J}}$ defined by (3.66) we obtain

$$\begin{aligned}\overline{g}_{\mathcal{I}_\mathcal{J}}(\boldsymbol{v}) &= \sup_{\boldsymbol{q}\in\mathbb{R}^{K_u+1}_{++}} \frac{q_{K_u+1}\cdot\mathcal{J}\Big(\frac{q_1}{q_{K_u+1}},\ldots,\frac{q_{K_u}}{q_{K_u+1}}\Big)}{\sum_{l=1}^{K_u+1} v_l q_l} \\ &= \sup_{\boldsymbol{q}\in\mathbb{R}^{K_u+1}_{++}} \frac{\mathcal{J}\Big(\frac{q_1}{q_{K_u+1}},\ldots,\frac{q_{K_u}}{q_{K_u+1}}\Big)}{\sum_{l=1}^{K_u} v_l \frac{q_l}{q_{K_u+1}} + v_{K_u+1}} \\ &= \sup_{\tilde{\boldsymbol{q}}\in\mathbb{R}^{K_u}_{++}} \frac{\mathcal{J}\Big(\tilde{q}_1,\ldots,\tilde{q}_{K_u}\Big)}{\sum_{l=1}^{K_u} v_l\tilde{q}_{K_u+1} + v_{K_u+1}} =: \overline{g}_\mathcal{J}(\boldsymbol{v})\,. \end{aligned} \tag{3.163}$$

Concave weakly standard interference functions are characterized by the following theorem.

Theorem 3.63. *A weakly standard interference function $\mathcal{J}$ is concave on $\mathbb{R}^{K_u}_+$ if and only if the following equivalent statements hold.*

- *The interference function $\mathcal{I}_\mathcal{J}$, as defined by (2.24), is concave.*
- *There exists a non-empty convex closed upward-comprehensive set $\mathcal{V}\subset\mathbb{R}^{K_u+1}_+$ such that for all $\boldsymbol{p}>0$*

$$\mathcal{J}(\boldsymbol{p}) = \min_{\boldsymbol{v}\in\mathcal{V}}\Big(\sum_{k\in\mathcal{K}_u} v_k p_k + v_{K_u+1}\Big)\,. \tag{3.164}$$

- *There is a function $\overline{g}_\mathcal{J}(\boldsymbol{v})$, as defined by (3.163), such that*

$$\mathcal{J}(\boldsymbol{p}) = \inf_{\substack{\boldsymbol{v}>0\\ \|\boldsymbol{v}\|_1=1}} \overline{g}_\mathcal{J}(\boldsymbol{v})\Big(\sum_{k\in\mathcal{K}_u} v_k p_k + v_{K_u+1}\Big)\,. \tag{3.165}$$

Proof. The proof is similar to the proof of Theorem 3.62. □

Similar to the functions $\mathcal{I}_1$ and $\mathcal{I}_2$ discussed in Subsections 3.4.4 and 3.3.4, respectively, we can show that the functions $1/\overline{g}_\mathcal{J}(\boldsymbol{v})$ and $1/\underline{g}_\mathcal{J}(\boldsymbol{v})$ are weakly standard interference functions. This implies that they are continuous on $\mathbb{R}^{K_u}_{++}$ (see Corolary 2.18).

3.7.3 The Structure of Convex and Concave Standard Interference Functions

Consider Theorem 3.62, which shows that a weakly standard interference function $\mathcal{J}$ is convex if and only if there exists a set $\mathcal{V}$ such that (3.156) is fulfilled. In this section we show corresponding results for the case that the function is *standard* instead of *weakly standard.*

Theorem 3.64. *$\mathcal{J}$ is a convex standard interference function if and only if for any $\boldsymbol{p} > 0$ the optimization problem (3.156) has a maximizer $\hat{\boldsymbol{v}} = \hat{\boldsymbol{v}}(\boldsymbol{p}) \in \mathcal{V}$ such that $\hat{v}_{K_u+1} > 0$.*

Proof. Suppose that such a maximizer always exists, then for any $\boldsymbol{p} > 0$ and any $\lambda > 1$ we have

$$\begin{aligned}\mathcal{J}(\lambda\boldsymbol{p}) &= \sum_{k=1}^{K_u} \hat{v}_k(\lambda)\cdot\lambda p_k + \hat{v}_{K_u+1}(\lambda)\\ &= \lambda\Big(\sum_{k=1}^{K_u} \hat{v}_k(\lambda)p_k + \frac{\hat{v}_{K_u+1}(\lambda)}{\lambda}\Big)\\ &< \lambda\Big(\sum_{k=1}^{K_u} \hat{v}_k(\lambda)p_k + \hat{v}_{K_u+1}(\lambda)\Big)\\ &\le \lambda\max_{v\in\mathcal{V}}\Big(\sum_{k=1}^{K_u} v_k p_k + v_{K_u+1}\Big) = \lambda\cdot\mathcal{J}(\boldsymbol{p})\,. \end{aligned} \tag{3.166}$$

That is, $\mathcal{J}$ is a standard interference function.

Conversely, assume that $\mathcal{J}$ is a standard interference function, and there is a $\boldsymbol{p} > 0$ such that for all maximizers $\hat{\boldsymbol{v}}$ (which is a convex subset of $\mathcal{V}$) we have $\hat{v}_{K_u+1} = 0$. Then

$$\mathcal{I}_{\mathcal{J}}(\boldsymbol{p},1) = \mathcal{J}(\boldsymbol{p}) = \max_{v\in\mathcal{V}}\Big(\sum_{k=1}^{K_u} v_k p_k + v_{K_u+1}\Big) = \sum_{k=1}^{K_u} \hat{v}_k p_k\,. \tag{3.167}$$

For $0 < \mu < 1$ we have

$$\begin{aligned}\mathcal{I}_{\mathcal{J}}(\boldsymbol{p},\mu) &= \max_{v\in\mathcal{V}}\Big(\sum_{k=1}^{K_u} v_k p_k + \mu v_{K_u+1}\Big)\\ &\ge \sum_{k=1}^{K_u} \hat{v}_k p_k + \mu\hat{v}_{K_u+1} = \sum_{k=1}^{K_u} \hat{v}_k p_k = \mathcal{I}_{\mathcal{J}}(\boldsymbol{p},1)\,.\end{aligned}$$

Thus, $\mathcal{I}_{\mathcal{J}}(\boldsymbol{p},\mu) = \mathcal{I}_{\mathcal{J}}(\boldsymbol{p},1)$ for all $0 < \mu < 1$. Consider an arbitrary $\lambda > 1$. We have

$$\mathcal{J}(\lambda\boldsymbol{p}) = \mathcal{I}_{\mathcal{J}}(\lambda\boldsymbol{p},1) = \lambda\mathcal{I}_{\mathcal{J}}(\boldsymbol{p},\tfrac{1}{\lambda}) = \lambda\mathcal{I}_{\mathcal{J}}(\boldsymbol{p},1) = \lambda\mathcal{J}(\boldsymbol{p})\,.$$

Thus, $\mathcal{J}$ is *not* a standard interference function, which is a contradiction. Thus, if $\mathcal{J}$ is a standard interference function then for any $\boldsymbol{p} > 0$ there is an optimizer $\hat{v} = \hat{v}(\boldsymbol{p})$ such that $\hat{v}_{K_u+1} > 0$. □

In a similar way, the corresponding result for concave standard interference functions is shown.

Theorem 3.65. *$\mathcal{J}$ is a concave standard interference function if and only if for any $\boldsymbol{p} > 0$ problem (3.164) has a minimizer $\hat{\boldsymbol{v}} = \hat{\boldsymbol{v}}(\boldsymbol{p}) \in \mathcal{V}$ such that $\hat{v}_{K_u+1} > 0$.*

3.7.4 The Structure of Log-Convex Standard Interference Functions

With the results of Section 2.4, we can characterize log-convex standard interference functions.

Log-convexity is defined with respect to the variable $\boldsymbol{s} = \log \boldsymbol{p}$ (componentwise). The second part of the following theorem builds on the function $\underline{f}_{\mathcal{I}}(\boldsymbol{w})$, as defind by (3.117), which was used for analyzing the structure of of log-convex interference functions. The function $\underline{f}_{\mathcal{I}}(\boldsymbol{w})$ is the conjugate of $\log \mathcal{I}(\exp \boldsymbol{s})$ [5].

Theorem 3.66. *Let $\mathcal{J}$ be a weakly standard interference function and $\mathcal{I}_{\mathcal{J}}$ be defined by (2.24). Then $\mathcal{J}(\exp \boldsymbol{s})$ is log-convex if and only if $\mathcal{I}_{\mathcal{J}}(\exp \underline{\boldsymbol{s}})$ is log-convex. In this case, we have*

$$\mathcal{J}(\boldsymbol{p}) = \sup_{\boldsymbol{w}: \underline{f}_{\mathcal{I}}(\boldsymbol{w}) > 0} \underline{f}_{\mathcal{I}}(\boldsymbol{w}) \prod_{l=1}^{K_u} (p_l)^{w_l} \cdot (1)^{w_{K_u+1}} . \tag{3.168}$$

Proof. If $\mathcal{I}_{\mathcal{J}}$ is log-convex then $\mathcal{J}$ must be log-convex as well. Conversely, we prove that $\mathcal{I}_{\mathcal{J}}$ is log-convex. To this end, consider two arbitrary $\hat{\underline{\boldsymbol{p}}}, \check{\underline{\boldsymbol{p}}} \in \mathbb{R}_{++}^{K_u+1}$. We define

$$\underline{\boldsymbol{s}}(\lambda) = (1-\lambda)\hat{\underline{\boldsymbol{s}}} + \lambda \check{\underline{\boldsymbol{s}}} , \quad \lambda \in (0,1) \tag{3.169}$$

$$\underline{\boldsymbol{p}}(\lambda) = \exp \underline{\boldsymbol{s}}(\lambda) = \hat{\underline{\boldsymbol{p}}}^{(1-\lambda)} \cdot \check{\underline{\boldsymbol{p}}}^{\lambda} . \tag{3.170}$$

Because $\mathcal{J}$ is log-convex by assumption, we have

$$
\begin{aligned}
\mathcal{I}_{\mathcal{J}}\big(\underline{\boldsymbol{p}}(\lambda)\big) = & \, \hat{\underline{p}}_{K_u+1}^{(1-\lambda)} \cdot \check{\underline{p}}_{K_u+1}^{\lambda} \quad \times \\
& \times \mathcal{J}\Big(\exp\big[(\hat{s}_1 - \hat{s}_{K_u+1})(1-\lambda) + (\check{s}_1 - \check{s}_{K_u+1})\lambda\big], \\
& \qquad \ldots, \exp\big[(\hat{s}_{K_u} - \hat{s}_{K_u+1})(1-\lambda) + (\check{s}_{K_u} - \check{s}_{K_u+1})\lambda\big]\Big) \\
\leq & \exp\big[\hat{\underline{s}}_{K_u+1}(1-\lambda) + \check{\underline{s}}_{K_u+1}\lambda\big] \quad \times \\
& \times \mathcal{J}\Big(\exp(\hat{s}_1 - \hat{s}_{K_u+1}), \ldots, \exp(\hat{s}_{K_u} - \hat{s}_{K_u+1})\Big)^{1-\lambda} \times \\
& \times \mathcal{J}\Big(\exp(\check{s}_1 - \check{s}_{K_u+1}), \ldots, \exp(\check{s}_{K_u} - \check{s}_{K_u+1})\Big)^{\lambda} \\
= & \Big(\mathcal{I}_{\mathcal{J}}(\hat{\underline{\boldsymbol{p}}})\Big)^{1-\lambda} \cdot \Big(\mathcal{I}_{\mathcal{J}}(\check{\underline{\boldsymbol{p}}})\Big)^{\lambda} .
\end{aligned}
$$

Thus, $\log \mathcal{I}_{\mathcal{J}}\big(\exp \underline{\boldsymbol{s}}\big)$ is convex on $\mathbb{R}^{K_u+1}$. □

Finally, we study the case of log-convex standard interference functions.

Theorem 3.67. *$\mathcal{J}$ is a log-convex standard interference function if and only if for any $\boldsymbol{p} > 0$ problem (3.168) has a maximizer $\hat{\boldsymbol{w}} = \hat{\boldsymbol{w}}(\boldsymbol{p})$ such that $\hat{w}_{K_u+1} > 0$.*

Proof. Assume that $\mathcal{J}$ is a log-convex standard interference function, and there is a $\boldsymbol{p} > 0$ such that for all maximizers $\hat{\boldsymbol{w}} = \hat{\boldsymbol{w}}(\boldsymbol{p})$ we always have $\hat{w}_{K_u+1}(\boldsymbol{p}) = 0$. For all $\lambda > 0$ we have $\mathcal{J}(\lambda \boldsymbol{p}) \leq \lambda \mathcal{J}(\boldsymbol{p})$. Exploiting $\sum_{l=1}^{K_u} \hat{w}_l = 1$, we have

$$
\begin{aligned}
\lambda \mathcal{J}(\boldsymbol{p}) &= \lambda \underline{f}_{\mathcal{I}}(\hat{\boldsymbol{w}}) \prod_{l=1}^{K_u} (p_l)^{\hat{w}_l} = \lambda^{\sum_{l=1}^{K_u} \hat{w}_l} \cdot \underline{f}_{\mathcal{I}}(\hat{\boldsymbol{w}}) \prod_{l=1}^{K_u} (p_l)^{\hat{w}_l} \\
&= \underline{f}_{\mathcal{I}}(\hat{\boldsymbol{w}}) \prod_{l=1}^{K_u} (\lambda p_l)^{\hat{w}_l} \leq \max_{\boldsymbol{w}} \underline{f}_{\mathcal{I}}(\boldsymbol{w}) \prod_{l=1}^{K_u} (\lambda p_l)^{w_l} \\
&= \mathcal{J}(\lambda \boldsymbol{p}) .
\end{aligned}
$$

Thus, $\mathcal{J}$ is not a standard interference function. This contradiction shows that $\hat{w}_{K_u+1} > 0$.

Conversely, assume that for any $\boldsymbol{p} > 0$ there is always a maximizer $\hat{\boldsymbol{w}} = \hat{\boldsymbol{w}}(\boldsymbol{p})$ such that $\hat{w}_{K_u+1} > 0$. For a given $\boldsymbol{p}$, we study $\lambda \boldsymbol{p}$, where $\lambda > 1$. The maximizer is $\hat{\boldsymbol{w}} = \hat{\boldsymbol{w}}(\lambda \boldsymbol{p})$, with $\hat{w}_{K_u+1} > 0$. With $\sum_{l=1}^{K_u} \hat{w}_l + \hat{w}_{K_u+1} = 1$ (see [5]), we have

$$
\begin{aligned}
\mathcal{J}(\lambda \boldsymbol{p}) &= \underline{f}_{\mathcal{I}}(\hat{\boldsymbol{w}}) \cdot \prod_{l=1}^{K_u} (\lambda p_l)^{\hat{w}_l} \\
&= (\lambda)^{\sum_{l=1}^{K_u} \hat{w}_l} \cdot \underline{f}_{\mathcal{I}}(\hat{\boldsymbol{w}}) \cdot \prod_{l=1}^{K_u} (p_l)^{\hat{w}_l} \\
&\leq \frac{\lambda}{\lambda^{\hat{w}_{K_u+1}}} \cdot \max_{\boldsymbol{w}} \underline{f}_{\mathcal{I}}(\boldsymbol{w}) \prod_{l=1}^{K_u} (p_l)^{w_l} \\
&= \frac{\lambda}{\lambda^{\hat{w}_{K_u+1}}} \cdot \mathcal{J}(\boldsymbol{p}) < \lambda \mathcal{J}(\boldsymbol{p}) , \qquad (3.171)
\end{aligned}
$$

because $\lambda > 1$ and $\hat{w}_{K_u+1} > 0$. Thus, $\mathcal{J}$ is standard. □

3.8 Convex and Concave Approximations

In the previous sections we have analyzed interference functions by means of minorants and majorants based on conjugate functions. In this section we discuss an alternative approach based on the sets $\mathcal{N}_0(\mathcal{I})$ and $\mathcal{W}_0(\mathcal{I})$, defined by (3.49) and (3.79), respectively.

3.8.1 Convex/Concave Bounds

One main result of Subsection 3.4.1 was to show that any convex interference function can be expressed as the maximum (3.81) of linear interference functions. Likewise, any concave interference function can be expressed as the minimum (3.51) of linear interference functions. This representation does not hold for arbitrary interference functions.

In this subsection it will be shown how *general* interference functions can be expressed as an optimum of linear functions. To this end, we introduce the following sets.

$$
\begin{aligned}
\mathcal{V}^{(1)} &= \{\tilde{\boldsymbol{p}} : \text{ there exists a } \hat{\boldsymbol{p}} \in \underline{L}(\mathcal{I}) \text{ and } \hat{\boldsymbol{p}} = 1/\tilde{\boldsymbol{p}} \ \} \\
\mathcal{V}^{(2)} &= \{\tilde{\boldsymbol{p}} : \text{ there exists a } \hat{\boldsymbol{p}} \in \overline{L}(\mathcal{I}) \text{ and } \hat{\boldsymbol{p}} = 1/\tilde{\boldsymbol{p}} \ \} .
\end{aligned}
$$

With Theorem 3.5, an arbitrary interference function $\mathcal{I}(\boldsymbol{p})$ can be expressed as

$$
\mathcal{I}(\boldsymbol{p}) = \min_{\tilde{\boldsymbol{p}} \in \mathcal{V}^{(1)}} \max_{k \in \mathcal{K}} (p_k \cdot \tilde{p}_k) = \max_{\tilde{\boldsymbol{p}} \in \mathcal{V}^{(2)}} \min_{k \in \mathcal{K}} (p_k \cdot \tilde{p}_k) .
$$

The set $\mathcal{V}^{(1)}$ can be rewritten as

$$
\begin{aligned}
\mathcal{V}^{(1)} &= \{\tilde{\boldsymbol{p}} > 0 : \mathcal{I}(1/\tilde{\boldsymbol{p}}) \leq 1\} \\
&= \{\tilde{\boldsymbol{p}} > 0 : 1 \leq 1/\mathcal{I}(1/\tilde{\boldsymbol{p}})\} = \overline{L}(\mathcal{I}_{inv}) ,
\end{aligned}
$$

where we have used the definition

$$\mathcal{I}_{inv}(\boldsymbol{p}) = \frac{1}{\mathcal{I}(1/\boldsymbol{p})}\,, \quad \text{for } \boldsymbol{p} > 0\,. \tag{3.172}$$

It can be verified that $\mathcal{I}_{inv}$ is an interference function: Property A2 follows from

$$\mathcal{I}_{inv}(\lambda\boldsymbol{p}) = \frac{1}{\mathcal{I}(1/\lambda\boldsymbol{p})} = \frac{1}{\frac{1}{\lambda}\cdot\mathcal{I}(1/\boldsymbol{p})} = \lambda\mathcal{I}_{inv}(\boldsymbol{p})\,.$$

Properties A1 and A3 are easily shown as well.

Defining $\mathcal{W} = \{\boldsymbol{w} > 0 : \|\boldsymbol{w}\|_1 = 1\}$, the following equation holds (see Lemma A.12 in the Appendix A.7).

$$\max_{k\in\mathcal{K}} p_k = \sup_{w\in\mathcal{W}} \sum_{k\in\mathcal{K}} w_k p_k\,, \quad \text{for any } \boldsymbol{p} > 0.$$

Hence, an arbitrary $\mathcal{I}(\boldsymbol{p})$ can be represented as

$$\mathcal{I}(\boldsymbol{p}) = \min_{\tilde{\boldsymbol{p}}\in\overline{L}(\mathcal{I}_{inv})} \sup_{\boldsymbol{w}\in\mathcal{W}} \sum_{k\in\mathcal{K}} w_k\tilde{p}_k\cdot p_k \tag{3.173}$$

$$= \max_{\tilde{\boldsymbol{p}}\in\underline{L}(\mathcal{I}_{inv})} \inf_{\boldsymbol{w}\in\mathcal{W}} \sum_{k\in\mathcal{K}} w_k\tilde{p}_k\cdot p_k\,. \tag{3.174}$$

It can be observed from (3.174) that this representation has a similar form as the convex function (3.87). For any given $\boldsymbol{w}$, a linear function is maximized over parameters $\tilde{\boldsymbol{p}}$. However, the interference function (3.174) is generally not convex because of the additional optimization with respect to $\boldsymbol{w}$, so the combined weights $w_k\tilde{p}_k$ are contained in a more general set. By choosing an arbitrary fixed $\boldsymbol{w}\in\mathcal{W}$, we obtain a convex upper bound:

$$\mathcal{I}(\boldsymbol{p}) \le \sup_{\tilde{\boldsymbol{p}}\in\underline{L}(\mathcal{I}_{inv})} \sum_{k\in\mathcal{K}} w_k\tilde{p}_k\cdot p_k =: \overline{\mathcal{I}}_{conv}(\boldsymbol{p},\boldsymbol{w})\,. \tag{3.175}$$

Note, that this convex upper bound can be trivial, i.e., the right hand side of (3.175) can tend to infinity. Inequality (3.175) holds for all $\boldsymbol{w}\in\mathcal{W}$, thus

$$\mathcal{I}(\boldsymbol{p}) \le \inf_{\boldsymbol{w}\in\mathcal{W}} \overline{\mathcal{I}}_{conv}(\boldsymbol{p},\boldsymbol{w})\,. \tag{3.176}$$

Similar results can be derived from (3.173), leading to a concave lower bound. This bound can also be trivial (i.e. zero).

Another interesting problem is the construction of a minorant $\hat{\mathcal{I}}$, such that $\hat{\mathcal{I}}(\hat{\boldsymbol{p}}) = \mathcal{I}(\hat{\boldsymbol{p}})$ for some point $\hat{\boldsymbol{p}}$, and $\hat{\mathcal{I}}(\boldsymbol{p}) \le \mathcal{I}(\boldsymbol{p})$ for all $\boldsymbol{p} > 0$. For general interference functions $\mathcal{I}(\boldsymbol{p})$, such a minorant is provided by the elementary interference function $\underline{\mathcal{I}}(\boldsymbol{p},\hat{\boldsymbol{p}})$, as shown in Section 3.1.4. For the special case of convex interference functions (3.87), another minorant is obtained by choosing an arbitrary $\boldsymbol{v}'\in\mathcal{V}$, for which we have $\mathcal{I}(\boldsymbol{p}) \ge \sum_k v'_k p_k$. However, such a linear minorant does not always exist, as shown by the following example.

Example 3.68. Consider the log-convex interference function

$$\mathcal{I}(\boldsymbol{p}) = C_1 \prod_{l \in \mathcal{K}} (p_l)^{w_l}, \quad \|\boldsymbol{w}\|_1 = 1,\ \boldsymbol{w} > 0,\ C_1 > 0\,. \tag{3.177}$$

We show by contradiction that no linear interference function can be a minorant of (3.177). Assume that there is a $\boldsymbol{w} > 0$ such that $\mathcal{I}(\boldsymbol{p}) \geq \sum_l p_l w_l$ for all $\boldsymbol{p} > 0$. Then we can construct a vector $\boldsymbol{p}(\rho) = (1, \ldots, 1, \rho, 1)$, where the rth component is set to some $\rho > 0$. The position r is chosen such that $w_r, v_r \neq 0$. By assumption, $\mathcal{I}(\boldsymbol{p}(\rho)) = C_1 \rho^{w_r} \geq v_r \rho + \sum_{l \neq r} v_l$. Dividing both sides by ρ we have $C_1 \rho^{w_r - 1} \geq v_r + \sum_{l \neq r} v_l / \rho$. Letting $\rho \to \infty$ leads to the contradiction $0 \geq v_r > 0$.

This discussion shows that in order to derive "good" minorants or majorants, it is important to exploit the structure of the interference function. Otherwise, trivial bounds can be obtained.

3.8.2 Least Concave Majorant and Greatest Convex Minorant

Next, we exploit the results of Sections 3.3 and 3.4 in order to derive best-possible convex minorants and best-possible concave majorants. We thereby complement the results from Subsection 3.3.4 and 3.4.4, where the minorant and majorant were derived in a different way.

Consider the sets $\mathcal{N}_0(\mathcal{I})$ and $\mathcal{W}_0(\mathcal{I})$ defined by (3.49) and (3.79), respectively. We know from Lemma 3.22 that for any $\boldsymbol{w} \in \mathcal{N}_0(\mathcal{I})$ we have $\mathcal{I}(\boldsymbol{p}) \leq \sum_l w_l p_l$, thus

$$\mathcal{I}(\boldsymbol{p}) \leq \min_{\boldsymbol{w} \in \mathcal{N}_0(\mathcal{I})} \sum_{l \in \mathcal{K}} w_l p_l \quad \text{for all } \boldsymbol{p} > 0. \tag{3.178}$$

This means that the function

$$\overline{\mathcal{I}}^{(v)}(\boldsymbol{p}) = \min_{\boldsymbol{w} \in \mathcal{N}_0(\mathcal{I})} \sum_{l \in \mathcal{K}} w_l p_l \tag{3.179}$$

is a concave majorant of $\mathcal{I}$. In a similar way, it follows from Lemma 3.34 that

$$\underline{\mathcal{I}}^{(x)}(\boldsymbol{p}) = \max_{\boldsymbol{w} \in \mathcal{W}_0(\mathcal{I})} \sum_{l \in \mathcal{K}} w_l p_l \leq \mathcal{I}(\boldsymbol{p}) \tag{3.180}$$

is a convex minorant.

Next, it will be shown that these approximations are best-possible.

Theorem 3.69. $\overline{\mathcal{I}}^{(v)}$ *is the least concave majorant of* $\mathcal{I}$*, and* $\underline{\mathcal{I}}^{(x)}$ *is the greatest convex minorant of* $\mathcal{I}$*.*

Proof. We prove the first statement by contradiction. The proof of the second statement follows in the same way. Suppose that there exists a concave interference function $\hat{\mathcal{I}}$, such that

$$\mathcal{I}(\boldsymbol{p}) \leq \hat{\mathcal{I}}(\boldsymbol{p}) \leq \overline{\mathcal{I}}^{(v)}(\boldsymbol{p}) \,, \quad \forall \boldsymbol{p} > 0 \,. \tag{3.181}$$

Both $\hat{\mathcal{I}}$ and $\overline{\mathcal{I}}^{(v)}$ are concave interference functions, thus we know from Theorem 3.23 that they can be represented as (3.51). If the conjugate of $\overline{\mathcal{I}}^{(v)}$ is finite for some $\boldsymbol{w} \geq 0$, i.e., $\underline{\mathcal{I}}^*(\boldsymbol{w}) > -\infty$, then it follows from inequality (3.181) that also the conjugates of $\hat{\mathcal{I}}$ and and $\mathcal{I}$ are finite. Thus,

$$\mathcal{N}_0(\overline{\mathcal{I}}^{(v)}) \subseteq \mathcal{N}_0(\hat{\mathcal{I}}) \subseteq \mathcal{N}_0(\mathcal{I}) \,. \tag{3.182}$$

The set $\mathcal{N}_0(\mathcal{I})$ is upward-comprehensive closed convex, as shown Section 3.3.2, so with Theorem 3.26 we have

$$\mathcal{N}_0(\mathcal{I}) = \mathcal{N}_0(\overline{\mathcal{I}}^{(v)}) \,. \tag{3.183}$$

Combining (3.182) and (3.183) we have $\mathcal{N}_0(\overline{\mathcal{I}}^{(v)}) = \mathcal{N}_0(\hat{\mathcal{I}})$. Hence, $\overline{\mathcal{I}}^{(v)}(\boldsymbol{p}) = \hat{\mathcal{I}}(\boldsymbol{p})$ for all $\boldsymbol{p} > 0$. □

In the next section, the convex minorant will be compared with the log-convex minorant.

3.8.3 Comparison of Convex and Log-Convex Minorants

In the previous sections it was shown that every general interference function $\mathcal{I}(\boldsymbol{p})$ has a greatest convex minorant $\underline{\mathcal{I}}^{(x)}(\boldsymbol{p})$. In Subsection 3.6.6 it was shown that $\mathcal{I}(\boldsymbol{p})$ also has a greatest log-convex minorant $\underline{\mathcal{I}}^{(lcnvx)}(\boldsymbol{p})$. Now, an interesting question is which class of functions provides the tightest minorant.

From Theorem 3.57 we know that $\underline{\mathcal{I}}^{(x)}(\mathrm{e}^{\boldsymbol{s}})$ is also log-convex. Thus, the set of log-convex interference functions is more general as the set of convex interference functions. That is, every convex interference function is log-convex, but not conversely. This means that the greatest log-convex minorant is better or as good as the greatest convex minorant, i.e.,

$$\underline{\mathcal{I}}^{(x)}(\boldsymbol{p}) \leq \underline{\mathcal{I}}^{(lcnvx)}(\boldsymbol{p}) \leq \mathcal{I}(\boldsymbol{p}), \quad \forall \boldsymbol{p} > 0 \,. \tag{3.184}$$

If the log-convex minorant $\underline{\mathcal{I}}^{(lcnvx)}$ is trivial, i.e., $\underline{\mathcal{I}}^{(lcnvx)}(\boldsymbol{p}) = 0$, $\forall \boldsymbol{p} > 0$, then also the convex minorant $\underline{\mathcal{I}}^{(x)}$ will be trivial. Conversely, if $\underline{\mathcal{I}}^{(x)}$ is trivial, then this does not imply that $\underline{\mathcal{I}}^{(lcnvx)}$ is trivial as well. This is shown by the following example.

Example 3.70. Consider the log-convex interference function

$$\mathcal{I}(\boldsymbol{p}) = \prod_{l \in \mathcal{K}} (p_l)^{w_l} \,, \quad \boldsymbol{w} \geq 0, \; \|\boldsymbol{w}\|_1 = 1 \,, \tag{3.185}$$

with the convex minorant

$$\underline{\mathcal{I}}^{(x)}(\boldsymbol{p}) = \max_{\boldsymbol{v} \in \mathcal{W}_0(\mathcal{I})} \sum_{l \in \mathcal{K}} v_l p_l \,. \tag{3.186}$$

It was already shown that $\underline{\mathcal{I}}^{(x)}(\boldsymbol{p}) \leq \mathcal{I}(\boldsymbol{p})$, $\forall \boldsymbol{p} > 0$. Suppose that there is a $\boldsymbol{v} \in \mathcal{W}_0(\mathcal{I})$ such that $v_r > 0$ for some index r. That is,

$$\prod_{l \in \mathcal{K}} (p_l)^{w_l} \geq \sum_{l \in \mathcal{K}} v_l p_l \geq v_r p_r > 0 \,, \quad \text{for all } \boldsymbol{p} > 0.$$

This would lead to the contradiction

$$0 = \lim_{p_r \to \infty} \frac{1}{p_r} \prod_{l \in \mathcal{K}} (p_l)^{w_l} \geq v_r > 0 \,.$$

Hence, $\mathcal{W}_0(\mathcal{I}) = 0$. The only convex minorant of the log-convex interference function (3.185) is the trivial function $\underline{\mathcal{I}}^{(x)}(\boldsymbol{p}) = 0$.

3.8.4 Convex and Concave Approximations of SIR Feasible Sets

The results can be applied to the SIR feasible region of a multi-user system. Consider K users with interference functions $\mathcal{I}_k(\boldsymbol{p}) > 0$ for all $k \in \mathcal{K}$. Certain SIR targets $\boldsymbol{\gamma} = [\gamma_1, \ldots, \gamma_K] > 0$ are said to be feasible if there exists a $\boldsymbol{p} > 0$ such that

$$\frac{p_k}{\mathcal{I}_k(\boldsymbol{p})} \geq \gamma_k - \epsilon, \quad \text{for all } \epsilon > 0 \text{ and } k \in \mathcal{K} \,.$$

That is, the SIR targets $\boldsymbol{\gamma}$ can be achieved, at least in an asymptotic sense. Whether or not this condition can be fulfilled depends on how the users are coupled by interference [2]. A point $\boldsymbol{\gamma}$ is feasible if and only if $C(\boldsymbol{\gamma}, \mathcal{I}) \leq 1$, where

$$C(\boldsymbol{\gamma}, \mathcal{I}) = \inf_{\boldsymbol{p} > 0} \Big(\max_{k \in \mathcal{K}} \frac{\gamma_k \mathcal{I}_k(\boldsymbol{p})}{p_k} \Big) \,. \tag{3.187}$$

The feasible region F is the sublevel set

$$F = \{ \boldsymbol{\gamma} > 0 : C(\boldsymbol{\gamma}, \mathcal{I}) \leq 1 \} \,. \tag{3.188}$$

If $\mathcal{I}_1(\mathrm{e}^{\boldsymbol{s}}), \ldots, \mathcal{I}_K(\mathrm{e}^{\boldsymbol{s}})$ are log-convex, then $C(\boldsymbol{\gamma}, \mathcal{I})$ is a log-convex interference function [2]. Thus, the sublevel set F is convex on a logarithmic scale. We will refer to such sets as "log-convex" in the following.

Now, consider general interference functions, with no further assumption on convexity or concavity. The corresponding region F need not be convex, which complicates the development of algorithms operating on the boundary of the region. However, with the results from the previous sections, we can derive convex and concave approximations.

For each $\mathcal{I}_k$, we have a log-convex minorant $\underline{\mathcal{I}}_k^{(lcnvx)}(\boldsymbol{p})$, as defined by (3.149). This leads to a region $\underline{F}^{(lcnvx)}$, characterized by $C(\boldsymbol{\gamma}, \underline{\mathcal{I}}^{(lcnvx)})$. Because $\mathcal{I}_k(\boldsymbol{p}) \geq \underline{\mathcal{I}}_k^{(lcnvx)}(\boldsymbol{p})$, for all $\boldsymbol{p} > 0$, we have $F \subseteq \underline{F}^{(lcnvx)}$. That is, the feasible region F is contained in the log-convex region $\underline{F}^{(lcnvx)}$. According to Theorem 3.59, this is the smallest region associated with log-convex interference functions. Moreover, the SIR region $\underline{F}$ has a useful property. For every mapping $QoS = \phi(\mathrm{SIR})$, with a log-convex inverse function $\phi^{[-1]}$, the resulting QoS region is log-convex [2].

Instead of approximating the underlying interference functions $\mathcal{I}(\boldsymbol{p})$, it is also possible to approximate the function $C(\boldsymbol{\gamma}) := C(\boldsymbol{\gamma}, \mathcal{I})$ directly. It can be verified that $C(\boldsymbol{\gamma})$ fulfills the axioms A1, A2, A3. Thus, the feasible SIR region F can also be regarded as a sublevel set of an interference function.

As shown in Section 3.8.2, we can construct the least concave majorant $\overline{C}$ and the greatest convex minorant $\underline{C}$. Consider the sublevel sets

$$\underline{F} = \{\boldsymbol{\gamma} > 0 : \underline{C}(\boldsymbol{\gamma}) \leq 1\}\,, \tag{3.189}$$

$$\overline{F} = \{\boldsymbol{\gamma} > 0 : \overline{C}(\boldsymbol{\gamma}) \leq 1\}\,. \tag{3.190}$$

Because $\underline{C}(\boldsymbol{\gamma}) \leq C(\boldsymbol{\gamma}) \leq \overline{C}(\boldsymbol{\gamma})$ for all $\boldsymbol{\gamma} > 0$, the resulting level sets fulfill

$$\underline{F} \supseteq F \supseteq \overline{F}\,. \tag{3.191}$$

Sublevel sets of convex interference functions are downward-comprehensive convex. Because $\underline{C}$ is the *greatest* convex minorant, the set $\underline{F}$ is the smallest closed downward-comprehensive convex subset of $\mathbb{R}_+^K$ containing F (the "convex comprehensive hull").

The other sublevel $\overline{F}$ is generally not convex, but it has a convex complementary set $\overline{F}^c = \{\boldsymbol{\gamma} > 0 : \overline{C}(\boldsymbol{\gamma}) > 1\}$. The complementary set $\overline{F}^c$ is a superlevel set of a concave interference functions, so it is upward-comprehensive convex. The set $\overline{F}$ is downward-comprehensive. Thus, $\overline{F}$ is the largest closed downward-comprehensive subset of F such that the complementary set $\overline{F}^c$ is convex.

These regions provide best-possible convex approximations of the original region. Of course, there can exist other bounds, which are non-convex, but tighter. For example, it is possible to construct a log-convex minorant $\underline{C}^l(\boldsymbol{\gamma})$, which fulfills $\underline{C}(\boldsymbol{\gamma}) \leq \underline{C}(\boldsymbol{\gamma})^l \leq C(\boldsymbol{\gamma})$. The resulting sublevel set

$$\underline{F}^l = \{\boldsymbol{\gamma} > 0 : \underline{C}(\boldsymbol{\gamma})^l(\boldsymbol{\gamma}) \leq 1\}$$

fulfills $F \subseteq \underline{F}^l \subseteq \underline{F}$. This is illustrated in Fig. 3.8. Note that these bounds need not be good. It can happen that only a trivial bound exists, as shown in Section 3.8.3.

Example 3.71. Consider the SIR supportable region $\mathcal{S}$ resulting from linear interference functions $\mathcal{I}_k(\boldsymbol{p}) = [\boldsymbol{V}\boldsymbol{p}]_k$, as defined by (1.21). For $K = 2$, we have a coupling matrix $\boldsymbol{V} = \left[\begin{smallmatrix} 0 & V_{12} \\ V_{21} & 0 \end{smallmatrix}\right]$. The closure of the non-supportable

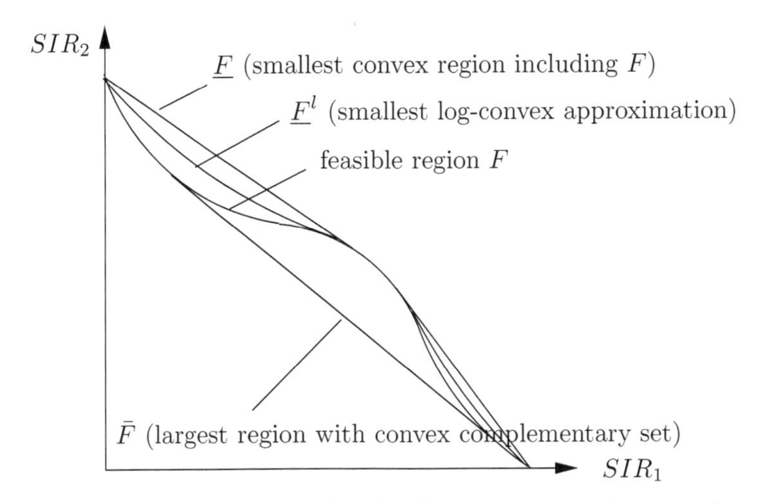

Fig. 3.8. Illustration: An arbitrary feasible SIR region F can be approximated by convex regions

region is the set $\{\boldsymbol{\gamma} : \rho(\boldsymbol{\Gamma V}) \geq 1\}$, where $\boldsymbol{\Gamma} = \text{diag}\{\boldsymbol{\gamma}\}$. It can be verified that the function $\rho(\boldsymbol{\gamma}) = \rho\big(\text{diag}\{\boldsymbol{\gamma}\}\boldsymbol{V}\big)$ fulfills the axioms A1, A2, A3, thus $\rho(\boldsymbol{\gamma})$ is an interference function. The spectral radius is

$$\rho(\boldsymbol{\gamma}) = \sqrt{\gamma_1\gamma_2 V_{12}V_{21}}\,, \tag{3.192}$$

thus $\rho(\boldsymbol{\gamma}) \geq 1$ if and only if $\gamma_2 \geq (V_{12}V_{21}\gamma_1)^{-1}$, which shows that the non-supportable region is convex. Perhaps interestingly, this set can be shown to be convex for $K < 4$ users [87]. However, this property does not extend to larger numbers $K \geq 4$, as shown in [88].

With the proposed theory, this problem can be understood in a more general context. This result shows that certain properties of the Perron root [87, 88] can be generalized to the min-max optimum $C(\boldsymbol{\gamma})$ for arbitrary convex/concave interference functions. The function $C(\boldsymbol{\gamma})$ is an indicator for feasibility of SIR targets $\boldsymbol{\gamma}$, and the level set (3.188) is the SIR region, i.e., the set of jointly feasible SIR.

3.8.5 Convex Comprehensive Level Sets

In the previous section we have discussed the SIR region, which is a comprehensive sublevel set of the interference function $C(\boldsymbol{\gamma}, \mathcal{I})$. This can be generalized to other level sets. It was shown in [3] that any closed downward-comprehensive subset of $\mathbb{R}^K_{++}$ can be expressed as a sublevel set of an interference function. Also, any closed upward-comprehensive subset of $\mathbb{R}^K_{++}$ can be expressed as a superlevel set of an interference function. Here, "closed" means *relatively closed on* $\mathbb{R}^K_{++}$ (see Definition 3.4 in Subsection 3.1.3).

In this section, we derive necessary and sufficient conditions for convexity. Consider an interference function $\mathcal{I}$ with the sublevel set

$$\underline{\mathcal{R}} = \{\boldsymbol{p} > 0 : \mathcal{I}(\boldsymbol{p}) \le 1\} , \qquad (3.193)$$

and the superlevel set

$$\overline{\mathcal{R}} = \{\boldsymbol{p} > 0 : \mathcal{I}(\boldsymbol{p}) \ge 1\} . \qquad (3.194)$$

Note, that the meaning of the vector $\boldsymbol{p}$ depends on the context. In the first part of the paper, $\boldsymbol{p}$ was introduced as a "power vector". However, $\boldsymbol{p}$ can stand for any other parameter, like the SIR vector $\boldsymbol{\gamma}$ used in the previous section.

Theorem 3.72. *The set $\overline{\mathcal{R}}$ is non-empty* upward-comprehensive closed convex *and $\overline{\mathcal{R}} \neq \mathbb{R}^K_{++}$ if and only if the interference function $\mathcal{I}$ is concave and there exists a $\mathbf{p} > 0$ such that $\mathcal{I}(\mathbf{p}) > 0$.*

Proof. Assume that the interference function $\mathcal{I}$ is concave. It was shown in [3] that the resulting superlevel set (3.194) is upward-comprehensive (this follows from axiom A3), closed (relatively on $\mathbb{R}^K_{++}$), and $\overline{\mathcal{R}} \neq \mathbb{R}^K_{++}$. The set $\overline{\mathcal{R}}$ is also convex since every superlevel set of a concave function is convex (see e.g. [23, p.75]).

Conversely, assume that the superlevel set $\overline{\mathcal{R}}$ is a upward-comprehensive closed convex set. It was shown in [3] that $\overline{\mathcal{R}} \neq \mathbb{R}^K_{++}$ implies the existence of a $\boldsymbol{p} > 0$ such that $\mathcal{I}(\boldsymbol{p}) > 0$. It remains to show that the interference function $\mathcal{I}(\boldsymbol{p})$ is concave. Consider arbitrary boundary points $\hat{\boldsymbol{p}}, \check{\boldsymbol{p}} \in \mathbb{R}^K_{++}$, such that $\mathcal{I}(\hat{\boldsymbol{p}}) = \mathcal{I}(\check{\boldsymbol{p}}) = 1$. Defining $\boldsymbol{p}(\lambda) = (1-\lambda)\hat{\boldsymbol{p}} + \lambda\check{\boldsymbol{p}}$, we have $\mathcal{I}\big(\boldsymbol{p}(\lambda)\big) \ge 1$ for all $\lambda \in (0,1)$. For arbitrary $\alpha, \beta > 0$ we define

$$1-\lambda = \frac{\alpha}{\alpha+\beta} \text{ and } \lambda = \frac{\beta}{\alpha+\beta} ,$$

which ensures the desired property $\lambda \in (0,1)$. With property A2, we have

$$1 \le \mathcal{I}\big(\tfrac{\alpha}{\alpha+\beta} \cdot \hat{\boldsymbol{p}} + \tfrac{\beta}{\alpha+\beta} \cdot \check{\boldsymbol{p}}\big) = \tfrac{1}{\alpha+\beta} \cdot \mathcal{I}\big(\alpha \cdot \hat{\boldsymbol{p}} + \beta \cdot \check{\boldsymbol{p}}\big) . \qquad (3.195)$$

Using $\mathcal{I}(\hat{\boldsymbol{p}}) = \mathcal{I}(\check{\boldsymbol{p}}) = 1$ and (3.195), we have

$$\alpha\mathcal{I}(\hat{\boldsymbol{p}}) + \beta\mathcal{I}(\check{\boldsymbol{p}}) = \alpha + \beta \le \mathcal{I}\big(\alpha \cdot \hat{\boldsymbol{p}} + \beta \cdot \check{\boldsymbol{p}}\big) . \qquad (3.196)$$

Next, consider arbitrary points $\hat{\boldsymbol{p}}', \check{\boldsymbol{p}}' \in \mathbb{R}^K_{++}$, from which we can construct boundary points $\hat{\boldsymbol{p}} = \hat{\boldsymbol{p}}'/\mathcal{I}(\hat{\boldsymbol{p}}')$ and $\check{\boldsymbol{p}} = \check{\boldsymbol{p}}'/\mathcal{I}(\check{\boldsymbol{p}}')$. It can be observed from A2 that $\mathcal{I}(\hat{\boldsymbol{p}}) = 1$ and $\mathcal{I}(\check{\boldsymbol{p}}) = 1$ holds. Defining $\hat{\alpha} = \alpha/\mathcal{I}(\hat{\boldsymbol{p}}')$ and $\check{\beta} = \beta/\mathcal{I}(\check{\boldsymbol{p}}')$, and using (3.196), we have

$$\hat{\alpha}\mathcal{I}(\hat{\boldsymbol{p}}') + \hat{\beta}\mathcal{I}(\check{\boldsymbol{p}}') \le \mathcal{I}\big(\hat{\alpha} \cdot \hat{\boldsymbol{p}}' + \check{\beta} \cdot \check{\boldsymbol{p}}'\big) . \qquad (3.197)$$

Inequality (3.197) holds for arbitrary $\hat{\alpha}, \check{\beta} > 0$ and $\hat{\boldsymbol{p}}', \check{\boldsymbol{p}}' \in \mathbb{R}^K_{++}$, thus implying concavity of $\mathcal{I}$. □

A similar result can be shown for the set $\underline{\mathcal{R}}$. The proof is similar to the proof of Theorem 3.72, but the directions of the inequalities are reversed.

Theorem 3.73. *The set* $\underline{\mathcal{R}}$ *is non-empty* downward-comprehensive closed convex *and* $\underline{\mathcal{R}} \neq \mathbb{R}^K_{++}$ *if and only if the interference function* $\mathcal{I}$ *is convex and there exists a* $\mathbf{p} > 0$ *such that* $\mathcal{I}(\mathbf{p}) > 0$.

Applying the result to the non-supportable SIR region inroduced in Example 3.71, it follows from Theorem 3.72 that the spectral radius $\rho(\boldsymbol{\gamma}) = \rho\big(\text{diag}\{\boldsymbol{\gamma}\}\boldsymbol{V}\big)$ needs to be concave in order for the non-supportable SIR region to be convex. It was shown [87] that $\rho(\mathrm{e}^{\boldsymbol{s}}\boldsymbol{V})$ is log-convex when using the substitution $\boldsymbol{\gamma} = \exp \boldsymbol{s}$. This does not imply that $\rho(\boldsymbol{\gamma})$ is concave.

Theorem 3.57 shows that every convex interference function $\mathcal{I}(\boldsymbol{p})$ is log-convex when we substitute $\boldsymbol{p} = \mathrm{e}^{\boldsymbol{s}}$. However, this does not mean that a concave function cannot be log-convex. For example, the function $\rho(\boldsymbol{\gamma})$, as defined by (3.192), is a concave interference function, even though $\rho(\mathrm{e}^{\boldsymbol{s}}\boldsymbol{V}) = \mathrm{e}^{s_1/2}\mathrm{e}^{s_2/2}\sqrt{V_{12}V_{21}}$ is log-convex.

The following example shows a case where an interference function $\mathcal{I}(\boldsymbol{p})$ is log-convex, but not concave. This discussion shows that log-convex interference functions need neither be convex nor concave. Both cases are possible, however.

Example 3.74. Consider two log-convex interference functions $\mathcal{I}_1$ and $\mathcal{I}_2$, where $\mathcal{I}_1(\boldsymbol{p})$ only depends on $p_1, \dots, p_r$ and $\mathcal{I}_2(\boldsymbol{p})$ only depends on $p_{r+1}, \dots, p_K$. We define

$$\mathcal{I}(\boldsymbol{p}) = \max\big(\mathcal{I}_1(\boldsymbol{p}), \mathcal{I}_2(\boldsymbol{p})\big) \tag{3.198}$$

The maximum of log-convex interference functions is a log-convex interference function. However, (3.198) is not concave. In order to show this, let $\boldsymbol{p}^{(1)} = [p_1^{(1)}, \dots, p_r^{(1)}, 0, \dots, 0]^T$ and $\boldsymbol{p}^{(2)} = [0, \dots, 0, p_{r+1}^{(1)}, \dots, p_K^{(1)}]^T$ be two arbitrary vectors such that $\mathcal{I}_1(\boldsymbol{p}^{(1)}) = 1$ and $\mathcal{I}_2(\boldsymbol{p}^{(2)}) = 1$. Defining $\boldsymbol{p}(\lambda) = (1-\lambda)\boldsymbol{p}^{(1)} + \lambda\boldsymbol{p}^{(2)}$, $\lambda \in (0,1)$, we have

$$\begin{aligned} \mathcal{I}_1\big(\boldsymbol{p}(\lambda)\big) &= (1-\lambda)\mathcal{I}_1\big(\boldsymbol{p}^{(1)}\big) = 1-\lambda \\ \mathcal{I}_2\big(\boldsymbol{p}(\lambda)\big) &= \lambda\mathcal{I}_2\big(\boldsymbol{p}^{(2)}\big) = \lambda \,. \end{aligned}$$

Thus,

$$\mathcal{I}\big(\boldsymbol{p}(\lambda)\big) = \max\big((1-\lambda), \lambda\big) < \mathcal{I}(\boldsymbol{p}^{(1)}) = \mathcal{I}(\boldsymbol{p}^{(2)}) = 1 \,.$$

The superlevel set $\{\boldsymbol{p} \geq 0 : \mathcal{I}(\boldsymbol{p}) \geq 1\}$ is not convex and $\mathcal{I}$ is not concave. This example shows that log-convex interference functions need not be concave.

The results can be further generalized by assuming a bijective mapping between a QoS vector $\boldsymbol{q}$ and the associated SIR values $\gamma(\boldsymbol{q}) = [\gamma_1(q_1), \dots, \gamma_K(q_K)]^T$. For a linear interference model with a coupling matrix $\boldsymbol{V}$, the QoS region is defined as

$$F_q = \{\boldsymbol{q} : \rho\big(\text{diag}\{\gamma(\boldsymbol{q})\}\boldsymbol{V}\big) \leq 1\} \,. \tag{3.199}$$

Under which condition is the QoS region $\mathcal{F}_q$ a convex set? This question is probably difficult and only partial answers exist. It was shown in [87] that if the function $\gamma(\boldsymbol{q})$ is log-convex, then $\rho\big(\text{diag}\{\gamma(\boldsymbol{q})\}\boldsymbol{V}\big)$ is convex for all irreducible $K \times K$ matrices $\boldsymbol{V}$. In this case, convexity of $\rho\big(\text{diag}\{\gamma(\boldsymbol{q})\}\boldsymbol{V}\big)$ implies convexity of the QoS feasible region $\mathcal{F}_q$. However, the converse is not true. That is, convexity of $\mathcal{F}_q$ does not imply convexity of $\rho\big(\text{diag}\{\gamma(\boldsymbol{q})\}\boldsymbol{V}\big)$. Note, that $\rho\big(\text{diag}\{\gamma(\boldsymbol{q})\}\boldsymbol{V}\big)$ is generally not an interference function with respect to $\boldsymbol{q}$ (except e.g. for the trivial case $\gamma = \boldsymbol{q}$), thus Theorem 3.73 cannot be applied.

4

Nash Bargaining and Proportional Fairness for Log-Convex Utility Sets

Interference calculus offers an analytical framework for modeling and optimizing utility tradeoffs between users (or players, agents). In this respect it is very similar to the game-theoretic approach. Game theory was originally introduced in the context of economics and social sciences. It is now a well-established tool for analyzing communication scenarios that would otherwise be too complex to be handled analytically [89–94]. Interference calculus complements existing concepts from game theory. It captures the essential properties of interference-coupled system, yet it is simple enough to be analytically tractable.

In this chapter we discuss how interference calculus and game theory can be combined in order to extend existing results. In particular, it will be shown how the well-known Nash Bargaining strategy [16, 18, 19] can be generalized to log-convex utility sets. The results appeared in [6, 10].

The achievable performances are commonly characterized by the *utility set* (utility region) $\mathcal{U}$. The utility set $\mathcal{U}$ is defined as the set of all achievable utility vectors $\boldsymbol{u} = [u_1, \dots, u_K]^T$, where $K \geq 2$ is the number of users. A particular utility set is the SIR-based QoS region introduced in Subsection 2.6.1. Some results of this chapter are specifically derived for such QoS regions, but other results are more general and hold as well for other utility sets. Examples of utility sets in the area of wireless communications are the QoS feasibility region for the vector broadcast channel [95] and the MISO interference channel [96, 97].

Game-theoretic strategies crucially depend on the structure of the set $\mathcal{U}$, thus a thorough understanding of the properties of $\mathcal{U}$ is important. Some often-made assumptions, which are discussed in this chapter, are

- comprehensiveness,
- convexity (here, a particular focus is on strict log-convexity)
- Pareto optimality,
- achievability of the boundary.

M. Schubert, H. Boche, *Interference Calculus*, Foundations in Signal Processing, Communications and Networking 7,

Comprehensiveness can be interpreted as free disposability of utility. Convexity facilitates the computation of a global optimum. Pareto optimality of the boundary means that no resources are wasted. The achievability of the boundary is an important prerequisite for the convergence of certain algorithms.

4.1 Nash Bargaining for Strictly Log-Convex Utility Sets

Strategies for distributing resources in a multiuser system are usually based on certain notions of "fairness" or "efficiency". In this section we focus on the game-theoretic strategy of *Nash bargaining* [16, 18, 19], which is closely related to concept of *proportional fairness* [98, 99]. Nash bargaining is a *cooperative* strategy, i.e., users (or players) unanimously agree on some solution outcome $\varphi(\mathcal{U})$, as illustrated in Fig. 4.1. This outcome is generally better

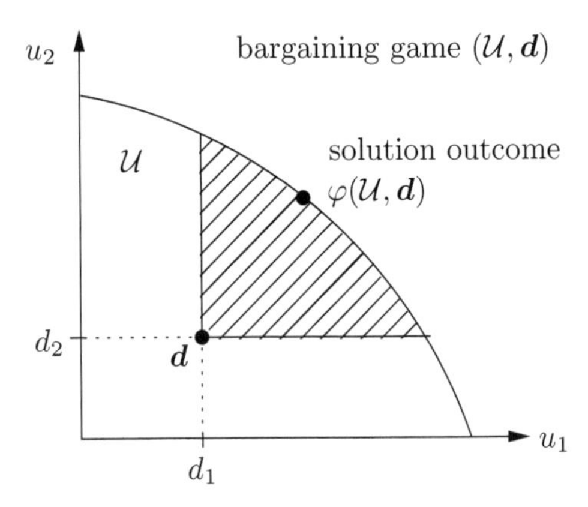

Fig. 4.1. Cooperative bargaining: given a utility region $\mathcal{U}$ and a disagreement point $\boldsymbol{d}$, the players agree on some solution outcome $\varphi(\mathcal{U})$, which is typically on the boundary of $\mathcal{U}$.

than the *Nash equilibrium* resulting from a non-cooperative approach. The gain from cooperation can be substantial (see e.g. [100, 101]). Nash bargaining was successfully applied to various problems in multi-user communication theory, e.g., [102–107].

The *Nash Bargaining Solution* (NBS) was introduced by Nash [16] and extended later (see, e.g., [18, 19, 108] and references therein). In its standard form, the NBS requires that the utility set $\mathcal{U}$ is *convex*. Nash bargaining for non-convex regions was studied, e.g. in [109–116]. However, these papers either deal with different types of regions (typically, only comprehensiveness is required, in which case uniqueness may be lost) or additional axioms are introduced in order to guarantee uniqueness. Also, most of this work was done in a context other than wireless communications.

In this section we will discuss how the original Nash bargaining framework can be generalized to certain non-convex sets [6, 10].

4.1.1 The Conventional Nash Bargaining Solution (NBS)

We begin by briefly reviewing the Nash Bargaining Solution.

Definition 4.1. *A bargaining game for K users is defined as a pair $(\mathcal{U}, \boldsymbol{d})$, where*

- $\mathcal{U}$ is a non-empty compact subset of $\mathbb{R}^K_+$.
- $\mathcal{U}$ is (downward)-comprehensive. *That is, for all $\boldsymbol{u} \in \mathcal{U}$ and $\boldsymbol{u}' \in \mathbb{R}^K_+$, the component-wise inequality $\boldsymbol{u}' \leq \boldsymbol{u}$ implies $\boldsymbol{u}' \in \mathcal{U}$.*
- $\boldsymbol{d} \in \{\boldsymbol{u} \in \mathcal{U} : \exists \boldsymbol{u}' > \boldsymbol{u}\}$ *is the* disagreement point, *which is the outcome in case that no agreement can be found.*

The class of sets with these properties is denoted by $\mathcal{D}^K$.

Definition 4.2. *Let $\mathcal{U} \in \mathcal{D}^K$ be convex, then the NBS is the unique (single-valued) solution that fulfills the following axioms.*

- Weak Pareto Optimality (WPO). *The users should not be able to collectively improve upon the solution outcome, i.e.,*

$$\varphi(\mathcal{U}) \in \{\boldsymbol{u} \in \mathcal{U} : \textit{there is no } \boldsymbol{u}' \in \mathcal{U} \textit{ with } \boldsymbol{u}' > \boldsymbol{u}\} .$$

- Symmetry (SYM). *If the game $(\mathcal{U}, \boldsymbol{d})$ is symmetric*[1]*, then the outcome does only depend on the employed strategies and not on the identities of the users, i.e., $\varphi_1(\mathcal{U}) = \cdots = \varphi_K(\mathcal{U})$. This does not mean that the utility set is symmetric, but rather that all users have the same priorities.*
- Independence of Irrelevant Alternatives (IIA). *If $\varphi(\mathcal{U})$ is the solution outcome of some utility set $\mathcal{U}$, then $\varphi(\mathcal{U})$ is also the solution outcome of every subset containing $\varphi(\mathcal{U})$, i.e.,*

$$\varphi(\mathcal{U}) \in \mathcal{U}', \textit{ with } \mathcal{U}' \subseteq \mathcal{U} \implies \varphi(\mathcal{U}') = \varphi(\mathcal{U}) .$$

- Scale Transformation Covariance (STC). *The optimization strategy is invariant with respect to a component-wise scaling of the region.*[2] *That is, for every $\mathcal{U} \in \mathcal{D}^K$, and all $\boldsymbol{a}, \boldsymbol{b} \in \mathbb{R}^K$ with $\boldsymbol{a} > 0$ and $(\boldsymbol{a} \circ \mathcal{U} + \boldsymbol{b}) \in \mathcal{D}^K$, we have*

$$\varphi(\boldsymbol{a} \circ \mathcal{U} + \boldsymbol{b}) = \boldsymbol{a} \circ \varphi(\mathcal{U}) + \boldsymbol{b} .$$

---

[1] A game $(\mathcal{U}, \boldsymbol{d})$ is said to be symmetric if $d_1 = \cdots = d_K$, and in addition, $\boldsymbol{u} = [u_1, \ldots, u_K] \in \mathcal{U} \Leftrightarrow \boldsymbol{u}' = [u_{\pi_1}, \ldots, u_{\pi_K}] \in \mathcal{U}$, for an arbitrary permutation $\boldsymbol{\pi}$.

[2] We use the component-wise Hadamard product $\circ$, and the notation $\boldsymbol{a} \circ \mathcal{U} = \{\boldsymbol{u} : \exists \boldsymbol{s} \in \mathcal{U} \text{ with } \boldsymbol{u} = \boldsymbol{a} \circ \boldsymbol{s}\}$.

If the utility set $\mathcal{U}$ is compact convex comprehensive, then the single-valued NBS fulfilling the four axioms is obtained by maximizing the product of utilities (Nash product).

$$\max_{\boldsymbol{u}\in\mathcal{U},\boldsymbol{u}\geq\boldsymbol{d}} \prod_{k\in\mathcal{K}} (u_k - d_k) . \tag{4.1}$$

Nash introduced the bargaining problem in [16] for convex compact sets and two players. Later, in [17] he extended this work by introducing the concept of a *disagreement point* (also known as *threat point*), which is the solution outcome in case that the players are unable to reach a unanimous agreement. Some "non-standard" variations of the Nash bargaining problem exist, including non-convex regions (see e.g. [109, 112–114]) and problem formulations without disagreement point (see e.g. [19] and the references therein).

In this chapter we study the Nash bargaining problem without disagreement point, i.e., $\boldsymbol{d} = \boldsymbol{0}$. Therefore, the axiom STC differs slightly from its common definition used in game-theoretical literature (e.g. [18]), where an additional invariance with respect to a translation of the region is required. Omitting the disagreement point is justified by the special structure of the problem under consideration. We are interested in utility sets for which the existence of a solution is always guaranteed. From a mathematical point of view, zero utilities must be excluded because of the possibility of singularities (SIR tending to infinity). However, from a technical perspective this corresponds to a bargaining game with disagreement point zero. The results are also relevant for certain games with non-zero disagreement point: if the zero of the utility scales does not matter then we can reformulate the game within a transformed coordinate system.

This leads to the following problem formulation, which is illustrated in Fig. 4.2.

$$\max_{\boldsymbol{u}\in\mathcal{U}} \prod_{k\in\mathcal{K}} u_k . \tag{4.2}$$

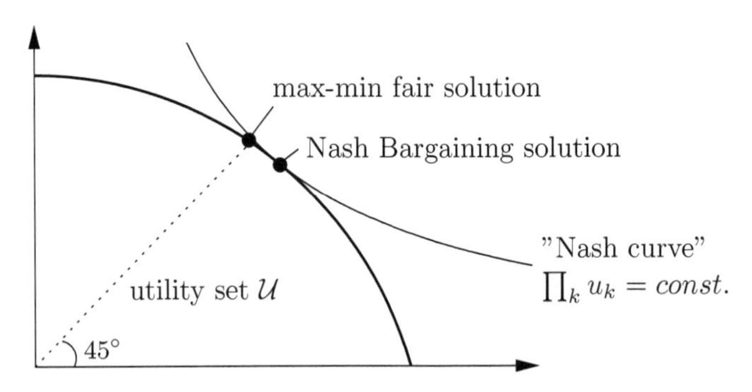

Fig. 4.2. Illustration of the Nash Bargaining solution

4.1.2 Proportional Fairness

Since $\log \max \prod_k u_k = \max \log \prod_k u_k = \max \sum_k \log u_k$, the optimum (4.2) can be found by solving

$$\max_{\boldsymbol{u} \in \mathcal{U}} \sum_{k \in \mathcal{K}} \log u_k \,. \tag{4.3}$$

In the following, we will refer to strategy (4.3) as *proportional fairness* (PF). In its original definition [98], a vector $\boldsymbol{u}^*$ is said to be proportionally fair if for any other feasible vector $\boldsymbol{u} \in \mathcal{U}$ the aggregated proportional change $\sum_k (u_k - u_k^*)/u_k^*$ is non-positive (see also [117]). For convex sets, this unique point is obtained as the optimizer of (4.3). In this case, Nash bargaining and proportional fairness are equivalent [98,102]. This relates the NBS to a known fairness criterion (see also [99,102,103,105,106,111]).

For every compact convex set from $\mathcal{D}^K$, the product maximizer (4.2) is the single-valued NBS characterized by the axioms WPO, SYM, IIA, and STC. However, convexity does not need to be fulfilled. An example is the SINR region discussed in Section 4.5.

A standard approach is to convexify the utility set based on randomization arguments (see e.g. [107, 114]), or by resource sharing. However, such a strategy is not always possible or even relevant. Again, the SINR region provides an example of a performance measure for which convexification is difficult to justify. Extensions and modifications of the NBS to non-convex utility sets have been studied in the literature, e.g., [109, 112–114]. However, the motivation of these papers is quite different from our approach.

In the following we discuss how Nash bargaining and proportional fairness can be generalized to certain non-convex sets that are log-convex.

4.1.3 Log-Convex Utility Sets

Consider the function $\log(\boldsymbol{u}) = [\log u_1, \ldots, \log u_K]^T$, where $\boldsymbol{u} \in \mathcal{U} \cap \mathbb{R}^K_{++}$. The image set of $\mathcal{U}$ is

$$\mathcal{L}og(\mathcal{U}) := \{\boldsymbol{q} = \log(\boldsymbol{u}) : \boldsymbol{u} \in \mathcal{U} \cap \mathbb{R}^K_{++}\} \,. \tag{4.4}$$

Definition 4.3. *We say that a set $\mathcal{U} \subseteq \mathbb{R}^K_+$ is a* log-convex *set if $\mathcal{L}og(\mathcal{U})$ is convex.*

The boundary of the utility set $\mathcal{U}$ is denoted by $\partial\mathcal{U}$. In the following, we focus on boundary points that are *Pareto optimal.* From a practical point of view, Pareto optimal means that it is not possible to improve the performance of one user without decreasing the performance of another user. A Pareto optimal operating point is "efficient" in the sense that the available system resources are fully utilized.

Definition 4.4. *A boundary point $\boldsymbol{u} \in \partial\mathcal{U}$ is said to be* Pareto optimal *if there is no $\hat{\boldsymbol{u}} \in \partial\mathcal{U}$ with $\hat{\boldsymbol{u}} \gneq \boldsymbol{u}$. The set of all Pareto optimal boundary points (the* Pareto boundary*) is denoted by $PO(\mathcal{U})$.*

Pareto optimality of certain SIR-based QoS sets will be studied later, in Section 4.5. In this chapter, we are interested in certain families of utility sets that are log-convex. These families $\mathcal{ST}$ and $\mathcal{ST}_c$ are specified as follows.

Definition 4.5. *By $\mathcal{ST}$ we denote the family of all closed downward-comprehensive utility sets $\mathcal{U} \subset \mathbb{R}_+^K$ such that the image set $\mathcal{Q} := \mathcal{L}og(\mathcal{U})$ is convex and the following additional property is fulfilled: For any $\hat{\boldsymbol{q}}, \check{\boldsymbol{q}} \in PO(\mathcal{Q})$, the connecting line $\boldsymbol{q}(\lambda) = (1-\lambda)\hat{\boldsymbol{q}} + \lambda\check{\boldsymbol{q}}$, with $\lambda \in (0,1)$, is contained in the interior of $\mathcal{Q}$. By $\mathcal{ST}_c$ we denote the family of all $\mathcal{U} \in \mathcal{ST}$, which are additionally bounded, thus compact.*

Definition 4.5 is illustrated in Fig. 4.3.

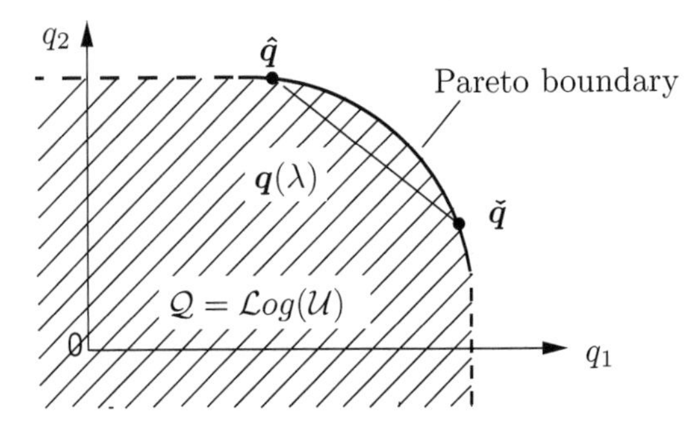

Fig. 4.3. Illustration of an image set $\mathcal{Q} := \mathcal{L}og(\mathcal{U})$ for $\mathcal{U} \in \mathcal{ST}_c$. The set is strictly convex with the exception of possible boundary segments parallel to the axes (dashed lines). These segments are irrelevant for the Nash solution.

In the following two subsections, we discuss how the Nash bargaining framework can be extended to certain non-convex utility sets, We begin by addressing the case of bounded sets $\mathcal{ST}_c$. The case of unbounded sets $\mathcal{ST}$ will be discussed later in Subsection 4.1.5.

4.1.4 Generalized Nash Bargaining for Compact Log-Convex Sets

We begin by observing that compactness and comprehensiveness are preserved by the log-transformation. That is, $\mathcal{U} \subset \mathbb{R}_+^K$ is compact comprehensive if and only if $\mathcal{L}og(\mathcal{U}) \subset \mathbb{R}^K$ is compact comprehensive. Every convex set from $\mathcal{D}^K$ is contained in $\mathcal{ST}_c$, but not conversely. Thus, $\mathcal{ST}_c$ is more general than the class of standard sets described by Definition 4.1. In the following we show that for any $\mathcal{U} \in \mathcal{ST}_c$, the product maximization (4.2) yields the single-valued NBS characterized by axioms WPO, SYM, IIA, and STC. This extends the classical Nash bargaining framework to certain non-convex sets $\mathcal{ST}_c$.

The properties of $\mathcal{ST}_c$ play an important role for the proof of uniqueness. We also exploit that the axioms WPO, SYM, IIA, and STC have direct counterparts for the image set $\mathcal{Q} := \mathcal{L}og(\mathcal{U})$. This is straightforward for axioms

WPO, SYM, and IIA, which are not affected by the logarithmic transformation. That is, axiom WPO in the utility set $\mathcal{U}$ corresponds directly to WPO in the image set $\mathcal{Q}$. The same holds for axioms SYM and IIA. We will denote the axioms associated with the image set by $\text{WPO}_\mathcal{Q}$, $\text{SYM}_\mathcal{Q}$, and $\text{IIA}_\mathcal{Q}$.

Axiom STC in the utility set $\mathcal{U} \in \mathcal{ST}_c$ also has a direct correspondence for the image set $\mathcal{Q} := \mathcal{L}og(\mathcal{U})$. Consider an arbitrary translation $\tilde{\boldsymbol{q}} \in \mathbb{R}^K$, leading to a translated set $\mathcal{Q}(\tilde{\boldsymbol{q}})$, defined as

$$\mathcal{Q}(\tilde{\boldsymbol{q}}) = \{\boldsymbol{q} \in \mathbb{R}^K : \exists \boldsymbol{q}_0 \in \mathcal{Q} \text{ with } \boldsymbol{q} = \boldsymbol{q}_0 + \tilde{\boldsymbol{q}}\} .$$

Also, let $\varphi_\mathcal{Q}$ be the log-transformed Nash bargaining solution, i.e., $\varphi_\mathcal{Q} = \log \varphi(\mathcal{U})$. Since the disagreement point is zero in our case, axiom STC becomes $\varphi(\boldsymbol{a} \circ \mathcal{U}) = \boldsymbol{a} \circ \varphi(\mathcal{U})$. In the log-transformed domain, this corresponds to

$$\varphi_\mathcal{Q}(\mathcal{Q}(\tilde{\boldsymbol{q}})) = \varphi_\mathcal{Q}(\mathcal{Q}) + \tilde{\boldsymbol{q}} . \tag{4.5}$$

We will refer to property (4.5) as $\text{STC}_\mathcal{Q}$. The following theorem shows that the transformed axioms are associated with a unique solution outcome $\varphi_\mathcal{Q}$ in the transformed set.

Theorem 4.6. *For an arbitrary set $\mathcal{U} \in \mathcal{ST}_c$, the solution outcome $\varphi_\mathcal{Q}$ in the transformed set $\mathcal{Q} = \mathcal{L}og(\mathcal{U})$ satisfies axioms WPO_Q, SYM_Q, STC_Q, and IIA_Q if and only if it is the unique maximizer*

$$\varphi_\mathcal{Q}(\mathcal{Q}) = \arg\max_{\boldsymbol{q} \in \mathcal{Q}} \sum_{k \in \mathcal{K}} q_k . \tag{4.6}$$

Proof. Non-Pareto-optimal boundary segments parallel to the axes can be safely excluded from the proof, since such points cannot be the solution of the product maximization (4.2). Thus, without loss of generality we can assume that $\mathcal{Q}$ is strictly convex. Given the properties of the region $\mathcal{U} \in \mathcal{ST}_c$ and its image set $\mathcal{L}og(\mathcal{U})$, it is clear that the solution (4.6) satisfies the axioms WPO_Q, SYM_Q, STC_Q, and IIA_Q.

It remains to show the converse. Consider a bargaining strategy on $\mathcal{Q} = \mathcal{L}og(\mathcal{U})$, that satisfies the axioms WPO_Q, SYM_Q, STC_Q, IIA_Q. We now show that these axioms are fulfilled by a unique solution, which is the optimizer of (4.6). This is illustrated by Figure 4.4.

Consider the set

$$\mathcal{Q}_1 := \{\boldsymbol{q} \in \mathbb{R}^K : \sum_{k \in \mathcal{K}} q_k \leq K\} .$$

Because of the $\text{STC}_\mathcal{Q}$ property (4.5), we know that the strategy is invariant with respect to a translation of the region. Thus, without loss of generality we can assume $\mathcal{Q} \subseteq \mathcal{Q}_1$, and

$$\hat{\boldsymbol{q}} = [1, \ldots, 1]^T = \arg\max_{\boldsymbol{q} \in \mathcal{Q}} \sum_{k \in \mathcal{K}} q_k .$$

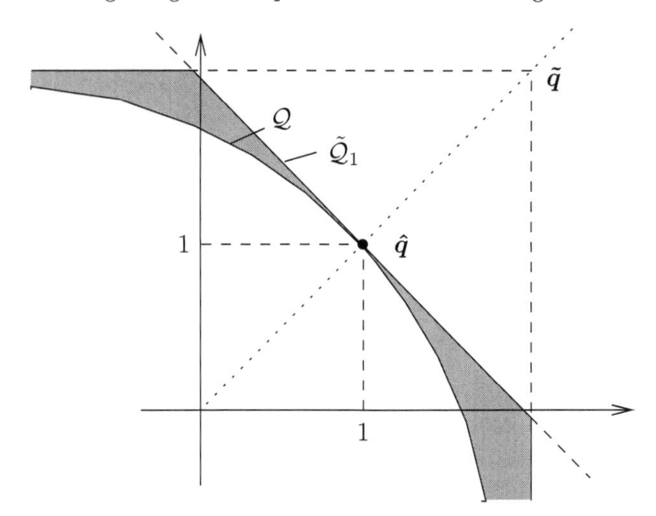

Fig. 4.4. Illustration of the proof of Theorem 4.6. The NBS in the transformed set $\mathcal{Q}$ is the unique solution that satisfies the transformed axioms.

That is, $\hat{\boldsymbol{q}}$ is the unique point which is on the boundaries of both sets $\mathcal{Q}$ and $\mathcal{Q}_1$. Since $\mathcal{Q}$ is upper-bounded by definition, there is a $\tilde{\boldsymbol{q}} \in \mathbb{R}^K$ such that

$$\tilde{\boldsymbol{q}} \geq \boldsymbol{q} \quad \text{for all } \boldsymbol{q} \in \mathcal{Q} \,.$$

Thus, $\mathcal{Q}$ is a sub-set of the set

$$\tilde{\mathcal{Q}}_1 = \{\boldsymbol{q} \in \mathcal{Q}_1 : \boldsymbol{q} \leq \tilde{\boldsymbol{q}}\} \,. \tag{4.7}$$

The set $\tilde{\mathcal{Q}}_1$ is symmetric and strictly convex. Let $\tilde{\mathcal{Q}}$ be the smallest symmetric and strictly convex closed set that fulfills

$$\tilde{\mathcal{Q}}_1 \supseteq \tilde{\mathcal{Q}} \supseteq \mathcal{Q} \,. \tag{4.8}$$

Since $\tilde{\mathcal{Q}}_1$ is upper-bounded, the set $\tilde{\mathcal{Q}}$ is compact. It is also strictly convex comprehensive, thus it is contained in $\mathcal{L}og(\mathcal{ST}_c)$, which is the class of all sets $\mathcal{L}og(\mathcal{U})$ such that $\mathcal{U} \in \mathcal{ST}_c$. Because of axiom SYM_Q, it follows that

$$\varphi_{\mathcal{Q}}(\tilde{\mathcal{Q}}) = \hat{\boldsymbol{q}} = [1, \ldots, 1]^T \,.$$

$\sum_k q_k = K$ describes a supporting hyperplane for $\tilde{\mathcal{Q}}$, i.e., $\hat{\boldsymbol{q}}$ is an optimizer of

$$\varphi_{\mathcal{Q}}(\tilde{\mathcal{Q}}) = \arg\max_{\boldsymbol{q} \in \tilde{\mathcal{Q}}} \sum_{k \in \mathcal{K}} q_k \,.$$

Now, $\mathcal{Q} \subseteq \tilde{\mathcal{Q}}$ and $\hat{\boldsymbol{q}} \in \mathcal{Q}$. Because of axiom $\text{IIA}_{\mathcal{Q}}$, we have

$$\varphi_{\mathcal{Q}}(\mathcal{Q}) = \varphi_{\mathcal{Q}}(\tilde{\mathcal{Q}}) = [1, \ldots, 1]^T = \arg\max_{\boldsymbol{q} \in \mathcal{Q}} \sum_{k \in \mathcal{K}} q_k \,, \tag{4.9}$$

which concludes the proof. □

Consequently, for all $\mathcal{U} \in \mathcal{ST}_c$ the optimization (4.9) in the transformed domain $\mathcal{Q} = \mathcal{L}og(\mathcal{U})$ leads to the unique optimum $\varphi_{\mathcal{Q}}(\mathcal{Q})$. Because of the strictly monotone logarithmic mapping between the sets $\mathcal{Q}$ and $\mathcal{U}$, we have the following result.

Corollary 4.7. *Let $\mathcal{U} \in \mathcal{ST}_c$. Then axioms WPO, SYM, STC, and IIA are satisfied by the unique solution*

$$\varphi(\mathcal{U}) = \arg\max_{\boldsymbol{u} \in \mathcal{U}} \prod_{k \in \mathcal{K}} u_k \,. \tag{4.10}$$

This result holds for arbitrary utility sets from $\mathcal{ST}_c$, including the conventional case of convex sets. An application example is the log-convex SIR region that will be discussed later in Section 4.4. Under certain conditions, the resulting QoS region is contained in $\mathcal{ST}_c$.

4.1.5 Generalized Nash Bargaining for Unbounded Sets

Sets from $\mathcal{ST}$ can be unbounded. As a consequence, the product maximization problem (4.2) does not need to have a solution. The following Theorem 4.8 provides a necessary and sufficient condition for the existence of a unique solution. Later, in Section 4.3.4 it will be discussed how the result can be applied in the context of SIR-based QoS regions.

We begin by introducing an auxiliary set

$$\underline{\mathcal{U}}(\lambda) = \mathcal{U} \cap \mathcal{G}(\lambda) \tag{4.11}$$

where

$$\mathcal{G}(\lambda) = \left\{ \boldsymbol{u} \in \mathbb{R}^K_{++} : \sum_{k \in \mathcal{K}} u_k \leq \lambda \right\}, \quad \lambda > 0 \,. \tag{4.12}$$

The set $\underline{\mathcal{U}}(\lambda)$ is illustrated in Fig. 4.5. Unlike $\mathcal{U}$, the set $\underline{\mathcal{U}}(\lambda)$ is always con-

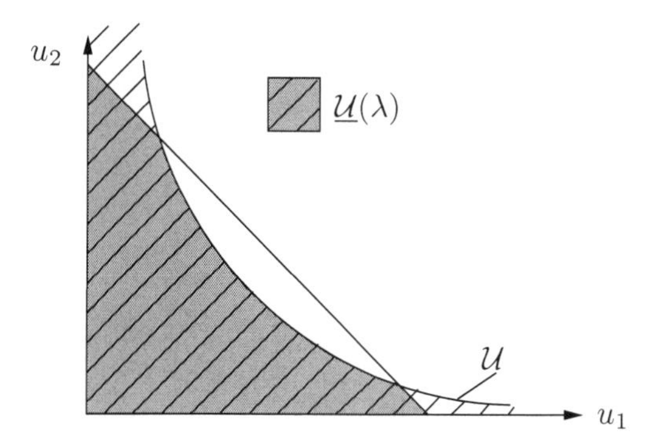

Fig. 4.5. The unbounded utility set $\mathcal{U}$ is approximated by a bounded set $\underline{\mathcal{U}}(\lambda) \in \mathcal{ST}_c$

tained in $\mathcal{ST}_c$. Thus, there is a unique Nash bargaining solution $\varphi(\underline{\mathcal{U}}(\lambda))$, given as the optimizer of the Nash product. The associated utilities are denoted by $\boldsymbol{u}(\lambda)$.

Theorem 4.8. *Let $\mathcal{U} \in \mathcal{ST}$. The product maximization problem (4.2) has a unique solution $\hat{\boldsymbol{u}}$ if and only if there exists a $\hat{\lambda}$ such that for all $\lambda \geq \hat{\lambda}$*

$$\varphi\big(\underline{\mathcal{U}}(\lambda)\big) = \arg\max_{\boldsymbol{u}\in\underline{\mathcal{U}}(\lambda)} \prod_{k\in\mathcal{K}} u_k = \boldsymbol{u}(\hat{\lambda}) \,. \tag{4.13}$$

Then, $\hat{\boldsymbol{u}} = \boldsymbol{u}(\hat{\lambda})$.

Proof. Assume that there is a $\hat{\lambda}$ such that (4.13) holds for any $\lambda \geq \hat{\lambda}$. Then, $\boldsymbol{u}(\hat{\lambda})$ is the solution of (4.2) for the set $\underline{\mathcal{U}}(\lambda)$. The solution is unique because $\underline{\mathcal{U}}(\lambda) \in \mathcal{ST}_c$. Thus, $\boldsymbol{u}(\hat{\lambda})$ is also the unique optimizer of the larger set $\mathcal{U}$.

With $\underline{\mathcal{U}}(\lambda) \subseteq \mathcal{U}$, we have

$$\max_{\boldsymbol{u}\in\underline{\mathcal{U}}(\lambda)} \prod_{k\in\mathcal{K}} u_k \leq \sup_{\boldsymbol{u}\in\mathcal{U}} \prod_{k\in\mathcal{K}} u_k =: C \,. \tag{4.14}$$

We show by contradiction that the supremum is finite. If $C = +\infty$, then for any $\mu > 0$ there is a $\boldsymbol{u}^{(\mu)} \in \mathcal{U}$ such that $\prod_k u_l^{(\mu)} > \mu$. There always exists a $\lambda \geq \hat{\lambda}$ such that $\boldsymbol{u}^{(\mu)} \in \underline{\mathcal{U}}(\lambda)$. Thus, the value $\max_{\boldsymbol{u}\in\underline{\mathcal{U}}(\lambda)} \prod_k u_k$ could become arbitrarily large, which contradicts the assumption that (4.13) holds for arbitrary $\lambda \geq \hat{\lambda}$. This implies $C < +\infty$. Inequality (4.14) is satisfied with equality for all $\lambda \geq \hat{\lambda}$. Since $\boldsymbol{u}(\hat{\lambda}) \in \mathcal{U}$, we have $\sup_{\boldsymbol{u}\in\mathcal{U}} \prod_k (u_k) = \prod_k (\hat{u}_k)$. That is, the maximum (4.2) is attained by $\boldsymbol{u}(\hat{\lambda})$.

Conversely, assume that $\hat{\boldsymbol{u}}$ is the solution of the product maximization (4.2). For any $\lambda > 0$, we have

$$\max_{\boldsymbol{u}\in\underline{\mathcal{U}}(\lambda)} \prod_{k\in\mathcal{K}} (u_k) \leq \max_{\boldsymbol{u}\in\mathcal{U}} \prod_{k\in\mathcal{K}} (u_k) = \prod_{k\in\mathcal{K}} (\hat{u}_k) \,. \tag{4.15}$$

There exists a $\hat{\lambda}$ for which this inequality is fulfilled with equality, with the maximizer $\boldsymbol{u}(\hat{\lambda}) = \hat{\boldsymbol{u}}$. This solution is also contained in any larger set $\underline{\mathcal{U}}(\lambda)$ where $\lambda \geq \hat{\lambda}$. □

Theorem 4.8 shows that the Nash bargaining framework outlined in Section 4.1.1 also holds for certain non-compact non-convex sets, provided that an optimizer exists. An example is the SIR region that will be discussed in Subsection 4.3.4).

4.1.6 Non-Symmetric Nash Solution and the Conjugate $\underline{f}_{\mathcal{I}}$

The Nash bargaining solution is symmetric, i.e., all users have the same priorities. But sometimes, a non-symmetric strategy is need. In [118], the connection between interference calculus and *non-symmetric Nash bargaining* was discussed. Given weighting factors $\boldsymbol{\alpha} = [\alpha_1, \dots, \alpha_K]^T$, with $\|\boldsymbol{\alpha}\|_1 = 1$, and a convex comprehensive utility set $\mathcal{V}$, the non-symmetric Nash solution is as follows.

$$\mathcal{N}(\boldsymbol{\alpha}) = \max_{\boldsymbol{v} \in \mathcal{V}} \prod_{k \in \mathcal{K}} (v_k)^{\alpha_k} . \tag{4.16}$$

This solution is characterized by a different set of axioms, as discussed in [118].

We can exploit that the utility set $\mathcal{V}$ is convex comprehensive. It was shown in Subsection 3.5 that any convex compact downward-comprehensive set from $\mathbb{R}^K_{++}$ can be expressed as a sublevel set of a convex interference function. From Theorem 3.45 we know that there is a convex interference function $\mathcal{I}_1$ such that

$$\mathcal{V} = \{\boldsymbol{v} > 0 : \mathcal{I}_1(\boldsymbol{v}) \leq 1\} .$$

The bargaining solution (4.16) is attained on the boundary of $\mathcal{V}$ being characterized by $\mathcal{I}_1(\boldsymbol{v}) = 1$. Thus, (4.16) can be rewritten as [118]

$$\mathcal{N}(\boldsymbol{\alpha}) = \max_{\{\boldsymbol{v}>0:\mathcal{I}_1(\boldsymbol{v})=1\}} \prod_{l \in \mathcal{K}} (v_l)^{\alpha_l} = \sup_{\boldsymbol{v}>0} \frac{\prod_{l \in \mathcal{K}} (v_l)^{\alpha_l}}{\mathcal{I}_1(\boldsymbol{v})} . \tag{4.17}$$

From Subsection 3.6.5 we know that every convex interference function is a log-convex interference function in the sense of Definition 1.4 (after the change of variable $\boldsymbol{s} = \log \boldsymbol{p}$). Thus, $\mathcal{I}_1$ can be expressed as (3.126). Comparing (4.17) with the function $\underline{f}_{\mathcal{I}_1}$, as defined by (3.117), we have

$$\mathcal{N}(\boldsymbol{\alpha}) = \frac{1}{\underline{f}_{\mathcal{I}_1}(\boldsymbol{\alpha})} . \tag{4.18}$$

This provides an interesting link between the Nash bargaining theory and the theory of (log-convex) interference functions. Problem (4.16) can also be interpreted as a *proportional fair* operating point [98] of a wireless system.

Note that there are other bargaining strategies which only rely on downward-comprehensive utility sets. Also in this case the set can be expressed as a sublevel set of an interference function, as shown by Theorem 3.11.

4.2 The SIR Region of Log-Convex Interference Functions

Consider the SIR region

$$\mathcal{S} = \{\boldsymbol{\gamma} \in \mathbb{R}^K_{++} : C(\boldsymbol{\gamma}) \leq 1\} , \tag{4.19}$$

which was introduced and motivated in Subsection 1.4.5. The structure of the region is determined by the min-max indicator function $C(\boldsymbol{\gamma})$. The boundary is characterized by $C(\boldsymbol{\gamma}) = 1$.

Resource allocation algorithms typically aim at finding a suitable operating point on the boundary. In this context, a crucial question is whether such a point is guaranteed to exist or not. Sometimes, the boundary can only be approached asymptotically, as discussed in Subsection 2.6.3. This can prevent the algorithm from converging. Thus, the achievability of boundary points is an important property of the SIR region.

The achievability of the boundary is closely connected with the existence of a fixed point $\boldsymbol{p}^*$ satisfying the following equation (see Subsection 2.6.3).

$$\boldsymbol{p} = \frac{1}{C(\boldsymbol{\gamma})} \operatorname{diag}\{\boldsymbol{\gamma}\}\boldsymbol{\mathcal{I}}(\boldsymbol{p}) .$$

Whether or not a solution exists for all boundary points $\boldsymbol{\gamma}$ depends on the structure of the underlying interference functions $\mathcal{I}_1, \ldots, \mathcal{I}_K$. In particular, it depends on how the interference functions are mutually coupled.

For linear interference functions, the existence of a fixed point is well understood. Boundary point are achievable if the coupling matrix is irreducible. This follows directly from the Perron-Frobenius theory of non-negative matrices. The entire boundary is achievable if the coupling matrix (1.9) is irreducible. However, achievability can be quite difficult to characterize for more general interference functions [2].

In this section, we focus on the class of log-convex interference functions. By exploiting the structure results from Section 3.6, it is possible to extend many of properties that are known for the linear case. Therefore, log-convex interference functions can be regarded as a natural extension of linear interference functions.

4.2.1 Existence of a Fixed Point for Constant $\boldsymbol{W}$

It was shown in Section 3.6 that every log-convex interference function can be represented as (3.126), based on coupling coefficients $\boldsymbol{w} \geq 0$, with $\|\boldsymbol{w}\|_1 = 1$. Now, we study the interactions between K log-convex interference functions. By $\boldsymbol{w}_k$ we denote a coefficient vector associated with user k. All coefficients are collected in a matrix

$$\boldsymbol{W} = [\boldsymbol{w}_1, \ldots, \boldsymbol{w}_K]^T \geq 0 , \quad \text{with } \|\boldsymbol{w}_k\|_1 = 1, \ \forall k \in \mathcal{K} .$$

Only in this subsection, it will be assumed that $\boldsymbol{W}$ is *constant*. This approach simplifies the analysis and reveals some characteristic properties. Arbitrary log-convex interference functions will be studied later in Section 4.2.4.

Because of the property $\sum_l w_{kl} = 1$, the matrix $\boldsymbol{W}$ is (row) stochastic. Let $\mathbf{1}$ be the all-one vector, then

$$\boldsymbol{W}\mathbf{1} = \mathbf{1} . \tag{4.20}$$

For arbitrary constants $f_k > 0$, we obtain interference functions

$$\mathcal{I}_k(\boldsymbol{p}, \boldsymbol{W}) = f_k \cdot \prod_{l \in \mathcal{K}} (p_l)^{w_{kl}} \,, \quad k \in \mathcal{K} \,. \tag{4.21}$$

The resulting min-max optimum for a constant $\boldsymbol{W}$ is

$$C(\boldsymbol{\gamma}, \boldsymbol{W}) = \inf_{\boldsymbol{p}>0} \Big(\max_{k \in \mathcal{K}} \frac{\gamma_k \mathcal{I}_k(\boldsymbol{p}, \boldsymbol{W})}{p_k} \Big) \,. \tag{4.22}$$

We are now interested in the existence of a fixed point $\boldsymbol{p}^* > 0$ fulfilling

$$C(\boldsymbol{\gamma}, \boldsymbol{W}) \, \boldsymbol{p}^* = \boldsymbol{\Gamma} \boldsymbol{\mathcal{I}}(\boldsymbol{p}^*, \boldsymbol{W}) \,. \tag{4.23}$$

The next lemma provides a necessary and sufficient condition for strict positivity of the fixed point. This basic property will be used later, e.g. in the proof of Theorem 4.14.

Lemma 4.9. *Let* $\boldsymbol{t} := (\gamma_1 f_1, \dots, \gamma_K f_K)^T$. *Equation (4.23) has a solution* $\boldsymbol{p}^* > 0$ *if and only if an additive translation of* $\log \boldsymbol{t}$ *(component-wise logarithm) lies in the range of the matrix* $\boldsymbol{I} - \boldsymbol{W}$. *That is, iff there exists a* $\mathcal{C} = \mathcal{C}(\boldsymbol{\gamma}, \boldsymbol{W}) = \log C(\boldsymbol{\gamma}, \boldsymbol{W}) \in \mathbb{R}$ *such that we can find an* $\boldsymbol{s}^* \in \mathbb{R}^K$ *with*

$$(\boldsymbol{I} - \boldsymbol{W})\boldsymbol{s}^* = \log \boldsymbol{t} - \mathcal{C}\boldsymbol{1} \,, \tag{4.24}$$

where $\boldsymbol{p}^* = \exp\{\boldsymbol{s}^*\}$ *(component-wise).*

Proof. Suppose there exists a $\boldsymbol{s}^* \in \mathbb{R}^K$ and a $\mathcal{C} \in \mathbb{R}$ such that (4.24) is fulfilled. Taking $\exp\{\cdot\}$ of both sides of (4.24), we have for all $k \in \mathcal{K}$,

$$\exp\{s_k^* - \sum_{l \in \mathcal{K}} w_{kl} s_l^*\} = \frac{p_k^*}{\prod_{l \in \mathcal{K}} (p_l^*)^{w_{kl}}} = \gamma_k f_k \frac{1}{C(\boldsymbol{\gamma}, \boldsymbol{W})} \,.$$

With (4.21) it follows that $\boldsymbol{p}^* = \exp\{\boldsymbol{s}^*\} > 0$ is a fixed point of (4.23), i.e., the infimum $C(\boldsymbol{\gamma}, \boldsymbol{W})$ is achieved.

Conversely, assume that there exists a solution $\boldsymbol{p}^* > 0$ such that (4.23) is fulfilled. By taking the logarithm of both sides we obtain (4.24). □

To conclude, if there exists a $\mathcal{C} \in \mathbb{R}$ such that $\log \boldsymbol{t} - \mathcal{C}\boldsymbol{1}$ lies in the range of $\boldsymbol{I} - \boldsymbol{W}$, then there is a $\boldsymbol{s}^* \in \mathbb{R}^K$ such that (4.24) holds. Thus, the existence of a fixed point $\boldsymbol{p}^* > 0$ depends on the subspace structure of $\boldsymbol{I} - \boldsymbol{W}$.

Corollary 4.10. *If there exists a* $\mathcal{C} \in \mathbb{R}$ *such that (4.24) holds, then* $\mathcal{C}$ *is unique.*

Proof. This follows from Lemma 2.21 and Lemma 4.9. □

Next, we show how the existence of a strictly positive fixed point depends on the structure of the non-negative square row stochastic matrix $\boldsymbol{W}$. We may assume, without loss of generality, that after simultaneous permutations of rows and columns, $\boldsymbol{W}$ is reduced to canonical form (4.85), with irreducible blocks along the main diagonal (see Appendix A.1 for a definition of irreducibility). The dimension of each square block $\boldsymbol{W}^{(n)} := \boldsymbol{W}^{(n,n)}$ on the main diagonal is equal or greater than two. This is a consequence of A1, which implies that each user is interfered by at least one other user. If $\boldsymbol{W}$ is irreducible, then it consists of one single block. Note, that the off-diagonal blocks need not be square.

Definition 4.11. *A diagonal block $\boldsymbol{W}^{(n)}$ is called* isolated *if $\boldsymbol{W}^{(n,m)} = \mathbf{0}$ for $m = 1, 2, \ldots, n-1$. We assume, without loss of generality, that the first i blocks are isolated.*

Definition 4.12. *A diagonal block is called* maximal *if its spectral radius equals the overall spectral radius $\rho(\boldsymbol{W})$.*

From the results of Section 3.6 we know that the matrix $\boldsymbol{W}$ is stochastic. There are some useful consequences, which are summarized by the following lemma.

Lemma 4.13. *If $\boldsymbol{W} \geq 0$ is stochastic then*

- *$\rho(\boldsymbol{W}) = 1$, which is a consequence of (4.20) and the Perron-Frobenius theorem. We have $\rho(\boldsymbol{W}) = \max_{1 \leq n \leq N} \rho(\boldsymbol{W}^{(n)}) = 1$.*
- *A diagonal block is maximal if and only if it is isolated. This follows from (4.20) and the results [81]. For all non-isolated blocks, we have $\rho(\boldsymbol{W}^{(n)}) < 1$.*
- *$\boldsymbol{I} - \boldsymbol{W}$ is singular, which becomes evident when rewriting (4.20) as $(\boldsymbol{I} - \boldsymbol{W})\mathbf{1} = \mathbf{0}$.*

We begin with the simple case where $\boldsymbol{W}$ consists of a single irreducible block.

Theorem 4.14. *Let $\boldsymbol{W} \geq 0$ be row-stochastic and irreducible, then there exists a unique (up to a scaling) fixed point $\mathbf{p}^* > 0$ fulfilling (4.23).*

Proof. The proof is given in the Appendix A.9. □

Next, we will address the more general case where $\boldsymbol{W}$ can be reducible. Without loss of generality, the canonical form (4.85) can be assumed. We exploit the special properties of stochastic matrices (cf. Lemma 4.13). In particular, each isolated block has a spectral radius one, and the non-isolated blocks have a spectral radius strictly less than one.

Let K_n denote the number of users belonging to the nth block $\boldsymbol{W}^{(n)}$, and $\mathcal{K}_n$ is the set of associated user indices. Also, $\boldsymbol{\gamma}^{(n)} \in \mathbb{R}_{++}^{K_n}$ is the vector of SIR targets associated with this block.

For each isolated block n, with $1 \leq n \leq i$, we define

$$C(\boldsymbol{\gamma}^{(n)}, \boldsymbol{W}^{(n)}) = \inf_{\boldsymbol{p} \in \mathbb{R}_{++}^{K}} \left(\max_{k \in \mathcal{K}_n} \frac{\gamma_k \mathcal{I}_k(\boldsymbol{p}, \boldsymbol{W})}{p_k} \right) \tag{4.25}$$

$$\leq C(\boldsymbol{\gamma}, \boldsymbol{W}) \, . \tag{4.26}$$

This inequality is a consequence of definition (4.22), where a larger set $\mathcal{K}$ is used instead of $\mathcal{K}_n$. Each isolated block n only depends on powers from the same block, so the users associated with this block form an independent subsystem.

The next lemma shows that $C(\boldsymbol{\gamma}, \boldsymbol{W})$ only depends on the isolated blocks. Inequality (4.26) is fulfilled with equality for at least one isolated block.

Lemma 4.15. *Let $\boldsymbol{W}$ be a row-stochastic matrix in canonical form (4.85), and $\boldsymbol{W}^{(1)}, \ldots, \boldsymbol{W}^{(i)}$ are the isolated irreducible blocks on the main diagonal, then*

$$C(\boldsymbol{\gamma}, \boldsymbol{W}) = \max_{1 \le n \le i} C(\boldsymbol{\gamma}^{(n)}, \boldsymbol{W}^{(n)}) . \tag{4.27}$$

Proof. The proof is given in the Appendix A.9. □

The proof of Lemma 4.15 shows that there always exists a vector $\hat{\boldsymbol{p}} > 0$ such that

$$\max_{k \in \mathcal{K}} \frac{\gamma_k \mathcal{I}_k(\hat{\boldsymbol{p}}, \boldsymbol{W})}{\hat{p}_k} = C(\boldsymbol{\gamma}, \boldsymbol{W}) . \tag{4.28}$$

That is, the infimum (4.22) is always achieved.

However, this alone does not guarantee the existence of a fixed point. A necessary and sufficient condition is provided by the following theorem:

Theorem 4.16. *There exists a fixed point $\boldsymbol{p}^* > 0$ satisfying (4.23) if and only if*

$$C(\boldsymbol{\gamma}, \boldsymbol{W}) = C(\boldsymbol{\gamma}^{(n)}, \boldsymbol{W}^{(n)}), \quad 1 \le n \le i . \tag{4.29}$$

Proof. Suppose that there exists a $\boldsymbol{p}^* > 0$ such that (4.23) holds. Then, for all isolated blocks $\boldsymbol{W}^{(n)}$, with $1 \le n \le i$, we have

$$\gamma_k \mathcal{I}_k(\boldsymbol{p}, \boldsymbol{W}) = C(\boldsymbol{\gamma}, \boldsymbol{W}) \cdot p_k , \quad k \in \mathcal{K}_n . \tag{4.30}$$

Because of uniqueness (Lemma 2.21, part 2) we know that $C(\boldsymbol{\gamma}, \boldsymbol{W}) = C(\boldsymbol{\gamma}^{(n)}, \boldsymbol{W}^{(n)})$ holds for all n with $1 \le n \le i$.

Conversely, assume that (4.29) holds. Then the proof of Lemma 4.15 shows that there is a $\boldsymbol{p}^* > 0$ such that (4.23) is fulfilled. For the isolated blocks, this follows from Theorem 4.14. For the non-isolated blocks, a vector can be constructed as in the proof of Lemma 4.15. □

The results show that the existence of a fixed point $\boldsymbol{p}^*$ only depends on the isolated blocks. However, $\boldsymbol{p}^*$ is generally not unique since different scalings are possible for the isolated blocks. Arbitrary SIR can be achieved by users with non-isolated blocks, as shown in the proof of Lemma 4.15.

4.2.2 Min-Max and Max-Min Balancing

In the previous section we have exploited that the min-max optimum $C(\boldsymbol{\gamma})$ characterizes the boundary of the SIR region. Now, an interesting question is whether an equivalent indicator function is obtained by max-min balancing, i.e.,

$$c(\boldsymbol{\gamma}) = \sup_{\boldsymbol{p}>0}\Big(\min_{k\in\mathcal{K}} \frac{\gamma_k \mathcal{I}_k(\boldsymbol{p})}{p_k}\Big) . \tag{4.31}$$

In general, we have [2]

$$c(\boldsymbol{\gamma}) \leq C(\boldsymbol{\gamma}) . \tag{4.32}$$

Note that (4.32) is not a simple consequence of Fan's minimax inequality since we do not only interchange the optimization order, but also the domain. Inequality (4.32) was derived in [2] by exploiting the special properties of interference functions. Even for simple linear interference functions, equality does not need to hold [2].

Now, we extend these results by showing special properties for log-convex interference functions.

Theorem 4.17. *Consider an arbitrary row-stochastic matrix* $\boldsymbol{W} \in \mathbb{R}_+^{K\times K}$ *with resulting log-convex interference functions* $\mathcal{I}_k(\boldsymbol{p}, \boldsymbol{W})$, $k \in \mathcal{K}$. *We have*

$$c(\boldsymbol{\gamma}, \boldsymbol{W}) = C(\boldsymbol{\gamma}, \boldsymbol{W}) \tag{4.33}$$

if and only if for all isolated blocks $n = 1, 2, \ldots, i$,

$$C(\boldsymbol{\gamma}^{(n)}, \boldsymbol{W}^{(n)}) = C(\boldsymbol{\gamma}, \boldsymbol{W}) . \tag{4.34}$$

Proof. If (4.34) holds, then it follows from Theorem 4.16 that there is a fixed point $\boldsymbol{p}^* > 0$ fulfilling (4.23), thus implying (4.33). Conversely, assume that (4.33) holds. With (4.26) we have $C(\boldsymbol{\gamma}^{(n)}, \boldsymbol{W}^{(n)}) \leq C(\boldsymbol{\gamma}, \boldsymbol{W})$ for all isolated blocks $1 \leq n \leq i$. In a similar way, we can use definition (4.31) in order to show $c(\boldsymbol{\gamma}^{(n)}, \boldsymbol{W}^{(n)}) \geq c(\boldsymbol{\gamma}, \boldsymbol{W})$. With (4.32) we have

$$c(\boldsymbol{\gamma}, \boldsymbol{W}) \leq C(\boldsymbol{\gamma}^{(n)}, \boldsymbol{W}^{(n)}) \leq C(\boldsymbol{\gamma}, \boldsymbol{W}) \,, \quad \forall n \in \{1, 2, \ldots, i\}.$$

With (4.33) this is fulfilled with equality, so (4.34) holds. □

The following corollary is a direct consequence of Theorems 4.16 and 4.17.

Corollary 4.18. *Consider an arbitrary row-stochastic matrix* $\boldsymbol{W} \in \mathbb{R}_+^{K\times K}$. *There exists a strictly positive fixed point* $\boldsymbol{p}^* > 0$ *satisfying (4.23) if and only if* $c(\boldsymbol{\gamma}, \boldsymbol{W}) = C(\boldsymbol{\gamma}, \boldsymbol{W})$.

Note that Corollary 4.18 is derived under the assumption of particular interference functions (4.21), where $\boldsymbol{W}$ and f_k are constant. The result cannot be transfered to general log-convex interference functions with adaptive $\boldsymbol{W}$. Even for simple linear interference functions (1.10), the condition $c(\boldsymbol{\gamma}) = C(\boldsymbol{\gamma})$

does not always ensure the existence of a fixed point (2.44), as shown in [119, 120].

In the next section we will study a more general class of log-convex interference functions where $\boldsymbol{W}$ is chosen adaptively. It will be shown (Theorem 4.19) that $c(\boldsymbol{\gamma}) = C(\boldsymbol{\gamma})$ holds if all possible $\boldsymbol{W}$ are irreducible.

4.2.3 Generalization to Adaptive $\boldsymbol{W}$

In the previous subsection we have considered a special class of log-convex interference functions (4.21), which depend on a fixed coefficient matrix $\boldsymbol{W}$. Now, the results will be extended by maximizing with respect to $\boldsymbol{W}$. The coefficients f_k are still assumed to be constant. General log-convex interference functions will be addressed later in Section 4.2.4.

Consider a coefficient set

$$\mathcal{W} = \left\{ \boldsymbol{W} = [\boldsymbol{w}_1, \ldots, \boldsymbol{w}_K]^T : \boldsymbol{w}_k \in \mathcal{L}_k, \ \forall k \in \mathcal{K} \right\} , \tag{4.35}$$

where $\mathcal{L}_k \subseteq \mathbb{R}^K_+$ is an arbitrary closed and bounded set such that any $\boldsymbol{w} \in \mathcal{L}_k$ fulfills $\|\boldsymbol{w}\|_1 = 1$. The set $\mathcal{W}$ is also closed and bounded.

Based on $\mathcal{W}$ and (4.21), we define log-convex interference functions

$$\mathcal{I}_k(\boldsymbol{p}) = \max_{\boldsymbol{W} \in \mathcal{W}} \mathcal{I}_k(\boldsymbol{p}, \boldsymbol{W}), \quad \forall k \in \mathcal{K} . \tag{4.36}$$

Note, that $\mathcal{I}_k(\boldsymbol{p}, \boldsymbol{W})$ only depends on $\boldsymbol{w}_k \in \mathcal{L}_k$, so we have K independent optimization problems. We will also use the vector notation

$$\boldsymbol{\mathcal{I}}(\boldsymbol{p}) = \begin{bmatrix} \max_{\boldsymbol{W} \in \mathcal{W}} \mathcal{I}_1(\boldsymbol{p}, \boldsymbol{W}) \\ \vdots \\ \max_{\boldsymbol{W} \in \mathcal{W}} \mathcal{I}_K(\boldsymbol{p}, \boldsymbol{W}) \end{bmatrix} . \tag{4.37}$$

Theorem 4.19. *Consider a set $\mathcal{W}$, as defined by (4.35), with the additional requirement that all elements $\boldsymbol{W} \in \mathcal{W}$ are irreducible, with resulting interference functions (4.37). Then $c(\boldsymbol{\gamma}) = C(\boldsymbol{\gamma})$ and there exists a fixed point $\boldsymbol{p}^* > 0$ satisfying (2.44).*

Proof. The proof is given in the Appendix A.9. □

The next theorem provides a necessary and sufficient condition for the existence of a strictly positive fixed point.

Theorem 4.20. *Let $\boldsymbol{\mathcal{I}}(\boldsymbol{p})$ be defined as by (4.37). A vector $\boldsymbol{p}^* > 0$ is a fixed point satisfying (2.44) if and only if there exists a stochastic matrix $\boldsymbol{W}^* \in \mathcal{W}$ and a $\mu > 0$ such that*

$$\mathcal{I}_k(\boldsymbol{p}^*) = \max_{\boldsymbol{W} \in \mathcal{W}} \mathcal{I}_k(\boldsymbol{p}^*, \boldsymbol{W}) = \mathcal{I}_k(\boldsymbol{p}^*, \boldsymbol{W}^*), \quad \forall k \in \mathcal{K} \tag{4.38}$$

$$\boldsymbol{\Gamma}\boldsymbol{\mathcal{I}}(\boldsymbol{p}^*, \boldsymbol{W}^*) = \mu \cdot \boldsymbol{p}^* . \tag{4.39}$$

Then,

$$\mu = C(\boldsymbol{\gamma}) = C(\boldsymbol{\gamma}, \boldsymbol{W}^*) . \tag{4.40}$$

Proof. If $\boldsymbol{p}^* > 0$ is a fixed point satisfying (2.44) then (4.38) and (4.39) are fulfilled. From (4.39) we know that $\boldsymbol{p}^*$ is also a fixed point of $\boldsymbol{\Gamma}\boldsymbol{\mathcal{I}}(\,\cdot\,,\boldsymbol{W}^*)$. Because $\boldsymbol{p}^* > 0$, we known from Lemma 2.21 (part 2) that (4.40) is fulfilled.

Conversely, assume that (4.39) and (4.38) are fulfilled. Then,

$$\begin{aligned}[\boldsymbol{\Gamma}\boldsymbol{\mathcal{I}}(\boldsymbol{p}^*)]_k &= \gamma_k \max_{\boldsymbol{W}\in\mathcal{W}} \mathcal{I}_k(\boldsymbol{p}^*,\boldsymbol{W}) = \gamma_k \mathcal{I}_k(\boldsymbol{p}^*,\boldsymbol{W}^*) \\ &= \mu[\boldsymbol{p}^*]_k \,. \end{aligned} \tag{4.41}$$

That is, $\boldsymbol{p}^* > 0$ is a fixed point of $\boldsymbol{\Gamma}\boldsymbol{\mathcal{I}}(\boldsymbol{p})$. Lemma 2.21 (part 2) yields (4.40). □

For the special case that all $\boldsymbol{W}$ are irreducible, we have the following result.

Theorem 4.21. *Consider a the set $\mathcal{W}$, as defined by (4.35), such that all $\boldsymbol{W} \in \mathcal{W}$ are irreducible. Then*

$$\max_{\boldsymbol{W}\in\mathcal{W}} C(\gamma,\boldsymbol{W}) = C(\gamma)\,, \tag{4.42}$$

and there is a $\boldsymbol{p}^* > 0$ such that $\boldsymbol{\Gamma}\boldsymbol{\mathcal{I}}(\boldsymbol{p}^*) = C(\gamma)\boldsymbol{p}^*$, where $\boldsymbol{\mathcal{I}}$ is defined by (4.37).

Proof. The proof is given in the Appendix A.9. □

4.2.4 Fixed Point Analysis for General Log-Convex Interference Functions

Finally, we will derive a condition under which the asymptotic matrix is irreducible. To this end we introduce the set

$$\mathcal{W}_{\mathcal{I}} = \left\{\boldsymbol{W} = [\boldsymbol{w}_1,\ldots,\boldsymbol{w}_K]^T : \boldsymbol{w}_k \in \mathcal{L}(\mathcal{I}_k),\ \forall k \in \mathcal{K}\right\}, \tag{4.43}$$

Note, that $\mathcal{W}_{\mathcal{I}}$ is based on the sets $\mathcal{L}(\mathcal{I}_k)$, as defined by (3.116). So it depends on the log-convex interference functions $\mathcal{I}_1,\ldots,\mathcal{I}_K$, which are arbitrary. In this respect it differs from the previously used set $\mathcal{W}$. Any $\boldsymbol{W} \in \mathcal{W}_{\mathcal{I}}$ is stochastic because of Lemma 3.49.

Theorem 4.22. *The asymptotic matrix $\boldsymbol{A}_{\mathcal{I}}$ (equivalently $\boldsymbol{D}_{\mathcal{I}}$) is irreducible if and only if there exists an irreducible stochastic matrix $\hat{\boldsymbol{W}} \in \mathcal{W}_{\mathcal{I}}$, and constants $C_1,\ldots,C_K > 0$, such that for all $\boldsymbol{p} > 0$,*

$$\mathcal{I}_k(\boldsymbol{p}) \ge C_k \prod_{l\in\mathcal{K}} (p_l)^{\hat{w}_{kl}}, \quad \forall k \in \mathcal{K}\,, \quad \forall \boldsymbol{p} > 0\,. \tag{4.44}$$

Proof. The proof is given in the Appendix A.9 □

Theorem 4.22 links irreducibility with the existence of non-zero lower bounds for the interference functions $\mathcal{I}_1, \ldots, \mathcal{I}_K$. This will be used in the next section.

In this section we will study the existence of a fixed point $\boldsymbol{p}^* > 0$ satisfying (2.44) for general log-convex interference functions as introduced in Definition 1.4. Consider the coefficient set $\mathcal{W}_{\mathcal{I}}$ as defined by (4.43). The first theorem shows that the existence of one irreducible coefficient matrix from $\mathcal{W}_{\mathcal{I}}$ is sufficient.

Theorem 4.23. *Let $\boldsymbol{\mathcal{I}} = [\mathcal{I}_1, \ldots, \mathcal{I}_K]^T$ be a vector of log-convex interference functions, such that there exists a stochastic irreducible matrix $\hat{\boldsymbol{W}} \in \mathcal{W}_{\mathcal{I}}$. Then for all $\gamma > 0$ there exists a fixed point $\mathbf{p}^* > 0$ such that*

$$\boldsymbol{\Gamma}\boldsymbol{\mathcal{I}}(\boldsymbol{p}^*) = C(\gamma)\boldsymbol{p}^* . \tag{4.45}$$

Proof. The proof is given in the Appendix A.9. □

In Theorem 4.23 we have required $\hat{\boldsymbol{W}} \in \mathcal{W}_{\mathcal{I}}$, which means that $\hat{\boldsymbol{W}}$ is stochastic and $\underline{f}_{\mathcal{I}_k}(\hat{\boldsymbol{w}}_k) > 0$ for all $k \in \mathcal{K}$. In this case, we know from (4.36) that

$$\mathcal{I}_k(\boldsymbol{p}) \geq \underline{f}_{\mathcal{I}_k}(\hat{\boldsymbol{w}}_k) \prod_{l \in \mathcal{K}} (p_l)^{\hat{w}_{kl}}, \quad \forall k \in \mathcal{K}, \ \forall \boldsymbol{p} > 0 . \tag{4.46}$$

Conversely, consider a stochastic matrix $\hat{\boldsymbol{W}}$ such that (4.44) is fulfilled for some $C_1, \ldots, C_K > 0$. Then,

$$\frac{\mathcal{I}_k(\boldsymbol{p})}{\prod_{l \in \mathcal{K}} (p_l)^{\hat{w}_{kl}}} \geq C_k > 0, \quad \forall k \in \mathcal{K}, \ \forall \boldsymbol{p} > 0 . \tag{4.47}$$

Thus, $\underline{f}_{\mathcal{I}_k}(\hat{\boldsymbol{w}}_k) > 0$, $\forall k \in \mathcal{K}$, which implies $\hat{\boldsymbol{W}} \in \mathcal{W}_{\mathcal{I}}$. Both conditions are equivalent, so Theorem 4.23 leads to the following corollary:

Corollary 4.24. *Assume there exist $C_1, \ldots, C_K > 0$ and a stochastic irreducible matrix $\hat{\mathbf{W}} \in \mathcal{W}_{\mathcal{I}}$ such that (4.44) holds, then for all $\gamma > 0$ there exists a fixed point $\mathbf{p}^* > 0$ such that (4.45) holds.*

With Theorem 4.22 we can reformulate this result as another corollary, which shows that irreducibility of the dependency matrix is always sufficient for the existence of a fixed point.

Corollary 4.25. *If the dependency matrix $\boldsymbol{D}_{\mathcal{I}}$ (equivalently $\boldsymbol{A}_{\mathcal{I}}$) is irreducible, then for all $\gamma > 0$ there exists a fixed point $\mathbf{p}^* > 0$ such that (4.45) holds.*

The next theorem addresses the case where the dependency matrix is *not irreducible*. Without loss of generality, we can choose the user indices such that $\boldsymbol{D}_{\mathcal{I}}$ has the canonical form (4.85). If an additional assumption is fulfilled, then there is at least one SIR vector which is not achievable:

Theorem 4.26. *Assume that the dependency matrix $\boldsymbol{D}_{\mathcal{I}}$ (equivalently $\boldsymbol{A}_{\mathcal{I}}$) is reducible, so it can be written in canonical form (4.85). Let $1, \ldots, l_1$ be the user indices associated with the isolated blocks. If*

$$\inf_{\boldsymbol{p}>0} \max_{k>l_1} \frac{\gamma_k \mathcal{I}_k(\boldsymbol{p})}{p_k} = \underline{C}_1(\boldsymbol{\gamma}) > 0 \,, \quad \forall \boldsymbol{\gamma} > 0 \,, \tag{4.48}$$

then there exists a $\boldsymbol{\gamma} > 0$ such that there is no fixed point $\mathbf{p}^* > 0$ fulfilling (4.45).

Proof. The proof is given in the Appendix A.9. □

Note that condition (4.48) in Theorem 4.26 is not redundant. In the remainder of this section we will discuss examples of log-convex interference functions with reducible $\boldsymbol{D}_{\mathcal{I}}$ where all $\boldsymbol{\gamma} > 0$ have a corresponding fixed point (4.45). But in these cases we have a trivial lower bound $\underline{C}_1(\boldsymbol{\gamma}) = 0$. In this sense, Theorem 4.26 is best possible.

A result corresponding to Theorem 4.26 is known from the theory of non-negative matrices [57], which is closely connected with the linear interference functions. For example, consider linear interference functions (1.10) based on a non-negative coupling matrix $\boldsymbol{V}$. Without loss of generality we can assume that $\boldsymbol{V}$ has canonical form (4.85). This is a special case of the log-convex interference model. We have $\boldsymbol{D}_{\mathcal{I}} = \boldsymbol{V}$. Let $\rho(\boldsymbol{\Gamma}^{(n)}\boldsymbol{V}^{(n)})$ be the spectral radius of the nth (weighted) block on the main diagonal, then it can be shown that

$$\underline{C}_1(\boldsymbol{\gamma}) = \max_{n>i} \rho(\boldsymbol{\Gamma}^{(n)}\boldsymbol{V}^{(n)}) \,, \tag{4.49}$$

where i is the number of isolated blocks.

Consider the example

$$\boldsymbol{\Gamma}\boldsymbol{V} = \operatorname{diag}[\gamma_1, \ldots, \gamma_K] \cdot \begin{bmatrix} 0 & 1 & 0 & 0 \\ 1 & 0 & 0 & 0 \\ 1 & 1 & 0 & 0 \\ 1 & 1 & 0 & 0 \end{bmatrix} , \tag{4.50}$$

The isolated block is zero, so $\underline{C}_1(\boldsymbol{\gamma}) = 0$. The overall spectral radius is $\rho(\boldsymbol{\Gamma}\boldsymbol{V}) = \sqrt{\gamma_1\gamma_2}$. It can easily be checked that for any $\boldsymbol{\Gamma}$ there is a $\boldsymbol{p}_{\Gamma} > 0$ such that $\boldsymbol{\Gamma}\boldsymbol{V}\boldsymbol{p}_{\Gamma} = \rho(\boldsymbol{\Gamma}\boldsymbol{V})\boldsymbol{p}_{\Gamma}$. This also follows from [81], where it was shown that an arbitrary $\boldsymbol{\gamma} > 0$ is associated with a positive fixed point $\boldsymbol{p}_{\Gamma} > 0$ if and only if the set of maximal blocks equals the set of isolated blocks, i.e.,

$$\rho(\boldsymbol{\Gamma}\boldsymbol{V}) = \rho(\boldsymbol{\Gamma}^{(n)}\boldsymbol{V}^{(n)}), \quad 1 \le n \le i \tag{4.51}$$

$$\text{and} \quad \rho(\boldsymbol{\Gamma}\boldsymbol{V}) > \rho(\boldsymbol{\Gamma}^{(n)}\boldsymbol{V}^{(n)}), \quad n > i \,. \tag{4.52}$$

These conditions are fulfilled for the example (4.50), because $\rho(\boldsymbol{\Gamma}^{(1)}\boldsymbol{V}^{(1)}) = \sqrt{\gamma_1\gamma_2}$ and $\rho(\boldsymbol{\Gamma}^{(2)}\boldsymbol{V}^{(2)}) = 0$.

With (4.51) and (4.52) we can also derive simple sufficient conditions for the non-existence of a fixed point. For example, we can choose a reducible matrix $\boldsymbol{\Gamma V}$ such that a non-isolated block $\boldsymbol{\Gamma}^{(n)}\boldsymbol{V}^{(n)}$, $n > i$, is maximal. Or we can choose γ such that an isolated block $\boldsymbol{\Gamma}^{(n)}\boldsymbol{V}^{(n)}$, $n \leq i$, is not maximal. In both cases there is no solution to the fixed point equation $\boldsymbol{\Gamma V p} = \rho(\boldsymbol{\Gamma V})\boldsymbol{p}$. Note that both cases require that at least one non-isolated block has a non-zero spectral radius, so $\underline{C}_1(\gamma) > 0$.

Discussing linear interference functions helps to better understand Theorem 4.26. However, the actual value of the theorem – as well as the other results – lies in its applicability to a broader class of interference functions. All results hold for arbitrary log-convex interference functions as introduced by Definition 1.4.

As a further illustration, consider the log-convex interference functions $\mathcal{I}_k(\boldsymbol{p}, \boldsymbol{W})$, as defined by (4.21), based on an arbitrary reducible stochastic matrix $\boldsymbol{W}$. We assume that there is at least one non-isolated block and a single isolated block. Every non-zero entry in $\boldsymbol{W}$ corresponds to a non-zero entry in $\boldsymbol{A}_{\mathcal{I}}$ and $\boldsymbol{D}_{\mathcal{I}}$ with the same position. From Lemma 4.15 and Theorem 4.16 we know that for any $\gamma > 0$ we have $C(\gamma, \boldsymbol{W}) = C(\gamma^{(1)}, \boldsymbol{W}^{(1)})$ and there is a fixed point $\boldsymbol{p}^* > 0$. This is a consequence of $\boldsymbol{W}$ having a single isolated block. Arbitrary γ_k can be achieved by the non-isolated users (see proof of Theorem 4.16), so $\underline{C}_1(\gamma) = 0$ for all $\gamma > 0$. That is, $\boldsymbol{D}_{\mathcal{I}}$ can be reducible and all $\gamma > 0$ are associated with a fixed point, but in this case $\underline{C}_1(\gamma) = 0$. This is another example showing that the requirement $\underline{C}_1(\gamma) > 0$ is generally important and cannot be omitted.

The results of this section show that the special properties of log-convex interference functions are very useful for the analysis of the fixed point equation (4.45), which is closely connected with the achievability of boundary points of the QoS region. In particular, the irreducibility of the dependency matrix $\boldsymbol{D}_{\mathcal{I}}$ is sufficient for the achievability of the *entire boundary*. This shows an interesting analogy to the theory of linear interference functions (Perron-Frobenius theory), where an irreducible "link gain matrix" is typically assumed to ensure the existence of a min-max optimal power vector. Linear interference functions are a special case of the axiomatic framework of log-convex interference functions. Note, that log-convexity is the key property which is exploited here. A similar characterization of the boundary can be more complicated for other classes of interference functions (see e.g. [2]). This is still an open problem for general interference functions being solely characterized by A1, A2, A3.

4.2.5 Fairness Gap

The min-max optimization (1.22) is one possible approach to fairness. In this definition, the value $C(\gamma)$ is the infimum over the weighted inverse SIR $\gamma_k \mathcal{I}_k(\boldsymbol{p})/p_k$. Note, that $\inf \max_k \mathrm{SIR}_k^{-1} = (\sup \min_k \mathrm{SIR}_k)^{-1}$. This optimization strategy is also referred to as *max-min fairness*.

An alternative approach is *min-max fairness*. This can also be formulated in terms of weighted inverse SIR, as the max-min optimization problem

$$c(\boldsymbol{\gamma}) = \sup_{\boldsymbol{p}>0}\Big(\min_{k\in\mathcal{K}} \frac{\gamma_k \mathcal{I}_k(\boldsymbol{p})}{p_k}\Big) \tag{4.53}$$

It is not obvious whether the max-min optimum $c(\boldsymbol{\gamma})$ and the min-max optimum $C(\boldsymbol{\gamma})$, as defined by (1.22), are identical. Both strategies can be regarded as fair. Note that we do not only interchange the optimization order, but also the domain, so Fan's minimax inequality cannot be applied here. Both values do not necessarily coincide. The difference is sometimes referred to as the *fairness gap* [42].

Example 4.27. Consider the linear interference model introduced in Section 1.4. If the coupling matrix $\boldsymbol{V}$ is irreducible, then $c(\boldsymbol{\gamma}) = C(\boldsymbol{\gamma})$ always holds. But this need not hold true for reducible coupling matrices. Consider the example

$$\boldsymbol{V} = \left[\begin{array}{c|c} \begin{matrix} 0 & 1 \\ 1 & 0 \end{matrix} & \begin{matrix} 0 & 0 \\ 0 & 0 \end{matrix} \\ \hline \boldsymbol{B} & \begin{matrix} 0 & \mu \\ \mu & 0 \end{matrix} \end{array}\right] \qquad \begin{array}{l} \text{with } \boldsymbol{B} \geq 0 \\ \quad 0 < \mu < 1\,. \end{array} \tag{4.54}$$

We have $C(\boldsymbol{\gamma}) = 1$, but $c(\boldsymbol{\gamma}) \leq 1$. If $\boldsymbol{B} = 0$, then we have two isolated subsystems with spectral radius 1 and μ. In this case, $C(\boldsymbol{\gamma}) = 1 > \mu = c(\boldsymbol{\gamma})$, which demonstrates that min-max fairness and max-min fairness are generally not equivalent. But also a different behavior can occur. For example, $C(\boldsymbol{\gamma}) = c(\boldsymbol{\gamma})$ is fulfilled if $\boldsymbol{B} > 0$.

In order to better understand these effects, notice that the function $c(\boldsymbol{\gamma})$ fulfills the properties A1, A2, A3. That is, $c(\boldsymbol{\gamma})$ is an interference function, so we can use Theorem 3.5 to analyze and compare both functions C and c.

The function C was already used in the definition (1.21) of the SIR region $\mathcal{S}$. With Theorem 3.5 it is clear that $\mathcal{S} = \underline{L}(C)$. Now, we will show some interesting analogies between $\underline{L}(C)$ and $\overline{L}(c)$, defined as

$$\overline{L}(c) = \{\boldsymbol{\gamma} > 0 : c(\boldsymbol{\gamma}) \geq 1\}\,. \tag{4.55}$$

From (4.53) we know that for every $\epsilon > 0$ there exists a $\boldsymbol{p}(\epsilon)$ such that

$$\min_{k\in\mathcal{K}} \frac{\gamma_k \mathcal{I}_k\big(\boldsymbol{p}(\epsilon)\big)}{p_k(\epsilon)} \geq c(\boldsymbol{\gamma}) - \epsilon\,.$$

If $c(\boldsymbol{\gamma}) \geq 1$, then

$$\gamma_k \geq (1-\epsilon)\cdot \frac{p_k(\epsilon)}{\mathcal{I}_k\big(\boldsymbol{p}(\epsilon)\big)}, \qquad \forall k \in \mathcal{K}\,. \tag{4.56}$$

This can be used for the following characterization

$$\begin{aligned} \overline{L}(c) = \{\boldsymbol{\gamma} > 0 :\ & \text{for every } \epsilon > 0 \text{ there exists a vector} \\ & \boldsymbol{p}(\epsilon) > 0 \text{ such that (4.56) is fulfilled}\}\,. \end{aligned}$$

With Theorem 3.5 we have

$$\sup_{\boldsymbol{p}>0}\Big(\min_{k\in\mathcal{K}}\frac{\gamma_k\mathcal{I}_k(\boldsymbol{p})}{p_k}\Big)=\max_{\hat{\boldsymbol{\gamma}}\in\overline{\mathcal{L}}(c)}\Big(\min_{k\in\mathcal{K}}\frac{\gamma_k}{\hat{\gamma}_k}\Big)=c(\boldsymbol{\gamma})\,. \tag{4.57}$$

Again, we can generally not replace the supremum by a maximum since the boundary of $\overline{\mathcal{L}}(c)$ cannot always be parametrized by $\boldsymbol{p}>0$.

It was shown in [2] that $c(\boldsymbol{\gamma})$ is always smaller than $C(\boldsymbol{\gamma})$. As mentioned before, this result is due to the specific properties A1, A2, A3, and does not follow from Fan's minimax inequality. Now, we can use the results of this book to show this property with a different approach, based on level sets.

Theorem 4.28. $c(\boldsymbol{\gamma})\le C(\boldsymbol{\gamma})$ *for all* $\boldsymbol{\gamma}>0$.

Proof. Consider an arbitrary $\tilde{\boldsymbol{\gamma}}$ from the interior of $\overline{\mathcal{L}}(c)$, i.e., $c(\tilde{\boldsymbol{\gamma}})\ge 1$. From (4.53), we know that there exists a $\tilde{\boldsymbol{p}}>0$ satisfying

$$\frac{\tilde{\gamma}_k\mathcal{I}_k(\tilde{\boldsymbol{p}})}{\tilde{p}_k}>1,\quad\forall k\in\mathcal{K}\,. \tag{4.58}$$

Now, we show that $\tilde{\boldsymbol{\gamma}}$ also lies in the interior of $\overline{\mathcal{L}}(C)$. From the definition of $C(\boldsymbol{\gamma})$, it follows that for all $\epsilon>0$ there exists a vector $\boldsymbol{p}(\epsilon)>0$ such that

$$\frac{\tilde{\gamma}_k\mathcal{I}_k\big(\boldsymbol{p}(\epsilon)\big)}{\tilde{p}_k(\epsilon)}\le C(\tilde{\boldsymbol{\gamma}})+\epsilon,\quad\forall k\in\mathcal{K}\,. \tag{4.59}$$

The ratio $\mathcal{I}(\boldsymbol{p})/p_k$ is invariant with respect to a scaling of $\boldsymbol{p}$, thus we can assume $\boldsymbol{p}(\epsilon)\ge\tilde{\boldsymbol{p}}$ without affecting (4.59). In addition, we can assume that there is an index $\hat{k}$ such that $p_{\hat{k}}(\epsilon)=\tilde{p}_{\hat{k}}$. With (4.58), (4.59), and property A3, we have

$$\begin{aligned}1<\frac{\tilde{\gamma}_{\hat{k}}\mathcal{I}_{\hat{k}}(\tilde{\boldsymbol{p}})}{\tilde{p}_{\hat{k}}}&=\frac{\tilde{\gamma}_{\hat{k}}\mathcal{I}_{\hat{k}}(\tilde{\boldsymbol{p}})}{p_{\hat{k}}(\epsilon)}\\&\le\frac{\tilde{\gamma}_{\hat{k}}\mathcal{I}_{\hat{k}}\big(\boldsymbol{p}(\epsilon)\big)}{p_{\hat{k}}(\epsilon)}\le C(\tilde{\boldsymbol{\gamma}})+\epsilon\,.\end{aligned} \tag{4.60}$$

This inequality holds for all $\epsilon>0$. Letting $\epsilon\to 0$, it follows that $C(\tilde{\boldsymbol{\gamma}})>1$, so $\tilde{\boldsymbol{\gamma}}$ is also contained in the interior of $\overline{\mathcal{L}}(C)$. Therefore,

$$c(\boldsymbol{\gamma})=\max_{\hat{\boldsymbol{\gamma}}\in\overline{\mathcal{L}}(c)}\Big(\min_{k\in\mathcal{K}}\frac{\gamma_k}{\hat{\gamma}_k}\Big)\le\max_{\hat{\boldsymbol{\gamma}}\in\overline{\mathcal{L}}(C)}\Big(\min_{k\in\mathcal{K}}\frac{\gamma_k}{\hat{\gamma}_k}\Big)=C(\boldsymbol{\gamma})\,. \tag{4.61}$$

Example 4.27 shows that strict inequality $c(\boldsymbol{\gamma})<C(\boldsymbol{\gamma})$ can actually occur. □

4.3 Proportional Fairness – Boundedness, Existence, and Strict Log-Convexity

In Subsection 4.1.2 we have introdced proportional fairness as the maximum sum of logarithmic utilities (4.3).

In this section, the utility set is the SIR region $\mathcal{S}$ based on general log-convex interference functions. Exploiting $-\sum_k \log \gamma_k = \sum_k \log \gamma_k^{-1}$, the problem can be formulated as

$$PF(\mathcal{I}) = \inf_{\gamma \in \mathcal{S}} \sum_{k \in \mathcal{K}} \log \gamma_k^{-1} . \tag{4.62}$$

Using the parametrization

$$\mathrm{SIR}_k(\boldsymbol{p}) = \frac{p_k}{\mathcal{I}_k(\boldsymbol{p})} , \quad k \in \mathcal{K} , \tag{4.63}$$

this can be rewritten as

$$PF(\mathcal{I}) = \inf_{\boldsymbol{p} \in \mathcal{P}} \sum_{k \in \mathcal{K}} \log \frac{\mathcal{I}_k(\boldsymbol{p})}{p_k} \tag{4.64}$$

where $\mathcal{P}$ is the set of power vectors. Since the SIR (4.63) is invariant with respect to a scaling of $\boldsymbol{p}$, we can define $\mathcal{P}$ as

$$\mathcal{P} = \{\boldsymbol{p} \in \mathbb{R}^K_{++} : \|\boldsymbol{p}\|_1 = 1\} . \tag{4.65}$$

Note, that the optimization (4.62) is over the SIR region directly, whereas (4.64) is over the set of power vectors. This approach allows to model the impact of the physical layer on the interference. For example, $\mathcal{I}(\boldsymbol{p})$ can depend on $\boldsymbol{p}$ in a nonlinear way. Some examples were given in Section 1.3.2.

Remark 4.29. For certain systems operating in a high-SIR regime, it is customary to approximate the data rate as $\log(1+\mathrm{SIR}) \approx \log(\mathrm{SIR})$ (see e.g. [43]). Then, our problem (4.64) can be interpreted as the maximization of the sum rate $\sum_k \log(1 + \mathrm{SIR}_k)$.

The SIR region $\mathcal{S}$ is generally non-convex and non-compact (because no power constraints are assumed), so it is not clear whether the frameworks of Nash bargaining and proportional fairness can be applied or not. It is even not clear whether the infimum (4.64) is actually attained.

Assuming log-convex interference functions, we will study the existence and uniqueness of a proportionally fair optimizer (4.64). We exploit that the interference coupling in the system can be characterized by a $K \times K$ *dependency matrix* $\boldsymbol{D}_{\mathcal{I}}$ (see Section 2.2.1) The following fundamental questions will be addressed:

1. *Boundedness:* When is $PF(\mathcal{I}) > -\infty$ fulfilled?

2. *Existence:* When does an optimizer $\hat{\boldsymbol{p}} > 0$ exist such that $PF(\mathcal{I}) = \sum_k \log \mathcal{I}_k(\hat{\boldsymbol{p}})/\hat{p}_k$?
3. *Uniqueness:* When is $\hat{\boldsymbol{p}} > 0$ the unique optimizer?

Property $PF(\mathcal{I}) > -\infty$ is necessary for the existence of $\hat{\boldsymbol{p}}$, but not sufficient. This justifies a separate treatment of problem 1) in Section 4.3.1. It is shown that $PF(\mathcal{I}) > -\infty$ implies the existence of a row or column permutation such that the dependency matrix $\boldsymbol{D}_{\mathcal{I}}$ has a strictly positive main diagonal. An additional condition is provided under which the converse holds as well.

In Subsection 4.3.2, the existence of an optimizer $\hat{\boldsymbol{p}} > 0$ is studied. Under certain monotonicity conditions, an optimizer exists, if and only if there exist row and column permutations such that the resulting matrix is block-irreducible [57] and its main diagonal is positive. Otherwise, no Pareto optimal operating point can be found.

In Subsection 4.3.3 we show that the uniqueness of an existing optimizer depends on the structure of the matrix $\boldsymbol{D}_{\mathcal{I}}\boldsymbol{D}_{\mathcal{I}}^T$. This extends recent results [85], which were carried out in the context of linear interference functions.

Finally, in Subsection 4.3.4 we study under which condition the feasible SIR set is strictly log-convex. If this is fulfilled, and if an optimizer exists, then it follows from the results of Section 4.1.3 that the proportionally fair operating point is obtained as the single-valued Nash bargaining solution.

4.3.1 Boundedness of the Cost Function

Having characterized the interference coupling, we are now in a position to study the existence of the proportionally fair infimum $PF(\mathcal{I})$ defined in (4.64). That is, we want to show under which conditions $PF(\mathcal{I}) > -\infty$. The following simple example shows that $PF(\mathcal{I})$ can be unbounded.

Example 4.30. Consider linear interference functions $\mathcal{I}_k(\boldsymbol{p}) = [\boldsymbol{V}\boldsymbol{p}]_k$, $k = 1,2,3$, with a coupling matrix

$$\boldsymbol{V} = \begin{bmatrix} 0 & 0 & 1 \\ 0 & 0 & 1 \\ 1 & 1 & 1 \end{bmatrix} . \tag{4.66}$$

Without loss of generality, we can scale $\boldsymbol{p}$ such that $\|\boldsymbol{p}\|_1 = p_1 + p_2 + p_3 = 1$. Then the cost function becomes

$$\sum_{k=1}^{3} \log \frac{\mathcal{I}_k(\boldsymbol{p})}{p_k} = \log\Big(\frac{p_3}{p_1 p_2}\Big) . \tag{4.67}$$

Choosing $p_2 = p_1$ and $p_3 = 1/n$, with $n > 1$. Since $\|\boldsymbol{p}\|_1 = 1$, we have $p_1 = \frac{1}{2} - \frac{1}{2n}$. Thus,

$$PF(\mathcal{I}) = \inf_{n>1} \log\Big(\frac{1}{n-1}\Big) = -\infty .$$

Before deriving the first result, we need to discuss an important property of our objective $\sum_k \log \mathcal{I}_k(\boldsymbol{p})/p_k$. Consider an arbitrary row permutation $\boldsymbol{\sigma} = [\sigma_1, \ldots, \sigma_K]$ applied to the matrix $\boldsymbol{D}_{\mathcal{I}}$. This corresponds to a reordering of the indices of $\mathcal{I}_1, \ldots, \mathcal{I}_K$, but without changing the indices of the transmission powers $p_1, \ldots, p_K$. Such a reordering does not affect the objective function in problem (4.64). For an arbitrary $\boldsymbol{p} > 0$ we have

$$\sum_{k\in\mathcal{K}} \log \frac{\mathcal{I}_k(\boldsymbol{p})}{p_k} = \sum_{k\in\mathcal{K}} \log \mathcal{I}_k(\boldsymbol{p}) - \sum_{k\in\mathcal{K}} \log p_k \tag{4.68}$$

$$= \sum_{k\in\mathcal{K}} \log \frac{\mathcal{I}_{\sigma_k}(\boldsymbol{p})}{p_k} . \tag{4.69}$$

This follows from the fact that the summands in (4.68) can be arranged and combined arbitrarily. Property (4.68) has an interesting interpretation in the context of user anonymity [121].

This means that the optimization problem (4.64) is invariant with respect to permutations of powers or interference functions. Defining arbitrary permutation matrices $\boldsymbol{P}^{(1)}$, $\boldsymbol{P}^{(2)}$, the permuted dependency matrix $\tilde{\boldsymbol{D}}_{\mathcal{I}} = \boldsymbol{P}^{(1)}\boldsymbol{D}_{\mathcal{I}}\boldsymbol{P}^{(2)}$ can equivalently be used in order to characterize the behavior of proportional fairness. This fundamental observation is the basis for the following results.

The next Lemma, which will be needed later for the proof of Theorem 4.33, shows a connection between boundedness and the structure of the dependency matrix $\boldsymbol{D}_{\mathcal{I}}$.

Definition 4.31. *We say that $K' \le K$ interference functions with indices $\sigma_1, \ldots, \sigma_{K'}$ depend on a power component with index l if at least one of these functions depends on this power, i.e., there exists a $k \in \{1, \ldots, K'\}$ such that $[\boldsymbol{D}_{\mathcal{I}}]_{\sigma_k, l} = 1$.*

Lemma 4.32. *If $PF(\mathcal{I}) > -\infty$, then for every $r \in \mathcal{K}$, arbitrary log-convex interference functions $\mathcal{I}_{\sigma_1}, \ldots, \mathcal{I}_{\sigma_r}$ depend on at least r components of the power vector $\boldsymbol{p}$.*

Proof. The proof is by contradiction. Assume that there is a number $\hat{r}$ and interference functions $\mathcal{I}_{k_1}, \ldots, \mathcal{I}_{k_{\hat{r}}}$, which only depend on powers $p_{l_1}, \ldots, p_{l_n}$, with $n < \hat{r}$. From (4.69) we know that interference functions and powers can be permuted such that $\mathcal{I}_1, \ldots, \mathcal{I}_{\hat{r}}$ only depend on $p_1, \ldots, p_n$, with $n < \hat{r}$. Consider the vector $\boldsymbol{p}(\delta)$, defined as

$$[\boldsymbol{p}(\delta)]_l = \begin{cases} \delta, & l = 1, \ldots, n \\ 1, & l = n+1, \ldots, K \end{cases}$$

where $0 < \delta \le 1$, i.e., $\boldsymbol{p}(\delta) \le \mathbf{1}$. Axiom A3 implies $\mathcal{I}_k\big(\boldsymbol{p}(\delta)\big) \le \mathcal{I}_k(\mathbf{1})$, so we have

$$\sum_{k=1}^{K}\log\frac{\mathcal{I}_k\big(\boldsymbol{p}(\delta)\big)}{p_k(\delta)} = \sum_{k=1}^{n}\log\frac{\delta\mathcal{I}_k(\mathbf{1})}{\delta} + \sum_{k=n+1}^{\hat{r}}\log\frac{\delta\mathcal{I}_k(\mathbf{1})}{1} + \\ + \sum_{k=\hat{r}+1}^{K}\log\frac{\mathcal{I}_k\big(\boldsymbol{p}(\delta)\big)}{1} \\ \leq \sum_{k=1}^{n}\log\mathcal{I}_k(\mathbf{1}) + \sum_{k=n+1}^{\hat{r}}\log\mathcal{I}_k(\mathbf{1}) + \\ + (\hat{r}-n)\log\delta + \sum_{k=\hat{r}+1}^{K}\log\mathcal{I}_k(\mathbf{1}) \,.$$

Therefore,

$$PF(\mathcal{I}) \leq \sum_{k=1}^{K}\log\frac{\mathcal{I}_k\big(\boldsymbol{p}(\delta)\big)}{p_k(\delta)} \leq \sum_{k=1}^{K}\log\mathcal{I}_k(\mathbf{1}) + (\hat{r}-n)\log\delta.$$

This holds for all δ, thus letting $\delta \to 0$, we obtain the contradiction $PF(\mathcal{I}) = -\infty$, thus concluding the proof. □

Necessary and Sufficient Condition for Boundedness

Using Lemma 4.32, the following result is shown.

Theorem 4.33. *Let $\mathcal{I}_1, \ldots, \mathcal{I}_K$ be arbitrary log-convex interference functions. If*

$$\inf_{\boldsymbol{p}>0}\sum_{k\in\mathcal{K}}\log\frac{\mathcal{I}_k(\boldsymbol{p})}{p_k} > -\infty \tag{4.70}$$

then there exists a row permutation $\boldsymbol{\sigma} = [\sigma_1, \ldots, \sigma_K]$ such that $[\boldsymbol{D}_{\mathcal{I}}]_{\sigma_k,k} > 0$ for all $k \in \mathcal{K}$. That is, the permuted matrix has a positive main diagonal.

Proof. Assume that (4.70) is fulfilled. Consider the function $\mathcal{I}_K$, which depends on L_K powers, with indices $\boldsymbol{k}^{(K)} = [k_1^{(K)}, \ldots, k_{L_K}^{(K)}]$. The trivial case $L_K = 0$ is ruled out by axiom A1. Consider the lth component $k_l^{(K)}$. The set $\mathcal{L}^{(K)}(k_l^{(K)})$ contains the indices $m \neq k_l^{(K)}$ on which $\mathcal{I}_1, \ldots, \mathcal{I}_{K-1}$ depend. More precisely, $\mathcal{L}^{(K)}(k_l^{(K)})$ is the set of indices $m \neq k_l^{(K)}$ such that there exists a $k \in \{1, 2, \ldots, K-1\}$ with $[\boldsymbol{D}_{\mathcal{I}}]_{km} \neq 0$. Let $\#\mathcal{L}^{(K)}(k_l^{(K)})$ denote the cardinality of this set. It follows from Lemma 4.32 that there exists at least one $\hat{l}$, $1 \leq \hat{l} \leq L_K$, such that

$$\#\mathcal{L}^{(K)}(k_{\hat{l}}^{(K)}) = K - 1 \,. \tag{4.71}$$

Otherwise, K interference functions could not depend on K powers. Note, that (4.71) need not be fulfilled for *all* indices $\boldsymbol{k}^{(K)}$. If (4.71) is fulfilled for multiple indices, then we can choose one. Because of (4.69) the powers can be arbitrarily permuted. Thus, we can choose a permutation $\boldsymbol{\sigma}$ such that $\sigma_K = k_{\hat{l}}^{(K)}$. That is, the interference function $\mathcal{I}_K$ depends on p_{σ_K}, thus $[\boldsymbol{D}_{\mathcal{I}}]_{K,\sigma_K} \neq 0$. This component σ_K is now kept fixed. It remains to consider the remaining functions $\mathcal{I}_1, \ldots, \mathcal{I}_{K-1}$ which depend on powers $p_{\sigma_1}, \ldots, p_{\sigma_{K-1}}$. These powers can still be permuted arbitrarily.

We continue with the interference function $\mathcal{I}_{K-1}$, which depends on $L_{K-1} > 0$ powers, with indices $\boldsymbol{k}^{(K-1)} = [k_1^{(K-1)}, \ldots, k_{L_{K-1}}^{(K-1)}]$. We denote by $\mathcal{L}^{(K-1)}(k_l^{(K-1)})$ the set of indices m (excluding σ_K and $k_l^{(K-1)}$) such that there exists a $k \in \{1, 2, \ldots, K-2\}$ with $[\boldsymbol{D}_{\mathcal{I}}]_{km} \neq 0$. There exists at least one $\hat{l}$, $1 \leq \hat{l} \leq L_{K-1}$ (no matter which one) such that

$$\#\mathcal{L}^{(K-1)}(k_{\hat{l}}^{(K-1)}) = K - 2 . \tag{4.72}$$

The remaining $K-1$ powers (except for σ_K) can still be permuted arbitrarily, so we can choose $\sigma_{K-1} = k_{\hat{l}}^{(K-1)}$. Thus, $[\boldsymbol{D}_{\mathcal{I}}]_{K-1,\sigma_{K-1}} \neq 0$. This component is also kept fixed, and we focus on the remaining functions $\mathcal{I}_1, \ldots, \mathcal{I}_{K-2}$ which depend on $p_{\sigma_1}, \ldots, p_{\sigma_{K-2}}$.

By repeating this procedure for all remaining interference functions, the result follows. □

Next, we are interested in the converse of Theorem 4.33. Under which condition does the existence of a permuted matrix with positive main diagonal imply the boundedness of $PF(\mathcal{I})$? In order to answer this question we introduce an additional property:

$$[\boldsymbol{D}_{\mathcal{I}}]_{k,l} > 0 \text{ implies } \mathcal{I}_k(\boldsymbol{e}_l) > 0 \text{ for any } k, l \in \mathcal{K} , \tag{4.73}$$

where $\boldsymbol{e}_k$ is defined in (2.4).

Theorem 4.34. *Under the additional property (4.73), the condition in Theorem 4.33 is necessary and sufficient.*

Proof. Assume that there exists a $\boldsymbol{\sigma}$ such that $[\boldsymbol{D}_{\mathcal{I}}]_{\sigma_k,k} > 0$ for all $k \in \mathcal{K}$. With (4.73) and properties A2, A3 we have

$$\mathcal{I}_{\sigma_k}(\boldsymbol{p}) \geq \mathcal{I}_{\sigma_k}(\boldsymbol{p} \circ \boldsymbol{e}_k) = p_k \cdot \mathcal{I}_{\sigma_k}(\boldsymbol{e}_k) = p_k \cdot C_k > 0 \tag{4.74}$$

for all $k \in \mathcal{K}$, where C_k are some positive values. The cost function is invariant with respect to a permutation of the indices of the interference functions, as can be seen from (4.68), so we have

$$\sum_{k \in \mathcal{K}} \log \frac{\mathcal{I}_k(\boldsymbol{p})}{p_k} \geq \sum_{k \in \mathcal{K}} \log C_k > -\infty ,$$

which completes the proof. □

Note, that property (4.73) is always fulfilled, e.g., for linear interference functions (1.10) or worst-case interference functions (1.26). However, there exist log-convex interference functions that do not fulfill (4.73). An example is the elementary log-convex interference function (3.118), for which $\mathcal{I}_k(\boldsymbol{e}_l) = 0$.

In the following it will be shown that the additional requirement (4.73) is justified. It is not possible to derive a sufficient condition for boundedness from $\boldsymbol{D}_\mathcal{I}$ alone, without further assumptions.

Elementary Log-Convex Interference Functions

It was shown in [5] that the elementary functions (3.118) play an important role in the analysis of log-convex interference functions. So in the remainder of this section we will study boundedness for this special case. For some given coefficient matrix $\boldsymbol{W}$, our cost function can be rewritten as

$$\sum_{k\in\mathcal{K}} \log \frac{\mathcal{I}_k(\boldsymbol{p})}{p_k} = \log\left(\frac{\prod_l (p_l)^{(\sum_k w_{kl})} \cdot \prod_k f_k}{\prod_k p_k}\right) . \tag{4.75}$$

The matrix $\boldsymbol{W}$ is row stochastic, i.e., $\boldsymbol{W}\mathbf{1} = \mathbf{1}$. This is an immediate consequence of axiom A2, as shown in [5]. The following theorem shows that in order for (4.75) to be bounded, $\boldsymbol{W}$ also needs to be *column stochastic.*

Theorem 4.35. *For interference functions (3.118), the infimum (4.64) is bounded if and only if* $\boldsymbol{W}$ *is doubly stochastic, i.e.,*

$$PF(\mathcal{I}) = \inf_{\boldsymbol{p}>0} \sum_{k\in\mathcal{K}} \log \frac{\mathcal{I}_k(\boldsymbol{p})}{p_k} > -\infty \quad \Leftrightarrow \quad \boldsymbol{W}^T\mathbf{1} = \mathbf{1} . \tag{4.76}$$

Proof. Assume $\boldsymbol{W}^T\mathbf{1} = \mathbf{1}$, i.e., $\sum_k w_{kl} = 1$ for all l. Then it can be observed from (4.75) that, independent of the choice of $\boldsymbol{p}$, we have

$$\sum_{k\in\mathcal{K}} \log \frac{\mathcal{I}_k(\boldsymbol{p})}{p_k} = \log\Big(\prod_{k\in\mathcal{K}} f_k\Big) > -\infty .$$

Conversely, assume that $PF(\mathcal{I}) > -\infty$. The proof is by contradiction: assume that $\boldsymbol{W}^T\mathbf{1} \neq \mathbf{1}$. Since $\boldsymbol{W}\mathbf{1} = \mathbf{1}$, we have $K = \sum_k(\sum_l w_{kl}) = \sum_l(\sum_k w_{kl})$. So $\boldsymbol{W}^T\mathbf{1} \neq \mathbf{1}$ implies the existence of a column index $\hat{l}$ such that $\sum_k w_{k\hat{l}} = \hat{w}_{\hat{l}} > 1$. Consider a sequence $\boldsymbol{p}(n) = [p_1(n), \ldots, p_K(n)]^T$, defined as

$$p_l(n) = \begin{cases} 1/n & , \; l = \hat{l} \\ \frac{1}{K-1}(1 - \frac{1}{n}) & , \; \text{otherwise.} \end{cases} \tag{4.77}$$

Using (4.75), (4.77), and $\sum_{l\neq\hat{l}} \sum_k w_{kl} = K - \hat{w}_{\hat{l}}$, we have

$$\sum_{k\in\mathcal{K}} \log \frac{\mathcal{I}_k(\boldsymbol{p}(n))}{p_k(n)}$$

$$= \log\Big(\big(\tfrac{1}{n}\big)^{\hat{w}_{\hat{l}}-1} \cdot \big(\tfrac{1}{K-1}(1-\tfrac{1}{n})\big)^{1-\hat{w}_{\hat{l}}} \cdot \prod_{k\in\mathcal{K}} f_k \Big) . \quad (4.78)$$

Letting $n \to \infty$, it can be observed that the argument of the log-function tends to zero, so (4.78) tends to $-\infty$. This contradicts the assumption, thus concluding the proof. □

Theorem 4.35 provides a necessary and sufficient condition for boundedness for a special log-convex interference function for which (4.73) is not fulfilled. It becomes apparent that in this case the boundedness does not depend on the structure of $\boldsymbol{D}_{\mathcal{I}}$. If $\boldsymbol{W}$ is chosen such that $\boldsymbol{W}^T\mathbf{1} \neq \mathbf{1}$, then the cost function is unbounded, even if $[\boldsymbol{D}_{\mathcal{I}}]_{kl} = 1$ for $k \neq l$. Hence, it is not possible to show the converse of Theorem 4.33 without additional assumptions. This is illustrated by a simple example.

Example 4.36. Consider log-convex interference functions (3.118) with a coefficient matrix

$$\boldsymbol{W} = \begin{bmatrix} 0 & 1 & 0 \\ \frac{1}{2} & 0 & \frac{1}{2} \\ \frac{1}{2} & \frac{1}{2} & 0 \end{bmatrix} . \quad (4.79)$$

We have $\boldsymbol{W}^T\mathbf{1} = [1\ \frac{3}{2}\ \frac{1}{2}]^T \neq \mathbf{1}$, so the condition in Theorem 4.35 is not fulfilled. With $\mathcal{I}_1(\boldsymbol{p}) = p_2$, $\mathcal{I}_2(\boldsymbol{p}) = (p_1)^{1/2} \cdot (p_3)^{1/2}$, and $\mathcal{I}_3(\boldsymbol{p}) = (p_1)^{1/2} \cdot (p_2)^{1/2}$, we have

$$\inf_{\boldsymbol{p}>0} \sum_{k=1}^{3} \log \frac{\mathcal{I}_k(\boldsymbol{p})}{p_k} = \inf_{\boldsymbol{p}>0} \log \frac{(p_2 p_3)^{1/2}}{p_3} = -\infty . \quad (4.80)$$

The infimum is not bounded, even though there exists a column permutation $\boldsymbol{P}^{(1)}$ such that the main diagonal of $\boldsymbol{D}_{\mathcal{I}}\boldsymbol{P}^{(1)}$ is non-zero.

4.3.2 Existence of a Proportionally Fair Optimizer

In the previous section it was shown that boundedness $PF(\mathcal{I}) > -\infty$ is connected with the positivity of the main diagonal of a permuted dependency matrix. Now, we investigate under which condition the infimum $PF(\mathcal{I}) > -\infty$ is actually attained by a power allocation $\boldsymbol{p} > 0$. The next example shows that this is not always fulfilled, even not for the simple linear interference functions (1.10).

Example 4.37. Consider linear interference functions $\mathcal{I}_k(\boldsymbol{p}) = [\boldsymbol{V}\boldsymbol{p}]_k$, $k = 1, 2, 3$, with a coupling matrix

$$\boldsymbol{V} = \begin{bmatrix} 0 & 1 & 0 \\ 1 & 0 & 1 \\ 1 & 1 & 0 \end{bmatrix} . \tag{4.81}$$

We have

$$\begin{aligned} PF(\mathcal{I}) = \inf_{\boldsymbol{p}>0} \sum_{k=1}^{3} \log \frac{\mathcal{I}_k(\boldsymbol{p})}{p_k} &= -\log\Big[\frac{p_1 \cdot p_3}{(p_1+p_3)(p_1+p_2)}\Big] \\ &\geq -\log\Big[\frac{p_1 \cdot p_3}{p_3 \cdot p_1}\Big] = 0 . \end{aligned} \tag{4.82}$$

Next, we will show that this inequality is fulfilled with equality. Choosing $p_1 = \lambda$, $p_2 = \lambda^2$, and $p_3 = 1 - \lambda - \lambda^2$, we have

$$\sum_{k=1}^{3} \log \frac{\mathcal{I}_k(\boldsymbol{p})}{p_k} = -\log \frac{(1-\lambda-\lambda^2)}{(1-\lambda^2)(1+\lambda)} .$$

This tends to zero as $\lambda \to 0$, Thus,

$$PF(\mathcal{I}) = \inf_{\boldsymbol{p}>0} \sum_{k=1}^{3} \log \frac{\mathcal{I}_k(\boldsymbol{p})}{p_k} \leq 0 . \tag{4.83}$$

Combining (4.82) and (4.83) it follows that $PF(\mathcal{I}) = 0 > -\infty$.

Now, we study whether this infimum is attained. Assume that there exists an optimizer $\boldsymbol{p}^* > 0$, then

$$\begin{aligned} 0 &= \log \frac{\mathcal{I}_1(\boldsymbol{p}^*)}{p_1^*} + \log \frac{\mathcal{I}_2(\boldsymbol{p}^*)}{p_2^*} + \log \frac{\mathcal{I}_3(\boldsymbol{p}^*)}{p_3^*} \\ &= -\log\Big[\frac{p_1^* \cdot p_3^*}{(p_1^*+p_3^*)(p_1^*+p_2^*)}\Big] \\ &> -\log\Big[\frac{p_1^* \cdot p_3^*}{p_3^* p_1^*}\Big] = 0 . \end{aligned} \tag{4.84}$$

This is a contradiction, so the infimum $PF(\mathcal{I}) = 0$ is not attained.

Now, consider arbitrary log-convex interference functions $\mathcal{I}_1, \ldots, \mathcal{I}_K$. The mutual coupling is characterized by the dependency matrix $\boldsymbol{D}_{\mathcal{I}}$ defined in (2.10). We may assume, without loss of generality, that $\boldsymbol{D}_{\mathcal{I}}$ is in canonical form [57, p. 75]

$$
\boldsymbol{D}_{\mathcal{I}} = \left[\begin{array}{ccc|ccc}
\boldsymbol{D}^{(1,1)} & & 0 & 0 & \dots & 0 \\
& \ddots & & \vdots & \dots & \vdots \\
0 & & \boldsymbol{D}^{(i,i)} & 0 & \dots & 0 \\
\hline
\boldsymbol{D}^{(i+1,1)} & \dots & \boldsymbol{D}^{(i+1,i)} & \boldsymbol{D}^{(i+1,i+1)} & & 0 \\
\vdots & \dots & \vdots & \vdots & \ddots & \\
\boldsymbol{D}^{(N,1)} & \dots & \boldsymbol{D}^{(N,i)} & \boldsymbol{D}^{(N,2)} & \dots & \boldsymbol{D}^{(N,N)}
\end{array}\right] . \tag{4.85}
$$

For any given dependency matrix $\boldsymbol{D}'_{\mathcal{I}}$ there always exists a permutation matrix $\boldsymbol{P}$ such that $\boldsymbol{D}_{\mathcal{I}} = \boldsymbol{P}\boldsymbol{D}'_{\mathcal{I}}\boldsymbol{P}^T$ has canonical form. This symmetric permutation preserves the relevant properties that will be exploited, so in the following we can simplify the discussion by assuming that $\boldsymbol{D}_{\mathcal{I}}$ has the form (4.85). The matrix $\boldsymbol{D}_{\mathcal{I}}$ has N *irreducible* blocks $\boldsymbol{D}^{(n)} := \boldsymbol{D}^{(n,n)}$ along its main diagonal (shaded in gray). Recall that $\boldsymbol{D}^{(n)}$ is irreducible if and only if its associated directed graph is *strongly connected* [57]. If $\boldsymbol{D}_{\mathcal{I}}$ is irreducible, then it consists of one single block. We say that $\boldsymbol{D}_{\mathcal{I}}$ is *block-irreducible* if

$$
\boldsymbol{D}_{\mathcal{I}} = \begin{bmatrix} \boldsymbol{D}^{(1)} & & 0 \\ & \ddots & \\ 0 & & \boldsymbol{D}^{(N)} \end{bmatrix},
$$

where all sub-blocks $\boldsymbol{D}^{(n)}$ are irreducible.

For the following theorem, recall the definition of strict monotonicity (Definition 2.7). Given this property, we can derive a necessary and sufficient condition for the existence of a proportionally fair optimizer.

Theorem 4.38. *Let $\mathcal{I}_1, \dots, \mathcal{I}_K$ be strictly monotone log-convex interference functions. We assume that (4.73) is fulfilled. There exists a proportionally fair optimizer $\hat{\mathbf{p}} > 0$ if and only if there exist permutation matrices $\boldsymbol{P}^{(1)}$, $\boldsymbol{P}^{(2)}$ such that $\hat{\boldsymbol{D}}_{\mathcal{I}} := \boldsymbol{P}^{(1)}\boldsymbol{D}_{\mathcal{I}}\boldsymbol{P}^{(2)}$ is block-irreducible and its main diagonal is strictly positive.*

Proof. See Appendix A.9 □

In the next section we will study whether the optimizer characterized by Theorem 4.38 is unique.

4.3.3 Uniqueness of the Solution

In the remainder of the paper, we assume that the interference functions $\mathcal{I}_1, \dots, \mathcal{I}_K$ are log-convex in the sense of Definition 1.4.

Assume that there exists an optimizer for the problem of proportional fairness (4.64). Is this optimizer unique or not? In order to answer this question, we analyze the cost function

$$G(\boldsymbol{s}) = \sum_{k\in\mathcal{K}} \log \frac{\mathcal{I}_k(\mathrm{e}^{\boldsymbol{s}})}{\mathrm{e}^{s_k}} \ , \ \text{on } \mathbb{R}^K \tag{4.86}$$

where we have used the substitution $\boldsymbol{p} = \mathrm{e}^{\boldsymbol{s}}$.

It is sufficient to show that the cost function $G(\boldsymbol{s})$ is strictly convex. Since $\boldsymbol{p} = \mathrm{e}^{\boldsymbol{s}}$ is a strictly monotone function, uniqueness of an optimizer $\boldsymbol{s}$ implies uniqueness of the original problem (4.64). Note, that it is not necessary to show strict convexity of the SIR region, this will be done later in Section 4.3.4.

We start with the following lemma, which will be needed later for Theorem 4.43.

Lemma 4.39. *The function $G(\boldsymbol{s})$ defined in (4.86) is strictly convex if and only if for arbitrary vectors $\hat{\boldsymbol{p}}, \check{\boldsymbol{p}} \in \mathbb{R}^K_{++}$, with $\hat{\boldsymbol{p}} \neq \mu\check{\boldsymbol{p}}$, $\mu \in \mathbb{R}_{++}$, there exists a $\lambda_0 \in (0,1)$ and at least one index k_0 such that*

$$\mathcal{I}_{k_0}(\boldsymbol{p}(\lambda_0)) < \big(\mathcal{I}_{k_0}(\hat{\boldsymbol{p}})\big)^{1-\lambda_0} \cdot \big(\mathcal{I}_{k_0}(\check{\boldsymbol{p}})\big)^{\lambda_0} . \tag{4.87}$$

Proof. Assume that (4.87) holds for k_0. With $\hat{\boldsymbol{p}} = \mathrm{e}^{\hat{\boldsymbol{s}}}$ and $\check{\boldsymbol{p}} = \mathrm{e}^{\check{\boldsymbol{s}}}$, we have

$$\begin{aligned}
G\big(\boldsymbol{s}(\lambda_0)\big) &= \sum_{k\in\mathcal{K}\backslash k_0} \log \frac{\mathcal{I}_k(\mathrm{e}^{\boldsymbol{s}(\lambda_0)})}{\mathrm{e}^{s_k(\lambda_0)}} + \log \frac{\mathcal{I}_{k_0}(\mathrm{e}^{\boldsymbol{s}(\lambda_0)})}{\mathrm{e}^{s_{k_0}(\lambda_0)}} \\
&\leq (1-\lambda_0) \sum_{k\in\mathcal{K}\backslash k_0} \log \frac{\mathcal{I}_k(\mathrm{e}^{\hat{\boldsymbol{s}}})}{\mathrm{e}^{\hat{s}_k}} + \\
&\quad + \lambda_0 \sum_{k\in\mathcal{K}\backslash k_0} \log \frac{\mathcal{I}_k(\mathrm{e}^{\check{\boldsymbol{s}}})}{\mathrm{e}^{\check{s}_k}} + \log \frac{\mathcal{I}_{k_0}(\mathrm{e}^{\boldsymbol{s}(\lambda_0)})}{\mathrm{e}^{s_{k_0}(\lambda_0)}} \\
&< (1-\lambda_0) \sum_{k\in\mathcal{K}} \log \frac{\mathcal{I}_k(\mathrm{e}^{\hat{\boldsymbol{s}}})}{\mathrm{e}^{\hat{s}_k}} + \lambda_0 \sum_{k\in\mathcal{K}} \log \frac{\mathcal{I}_k(\mathrm{e}^{\check{\boldsymbol{s}}})}{\mathrm{e}^{\check{s}_k}} \\
&= (1-\lambda_0) G(\hat{\boldsymbol{s}}) + \lambda_0 G(\check{\boldsymbol{s}}) \ ,
\end{aligned} \tag{4.88}$$

where the first inequality follows from the convexity of $G\big(\boldsymbol{s}(\lambda_0)\big)$ [5], and the second strict inequality is due to (4.87).

Conversely, assume that G is strictly convex. The proof is by contradiction: suppose that there is $\hat{\boldsymbol{s}}, \check{\boldsymbol{s}} \in \mathbb{R}^K$ and $\lambda_0 \in (0,1)$, such that for all $k \in \mathcal{K}$,

$$\mathcal{I}_k\big(\mathrm{e}^{\boldsymbol{s}(\lambda_0)}\big) = \big(\mathcal{I}_k(\hat{\boldsymbol{s}})\big)^{1-\lambda_0} \cdot \big((\mathcal{I}_k(\check{\boldsymbol{s}})\big)^{\lambda_0} . \tag{4.89}$$

With (4.89), we have

$$\begin{aligned}
G\big(\boldsymbol{s}(\lambda_0)\big) &= \sum_{k\in\mathcal{K}} \log \frac{\big(\mathcal{I}_k(\mathrm{e}^{\hat{\boldsymbol{s}}})\big)^{1-\lambda_0} \cdot \big(\mathcal{I}_k(\mathrm{e}^{\check{\boldsymbol{s}}})\big)^{\lambda_0}}{\mathrm{e}^{(1-\lambda_0)\hat{s}_k} \cdot \mathrm{e}^{(\lambda_0)\check{s}_k}} \\
&= (1-\lambda_0) G(\hat{\boldsymbol{s}}) + \lambda_0 G(\check{\boldsymbol{s}}) \ ,
\end{aligned} \tag{4.90}$$

which contradicts the assumption of strict convexity, thus concluding the proof. □

Note that, if (4.87) holds for a $\lambda_0 \in (0,1)$, then it holds for *all* $\lambda \in (0,1)$. This is a direct consequence of log-convexity (1.5).

In order to show the next Theorem 4.43, we need the following three Lemmas 4.40, 4.41, and 4.42. We also need the strict-logconvexity (Definition 2.8 from Section 2.3). We have the following result.

Lemma 4.40. *Let $\mathcal{I}_k$ be a strictly log-convex interference function in the sense of Definition 2.8. For all $\lambda \in (0,1)$, we have*

$$\mathcal{I}_k\big(\boldsymbol{p}(\lambda)\big) = \big(\mathcal{I}_k(\hat{\boldsymbol{p}})\big)^{1-\lambda} \cdot \big(\mathcal{I}_k(\check{\boldsymbol{p}})\big)^{\lambda} \tag{4.91}$$

if and only if for all $l \in \mathsf{L}(k)$,

$$\hat{p}_l = \mu \check{p}_l \,, \quad \mu > 0 \,. \tag{4.92}$$

Proof. Assume that (4.92) holds. We have

$$p_l(\lambda) = \hat{p}_l^{1-\lambda} \cdot \check{p}_l^{\lambda} = \mu^{1-\lambda} \cdot \check{p}_l \,, \quad \forall l \in \mathsf{L}(k) \,, \tag{4.93}$$

and thus

$$\mathcal{I}_k\big(\boldsymbol{p}(\lambda)\big) = \mu^{1-\lambda} \cdot \mathcal{I}_k(\check{\boldsymbol{p}}) \,. \tag{4.94}$$

With $\mathcal{I}_k(\hat{\boldsymbol{p}}) = \mu \mathcal{I}_k(\check{\boldsymbol{p}})$, we have

$$\begin{aligned}\mathcal{I}_k\big(\boldsymbol{p}(\lambda)\big) &= \big(\mathcal{I}_k(\hat{\boldsymbol{p}})\big)^{1-\lambda} \cdot \frac{\mathcal{I}_k(\check{\boldsymbol{p}})}{\big(\mathcal{I}_k(\check{\boldsymbol{p}})\big)^{1-\lambda}} \\ &= \big(\mathcal{I}_k(\hat{\boldsymbol{p}})\big)^{1-\lambda} \cdot \big(\mathcal{I}_k(\check{\boldsymbol{p}})\big)^{\lambda} \,.\end{aligned} \tag{4.95}$$

Conversely, assume that (4.91) is fulfilled. Then strict log-convexity implies $\hat{p}_l = \mu \check{p}_l$ for all $l \in \mathsf{L}(k)$. □

Based on Lemma 4.40 we can show the following result.

Lemma 4.41. *Let $\mathcal{I}_1, \ldots, \mathcal{I}_K$ be strictly log-convex interference functions. Assume that $\boldsymbol{D}_{\mathcal{I}} \boldsymbol{D}_{\mathcal{I}}^T$ is irreducible. For arbitrary $\hat{\boldsymbol{p}}, \check{\boldsymbol{p}} \in \mathbb{R}^K_{++}$ and $\lambda_0 \in (0,1)$, the equality*

$$\mathcal{I}_k\big(\boldsymbol{p}(\lambda_0)\big) = \big(\mathcal{I}_k(\hat{\boldsymbol{p}})\big)^{1-\lambda_0} \cdot \big(\mathcal{I}_k(\check{\boldsymbol{p}})\big)^{\lambda_0} \,, \tag{4.96}$$

holds for all $k \in \mathcal{K}$, if and only if there exists a $\mu \in \mathbb{R}_{++}$ such that

$$\hat{\boldsymbol{p}} = \mu \check{\boldsymbol{p}} \,. \tag{4.97}$$

Proof. If (4.97) is fulfilled, then (4.96) is fulfilled for all $k \in \mathcal{K}$.

Conversely, assume that (4.96) is fulfilled, then it follows from Lemma 4.40 that

$$\hat{p}_l = \mu^{(k)} \cdot \check{p}_l \,, \quad \forall l \in \mathsf{L}(k) \,, \tag{4.98}$$

where $\mu^{(k)} \in \mathbb{R}$ is associated with the kth user. If $l \in \mathsf{L}(k_1) \cap \mathsf{L}(k_2)$, then (4.98) is fulfilled for both k_1 and k_2, i.e.,

$$\mu^{(k_1)} = \mu^{(k_2)} .$$

Since $\boldsymbol{D}_{\mathcal{I}} \boldsymbol{D}_{\mathcal{I}}^T$ is irreducible, for each k there is a sequence of indices k_0 to k_r, with $k_0 = 1$ and $k_r = k$, such that

$$\mathsf{L}(k_s) \cap \mathsf{L}(k_{s+1}) \neq \emptyset , \quad s = 0, \dots, r-1 . \tag{4.99}$$

It can be concluded that

$$\mu^{(1)} = \mu^{(k_1)} = \cdots = \mu^{(k)} , \tag{4.100}$$

which shows (4.97). □

With Lemma 4.41 we can show the following result.

Lemma 4.42. *Let $\mathcal{I}_1, \dots, \mathcal{I}_K$ be strictly log-convex interference functions. There is at least one $k_0 \in \mathcal{K}$ such that the strict inequality (4.87) is fulfilled for $\hat{\boldsymbol{p}} \neq \mu \check{\boldsymbol{p}}$, if and only if $\boldsymbol{D}_{\mathcal{I}} \boldsymbol{D}_{\mathcal{I}}^T$ is irreducible.*

Proof. From Lemma 4.41 we know that if $\boldsymbol{D}_{\mathcal{I}} \boldsymbol{D}_{\mathcal{I}}^T$ is irreducible, and $\hat{\boldsymbol{p}} \neq \mu \check{\boldsymbol{p}}$, for arbitrary $\hat{\boldsymbol{p}}, \check{\boldsymbol{p}} \in \mathbb{R}^K_{++}$, then there exists a $k_0 \in \mathcal{K}$ and a λ_0 such that (4.87) holds.

Conversely, assume that (4.87) is fulfilled. The proof is by contradiction. Suppose that $\boldsymbol{D}_{\mathcal{I}} \boldsymbol{D}_{\mathcal{I}}^T$ is not irreducible. Then there are at least two indices $k_1, k_2 \in \mathcal{K}$, which are not connected (see Definition 4 and Theorem 3 in [85]). Let $\mathcal{K}^{(1)}$ and $\mathcal{K}^{(2)}$ denote the sets of indices connected with k_1 and k_2, respectively. We have $\mathcal{K}^{(1)} \cap \mathcal{K}^{(2)} = \emptyset$. All other indices are collected in the (possibly) non-empty set $\mathcal{K}^{(3)} = \mathcal{K} \backslash (\mathcal{K}^{(1)} \cup \mathcal{K}^{(2)})$.

Consider a vector $\boldsymbol{p}^{(1)}$, and positive scalars $c^{(1)}$, $c^{(2)}$, where $c^{(1)} \neq c^{(2)}$. We define a vector $\boldsymbol{p}^{(2)}$ such that

$$p_k^{(2)} = \begin{cases} p_k^{(1)} & \text{if } k \in \mathcal{K}^{(3)} \\ c^{(1)} p_k^{(1)} & \text{if } k \in \mathcal{K}^{(1)} \\ c^{(2)} p_k^{(1)} & \text{if } k \in \mathcal{K}^{(2)} . \end{cases} \tag{4.101}$$

Since $c^{(1)} \neq c^{(2)}$, we have $\boldsymbol{p}^{(1)} \neq \boldsymbol{p}^{(2)}$. Now, consider

$$p_k(\tfrac{1}{2}) := (p_k^{(1)})^{1/2} \cdot (p_k^{(2)})^{1/2} , \quad \forall k \in \mathcal{K} . \tag{4.102}$$

For $k \in \mathcal{K}^{(3)}$ we have $\mathsf{L}(k) \cap \mathcal{K}^{(1)} = \emptyset$ and $\mathsf{L}(k) \cap \mathcal{K}^{(2)} = \emptyset$. Therefore, $\mathcal{I}_k(\boldsymbol{p}^{(1)}) = \mathcal{I}_k(\boldsymbol{p}^{(2)})$, and thus

$$\mathcal{I}_k(\boldsymbol{p}(\tfrac{1}{2})) = \big(\mathcal{I}_k(\boldsymbol{p}^{(1)})\big)^{1/2} \cdot \big(\mathcal{I}_k(\boldsymbol{p}^{(2)})\big)^{1/2} . \tag{4.103}$$

For $k \in \mathcal{K}^{(1)}$ we have $p_l^{(2)} = c^{(1)} p_l^{(1)}$ for all $l \in \mathsf{L}(k)$, thus

$$\mathcal{I}_k\big(\boldsymbol{p}(\tfrac{1}{2})\big) = \big(\mathcal{I}_k(\boldsymbol{p}^{(1)})\big)^{1/2} \cdot \big(\mathcal{I}_k(\boldsymbol{p}^{(2)})\big)^{1/2} . \tag{4.104}$$

The corresponding result can be shown for $k \in \mathcal{K}^{(2)}$. Thus, (4.104) holds for *all* $k \in \mathcal{K}$. However, this contradicts the assumed strict convexity of the interference function. Hence, $\boldsymbol{D}_{\mathcal{I}}\boldsymbol{D}_{\mathcal{I}}^T$ must be irreducible. □

This leads to the following result.

Theorem 4.43. *Let $\mathcal{I}_1, \ldots, \mathcal{I}_K$ be strictly log-convex interference functions. The cost function $G(\boldsymbol{s})$ defined in (4.86) is strictly convex if and only if $\boldsymbol{D}_{\mathcal{I}}\boldsymbol{D}_{\mathcal{I}}^T$ is irreducible.*

Proof. This follows from Lemma 4.39 and Lemma 4.42. □

Hence, if a proportionally fair optimizer exists, and if $\boldsymbol{D}_{\mathcal{I}}\boldsymbol{D}_{\mathcal{I}}^T$ is irreducible then we know from Theorem 4.43 that the solution is unique. However, $\boldsymbol{D}_{\mathcal{I}}\boldsymbol{D}_{\mathcal{I}}^T$ alone is not sufficient for the existence of an optimizer. This is shown by the next example.

Example 4.44. Consider the coupling matrix $\boldsymbol{V}$ defined in (4.81). The matrix $\boldsymbol{V}$ is irreducible. The product

$$\boldsymbol{V}\boldsymbol{V}^T = \begin{bmatrix} 1 & 0 & 1 \\ 0 & 2 & 1 \\ 1 & 1 & 2 \end{bmatrix}$$

is irreducible as well. The function $\sum_{k=1}^{3} \log \frac{[\boldsymbol{V}\hat{\boldsymbol{p}}]_k}{\hat{p}_k}$ is strictly convex if we substitute $\boldsymbol{p} = \mathrm{e}^{\boldsymbol{s}}$. The resulting SIR region is strictly log-convex according to Theorem 4.43. However, the previous Example 4.37 shows that no optimizer exists. This is because the requirements in Theorem 4.38 are not satisfied.

Lemma 4.45. *Consider an arbitrary dependency matrix $\hat{\boldsymbol{D}}_{\mathcal{I}}$ with a positive main diagonal. If $\hat{\boldsymbol{D}}_{\mathcal{I}}$ is irreducible then $\hat{\boldsymbol{D}}_{\mathcal{I}}\hat{\boldsymbol{D}}_{\mathcal{I}}^T$ is irreducible too.*

Proof. Defining $\hat{\boldsymbol{D}}_{\mathcal{I}}' := \hat{\boldsymbol{D}}_{\mathcal{I}}\hat{\boldsymbol{D}}_{\mathcal{I}}^T$, we have

$$[\hat{\boldsymbol{D}}_{\mathcal{I}}']_{kl} = \sum_{n=1}^{K} [\hat{\boldsymbol{D}}_{\mathcal{I}}]_{kn} [\hat{\boldsymbol{D}}_{\mathcal{I}}^T]_{nl} = \sum_{n=1}^{K} [\hat{\boldsymbol{D}}_{\mathcal{I}}]_{kn} [\hat{\boldsymbol{D}}_{\mathcal{I}}]_{ln} . \tag{4.105}$$

Consider the summand $n = l$. We have $[\hat{\boldsymbol{D}}_{\mathcal{I}}']_{kl} \geq [\hat{\boldsymbol{D}}_{\mathcal{I}}]_{kl}[\hat{\boldsymbol{D}}_{\mathcal{I}}]_{ll} \geq 0$. By assumption of a positive main diagonal, we have $[\hat{\boldsymbol{D}}_{\mathcal{I}}]_{ll} > 0$. Thus, $[\hat{\boldsymbol{D}}_{\mathcal{I}}]_{kl} > 0$ implies that $[\hat{\boldsymbol{D}}_{\mathcal{I}}']_{kl} > 0$ for an arbitrary choice of indices k, l. Hence, irreducibility of $\hat{\boldsymbol{D}}_{\mathcal{I}}$ implies irreducibility of $\hat{\boldsymbol{D}}_{\mathcal{I}}\hat{\boldsymbol{D}}_{\mathcal{I}}^T$. □

Lemma 4.45 leads to the following Theorem 4.46, which complements Theorem 4.38. It provides a necessary and sufficient condition for the existence of a *unique* optimizer.

Theorem 4.46. *Let $\mathcal{I}_1, \dots, \mathcal{I}_K$ be strictly monotone log-convex interference functions. We assume that (4.73) is fulfilled. Then problem (4.64) has a unique optimizer $\hat{\boldsymbol{p}} > 0$, $\|\hat{\boldsymbol{p}}\|_1 = 1$, if and only if there exist permutation matrices $\boldsymbol{P}^{(1)}$, $\boldsymbol{P}^{(2)}$ such that $\hat{\boldsymbol{D}}_\mathcal{I} = \boldsymbol{P}^{(1)} \boldsymbol{D}_\mathcal{I} \boldsymbol{P}^{(2)}$ is irreducible and its main diagonal is strictly positive.*

Proof. Assume that a unique optimizer $\hat{\boldsymbol{p}} > 0$ exists. Theorem 4.38 implies the existence of permutations such that $\hat{\boldsymbol{D}}_\mathcal{I}$ is block-irreducible with strictly positive main diagonal. That is, $\hat{\boldsymbol{D}}_\mathcal{I}$ is block-diagonal with $r \geq 1$ irreducible blocks $\hat{\boldsymbol{D}}_\mathcal{I}^{(1)}, \dots, \hat{\boldsymbol{D}}_\mathcal{I}^{(r)}$. The optimization $\inf_{\boldsymbol{p}>0} \sum_k \log(\mathcal{I}_k(\boldsymbol{p})/p_k)$ is reduced to r independent sub-problems with the respective dependency matrices. This leads to proportionally fair power allocations $\hat{\boldsymbol{p}}^{(1)}, \dots, \hat{\boldsymbol{p}}^{(r)}$. Uniqueness of $\hat{\boldsymbol{p}}$ implies $r = 1$, i.e. $\hat{\boldsymbol{D}}_\mathcal{I}$ consists of a single irreducible block. To show this, suppose that $r > 1$. Since each power vector can be arbitrarily scaled, every vector

$$\hat{\boldsymbol{p}} = \begin{bmatrix} \mu_1 \cdot \hat{\boldsymbol{p}}^{(1)} \\ \vdots \\ \mu_r \cdot \hat{\boldsymbol{p}}^{(r)} \end{bmatrix}, \quad \text{with } \mu_1, \dots, \mu_r > 0$$

is proportionally fair. Thus, $\hat{\boldsymbol{p}}$ is not unique. This contradicts the hypothesis and implies irreducibility.

Conversely, assume that there is an irreducible matrix $\hat{\boldsymbol{D}}_\mathcal{I}$ with a positive main diagonal. Since the requirements of Theorem 4.38 are fulfilled, we know that problem (4.64) has an optimizer $\hat{\boldsymbol{p}} > 0$. It remains to show that $\hat{\boldsymbol{p}} > 0$, with $\|\hat{\boldsymbol{p}}\|_1 = 1$, is unique. From Lemma 4.45, we know that $\hat{\boldsymbol{D}}_\mathcal{I} \hat{\boldsymbol{D}}_\mathcal{I}^T$ is irreducible. We have

$$\begin{aligned} \hat{\boldsymbol{D}}_\mathcal{I} \hat{\boldsymbol{D}}_\mathcal{I}^T &= \boldsymbol{P}^{(1)} \boldsymbol{D}_\mathcal{I} \boldsymbol{P}^{(2)} (\boldsymbol{P}^{(2)})^T \boldsymbol{D}_\mathcal{I}^T (\boldsymbol{P}^{(1)})^T \\ &= \boldsymbol{P}^{(1)} \boldsymbol{D}_\mathcal{I} \boldsymbol{D}_\mathcal{I}^T (\boldsymbol{P}^{(1)})^T . \end{aligned}$$

Thus, $\boldsymbol{D}_\mathcal{I} \boldsymbol{D}_\mathcal{I}^T$ is irreducible as well. It follows from Theorem 4.43 that the cost function $G(\boldsymbol{s})$ defined in (4.86) is strictly convex. Since the function $\exp\{\cdot\}$ is strictly monotonic, it can be concluded that the optimizer $\hat{\boldsymbol{p}}$ is unique. □

4.3.4 Equivalence of Nash Bargaining and Proportional Fairness

In the previous section we have studied the existence and uniqueness of a proportionally fair optimizer directly, without analyzing the underlying SIR region.

In this section, we use the results of Section 4.1.3, where the Nash bargaining theory was extended to the class of non-compact sets $\mathcal{ST}$. We investigate conditions under which the SIR region is contained in $\mathcal{ST}$. If this is fulfilled, and if an optimizer exists, then we know that it is the unique NBS.

For the problem at hand, boundary points $\hat{\gamma}$ with $C(\hat{\gamma}) = 1$ need not be achievable. In order to guarantee the existence of a $\hat{\boldsymbol{p}} > 0$ such that

$$1 = C(\hat{\gamma}) = \frac{\hat{\gamma}_k \mathcal{I}_k(\hat{\boldsymbol{p}})}{\hat{p}_k}, \tag{4.106}$$

we need the additional requirement that $\boldsymbol{D}_{\mathcal{I}}$ is irreducible. This ensures the existence of a power allocation $\boldsymbol{p} > 0$ such that (4.106) is fulfilled [5]. Note, that this solution is not required to be unique. An SIR boundary point may be associated with different power vectors. However, different SIR boundary points will always be associated with different power vectors.

Theorem 4.47. *Let $\mathcal{I}_1, \ldots, \mathcal{I}_K$ be strictly log-convex and strictly monotone interference functions. If $\boldsymbol{D}_{\mathcal{I}}$ and $\boldsymbol{D}_{\mathcal{I}}\boldsymbol{D}_{\mathcal{I}}^T$ are irreducible then the SIR region $\mathcal{S}$ defined in (4.19) is contained in $\mathcal{ST}$.*

Proof. Consider arbitrary boundary points $\hat{\gamma}$, $\check{\gamma}$ with $\hat{\gamma} \neq \check{\gamma}$ (at least one component). Since $\boldsymbol{D}_{\mathcal{I}}$ is irreducible, the points $\hat{\gamma}$, $\check{\gamma}$ are attained by power vectors $\hat{\boldsymbol{p}}$, $\check{\boldsymbol{p}}$, with $\hat{\boldsymbol{p}} \neq c\check{\boldsymbol{p}}$ for all $c > 0$, such that (4.106) is fulfilled. Next, consider $\boldsymbol{p}(\lambda)$ defined by (1.4). Defining $\boldsymbol{\gamma}(\lambda) = \hat{\gamma}^{1-\lambda} \cdot \check{\gamma}^{\lambda}$, we have [5]

$$\gamma_k(\lambda) \leq \frac{p_k(\lambda)}{\mathcal{I}_k(\boldsymbol{p}(\lambda))}, \quad \forall k \in \mathcal{K}. \tag{4.107}$$

It can be observed that $\boldsymbol{\gamma}(\lambda)$ is feasible, i.e., $C(\boldsymbol{\gamma}(\lambda)) \leq 1$. Next, consider the image set $\mathcal{L}og(\mathcal{S})$, with boundary points $\log \check{\gamma}$ and $\log \hat{\gamma}$. Since $\boldsymbol{\gamma}(\lambda)$ is contained in $\mathcal{S}$, it follows that all convex combinations $\log \boldsymbol{\gamma}(\lambda) = (1-\lambda)\log \hat{\gamma} + \lambda \log \check{\gamma}$ are contained in $\mathcal{L}og(\mathcal{S})$. Thus, $\mathcal{S}$ is log-convex. It remains to show strictness.

From Lemma 4.42 we know that there is at least one k_0 for which inequality (4.107) is strict. Following the same reasoning as in [85], we can successively reduce the powers of users for which strict inequality holds. Since $\boldsymbol{D}_{\mathcal{I}}\boldsymbol{D}_{\mathcal{I}}^T$ is irreducible, this reduces interference of other users, which in turn can reduce their power. The irreducibility of $\boldsymbol{D}_{\mathcal{I}}\boldsymbol{D}_{\mathcal{I}}^T$ ensures that all users benefit from this approach, so after a finite number of steps, we find a power vector $\tilde{\boldsymbol{p}} > 0$ such that

$$\gamma_k(\lambda) < \frac{\tilde{p}_k}{\mathcal{I}_k(\tilde{\boldsymbol{p}})}, \quad \forall k \in \mathcal{K}. \tag{4.108}$$

Thus, $C(\boldsymbol{\gamma}(\lambda)) < 1$, which proves strict log-convexity. □

Note that strict convexity of the SIR set does not imply that the PF problem (4.64) has an optimizer $\boldsymbol{p}^* > 0$. Example 4.44 in the previous section shows that $\boldsymbol{D}_{\mathcal{I}}$ and $\boldsymbol{D}_{\mathcal{I}}\boldsymbol{D}_{\mathcal{I}}^T$ can both be irreducible, however no optimizer exists if the conditions in Theorem 4.38 are not fulfilled.

The following theorem links the previous results on the existence and uniqueness of a proportional fair optimizer with the Nash bargaining framework derived in Section 4.1.3.

Corollary 4.48. *Let $\mathcal{I}_1, \ldots, \mathcal{I}_K$ be strictly log-convex and strictly monotone interference functions, and let $\boldsymbol{D}_{\mathcal{I}}$ and $\boldsymbol{D}_{\mathcal{I}}\boldsymbol{D}_{\mathcal{I}}^T$ be irreducible. There is a unique* *optimizer* $\hat{\boldsymbol{p}} > 0$ *to the problem of proportional fairness (4.64), with an associated SIR vector* $\hat{\boldsymbol{\gamma}}$*, if and only if there is a single-valued solution outcome* φ *satisfying the Nash axioms WPO, SYM, IIA, STC, and* $\varphi = \hat{\boldsymbol{\gamma}}$*.*

Proof. This follows from Theorems 4.8 and 4.47. □

4.3.5 Weighted Utility and Cost Optimization

In this section we consider another application example for the framework of log-convex interference functions. Assume that the SIR is related to the QoS by a function $\phi(x) = g(1/x)$, i.e.,

$$QoS = g(1/\text{SIR}) .$$

The function g is assumed to be monotone increasing and $g(\mathrm{e}^x)$ is convex with respect to x, like $g(x) = x$ or $g(x) = \log x$. We are interested in the optimization problem

$$\inf_{\boldsymbol{s} \in \mathbb{R}^K} \sum_{k \in \mathcal{K}} \alpha_k \, g\big(\mathcal{I}_k(\mathrm{e}^{\boldsymbol{s}})/\mathrm{e}^{s_k}\big) \quad \text{s.t.} \quad \|\mathrm{e}^{\boldsymbol{s}}\|_1 \le P_{\max} , \tag{4.109}$$

where $\mathcal{I}_k(\mathrm{e}^{\boldsymbol{s}})$ is a log-convex interference function. The weights $\boldsymbol{\alpha} = [\alpha_1, \ldots, \alpha_K] > 0$ can model individual user requirements and possibly depend on system parameters like priorities, queue lengths, etc. By appropriately choosing $\boldsymbol{\alpha}$ it is possible to trade off overall efficiency against fairness.

The next theorem, which is proven in the Appendix A.9, shows conditions for convexity.

Theorem 4.49. *Suppose that* $\mathcal{I}_k(\mathrm{e}^{\boldsymbol{s}})$ *is log-convex for all* $k \in \mathcal{K}$ *and* g *is monotone increasing. Then problem (4.109) is convex if and only if* $g(\mathrm{e}^x)$ *is convex on* $\mathbb{R}$*.*

If the optimization problem (4.109) is convex, then it can be solved by standard convex optimization techniques. Note, that the optimization is over the non-compact set $\mathbb{R}^K$, thus even if the problem is convex, it is not obvious that the optimum is achieved (e.g. $\boldsymbol{s} \to -\infty$ might occur). However, this case can be ruled out for a practical system with receiver noise $\sigma_n^2 > 0$, in which case $\mathrm{e}^{s_k} \to 0$ can never happen, since otherwise the objective would tend to infinity, away from the minimum. Without noise, however, it can happen that one or more power components tend to zero, in which case the infimum is not achieved (see e.g. the discussion in [2]).

A special case of problem (4.109) is (weighted) *proportional fairness* [98].

$$\sup_{\boldsymbol{p} > 0} \Big(- \sum_{k \in \mathcal{K}} \alpha_k \log \frac{\mathcal{I}_k(\boldsymbol{p})}{p_k} \Big) = \sup_{\boldsymbol{p} > 0} \Big(\sum_{k \in \mathcal{K}} \alpha_k \log \frac{p_k}{\mathcal{I}_k(\boldsymbol{p})} \Big) . \tag{4.110}$$

Note, that this problem (4.110) is also related to the problem of throughput maximization (see e.g. [43, 122]). In the high SIR regime, we can approximate

$\log(1 + SIR) = \log(SIR)$, so (4.110) can be interpreted as the weighted sum throughput of the system.

Similar to the cost minimization problem (4.109), we formulate a utility maximization problem.

$$\sup_{s \in \mathbb{R}^K} \sum_{k \in \mathcal{K}} \alpha_k \, g\big(\mathcal{I}_k(e^{s})/e^{s_k}\big) \quad \text{s.t.} \quad \|e^{s}\|_1 \leq P_{\max} \,. \tag{4.111}$$

In this case, the function $g(e^x)$ is required to be monotone decreasing instead of increasing. As in Theorem 4.49, convexity of $g(e^x)$ can be shown to be necessary and sufficient for (4.111) to be convex.

Notice that the supremum (4.111) can be written as a convex function $u(\boldsymbol{\alpha})$ of the weights $\boldsymbol{\alpha} = [\alpha_1, \ldots, \alpha_K]$. Moreover, $u(\boldsymbol{\alpha})$ fulfills the properties A1, A2, A3, so it can be regarded as an "interference function". Using a substitution $\boldsymbol{\alpha} = \exp \boldsymbol{\beta}$, the function $u(\boldsymbol{\alpha})$ is a *log-convex interference function* in the sense of Definition 1.4. This is a further example, which shows that log-convex interference functions arise naturally in many different contexts. Even though our discussion is motivated by power control, the proposed theoretical framework provides a general tool, which is not limited to interference in a physical sense.

Also, (4.111) provides another example for a combination of log-convex interference functions resulting in a log-convex interference function. Again, it can be observed that certain operations are closed within the framework of log-convex interference functions.

4.4 SINR Region under a Total Power Constraint

In the previous section we have discussed general SIR regions based on log-convex interference functions. Conditions were derived under which the SIR region is contained in $\mathcal{ST}$ and a unique optimizer exists.

The situation is much simpler if the log-convex interference functions is standard. Under the assumption of a sum power constraint, the interference functions are not only coupled by interference, but also by the limited power budget. This simplifies the analysis. The more complicated case of individual power constraints wil be addressed in the following Section 4.5.

Consider the sum-power-constrained SINR region $\mathcal{S}(P_{\max})$, as defined in (2.47). The sum of all transmission powers is limited by $P_{\max}$. The next theorem shows that the resulting SINR set is strictly convex after a logarithmic transformation.

Theorem 4.50. *Let $\mathcal{I}_1, \ldots, \mathcal{I}_{K_u}$ be arbitrary log-convex interference functions. Then for all $0 < P_{\max} < +\infty$ the logarithmic transformation of the SINR region $\mathcal{L}og\big(\mathcal{S}(P_{\max})\big)$ is strictly convex, the entire boundary of $\mathcal{S}(P_{\max})$ is Pareto optimal, and $\mathcal{S}(P_{\max}) \in \mathcal{ST}_c$.*

Proof. In order to show strict convexity, consider arbitrary points $\hat{\boldsymbol{q}}, \check{\boldsymbol{q}}$, with $\hat{\boldsymbol{q}} \neq \check{\boldsymbol{q}}$, from the boundary of $\mathcal{Log}\big(\mathcal{S}(P_{\max})\big)$. This set is strictly convex if the line segment $\boldsymbol{q}(\lambda) = (1-\lambda)\hat{\boldsymbol{q}} + \lambda\check{\boldsymbol{q}}$, with $\lambda \in (0,1)$, is in the interior of the region. This is shown in the SINR domain, where $\hat{\boldsymbol{\gamma}} = \exp \hat{\boldsymbol{q}}$ and $\check{\boldsymbol{\gamma}} = \exp \check{\boldsymbol{q}}$ are the corresponding boundary points, with $\hat{\boldsymbol{\gamma}} \neq \check{\boldsymbol{\gamma}}$. The line segment is transformed to the curve (all operations are component-wise)

$$\boldsymbol{\gamma}(\lambda) = \exp \boldsymbol{q}(\lambda) = (\hat{\boldsymbol{\gamma}})^{1-\lambda} \cdot (\check{\boldsymbol{\gamma}})^{\lambda} \,. \tag{4.112}$$

A point $\boldsymbol{q}(\lambda)$ on the line segment is in the interior of $\mathcal{Log}\big(\mathcal{S}(P_{\max})\big)$ if and only if $C(\boldsymbol{\gamma}(\lambda), P_{\max}) < 1$. We exploit that for any $\boldsymbol{\gamma} > 0$ there exists a unique power vector $\boldsymbol{p}(\boldsymbol{\gamma}) > 0$ such that

$$\gamma_k \frac{1}{C(\boldsymbol{\gamma}, P_{\max})} = \frac{p_k(\boldsymbol{\gamma})}{\mathcal{J}_k\big(\boldsymbol{p}(\boldsymbol{\gamma})\big)} \,, \quad \forall k \in \mathcal{K}_u \,. \tag{4.113}$$

This can be shown in a similar way as in [1], by exploiting strict monotonicity (2.22), and the fact that $\gamma_k / C(\boldsymbol{\gamma}, P_{\max})$ is a boundary point. Let us define $\boldsymbol{p}(\lambda)$, where $p_k(\lambda) = (\hat{p}_k)^{1-\lambda} \cdot (\check{p}_k)^{\lambda}$, and $\hat{\boldsymbol{p}} := \boldsymbol{p}(\hat{\boldsymbol{\gamma}})$, $\check{\boldsymbol{p}} := \boldsymbol{p}(\check{\boldsymbol{\gamma}})$ are the power vectors that achieve the boundary points $\hat{\boldsymbol{p}}$ and $\check{\boldsymbol{p}}$, respectively. Because of uniqueness, $\hat{\boldsymbol{\gamma}} \neq \check{\boldsymbol{\gamma}}$ implies $\hat{\boldsymbol{p}} \neq \check{\boldsymbol{p}}$. By exploiting log-convexity of the interference functions $\mathcal{I}_1, \ldots, \mathcal{I}_{K_u}$, we have

$$\frac{\gamma_k(\lambda) \cdot \mathcal{J}_k\big(\boldsymbol{p}(\lambda)\big)}{p_k(\lambda)} \leq \left(\frac{\hat{\gamma}_k \cdot \mathcal{J}_k(\hat{\boldsymbol{p}})}{\hat{p}_k}\right)^{1-\lambda} \cdot \left(\frac{\check{\gamma}_k \cdot \mathcal{J}_k(\check{\boldsymbol{p}})}{\check{p}_k}\right)^{\lambda} \tag{4.114}$$

for all $k \in \mathcal{K}_u$. Combining (2.48) and (4.114), we have

$$C\big(\boldsymbol{\gamma}(\lambda), \boldsymbol{\mathcal{I}}, P_{\max}\big) \leq \big[C\big(\hat{\boldsymbol{\gamma}}, \boldsymbol{\mathcal{I}}, P_{\max}\big)\big]^{1-\lambda} \cdot \big[C\big(\check{\boldsymbol{\gamma}}, \boldsymbol{\mathcal{I}}, P_{\max}\big)\big]^{\lambda} \,.$$

Since $\hat{\boldsymbol{\gamma}}$ and $\check{\boldsymbol{\gamma}}$ are boundary points, we have $C\big(\hat{\boldsymbol{\gamma}}, \boldsymbol{\mathcal{I}}, P_{\max}\big) = C\big(\check{\boldsymbol{\gamma}}, \boldsymbol{\mathcal{I}}, P_{\max}\big) = 1$, and thus

$$C\big(\boldsymbol{\gamma}(\lambda), \boldsymbol{\mathcal{I}}, P_{\max}\big) \leq 1 \,. \tag{4.115}$$

It remains to show that inequality (4.115) is strict. Since $\hat{\boldsymbol{p}} \neq \check{\boldsymbol{p}}$, Hölder's inequality leads to

$$\sum_{k \in \mathcal{K}_u} p_k(\lambda) < \Big(\sum_{k \in \mathcal{K}_u} (\hat{p}_k)^{(1-\lambda)n}\Big)^{\frac{1}{n}} \cdot \Big(\sum_{k \in \mathcal{K}_u} (\check{p}_k)^{\lambda m}\Big)^{\frac{1}{m}} \tag{4.116}$$

where $1 = 1/n + 1/m$. This expression is simplified by choosing $n = 1/(1-\lambda)$ and $m = 1/\lambda$. Since the sum-power constraint is active for points on the boundary, we have $P_{\max} = \sum_k \hat{p}_k = \sum_k \check{p}_k$. Thus,

$$\sum_{k \in \mathcal{K}_u} p_k(\lambda) < \Big(\sum_{k \in \mathcal{K}_u} \hat{p}_k\Big)^{1-\lambda} \cdot \Big(\sum_{k \in \mathcal{K}_u} \check{p}_k\Big)^{\lambda} = P_{\max} \,. \tag{4.117}$$

Since inequality (4.117) is strict, there exists a $\mu > 1$, and a new vector $\boldsymbol{p}'(\lambda) = \mu\boldsymbol{p}(\lambda)$ that also fulfills the inequality. By exploiting axioms A2 and strict monotonicity (2.22), we have

$$\frac{\gamma_k(\lambda) \cdot \mathcal{J}_k\big(\boldsymbol{p}'(\lambda)\big)}{p'_k(\lambda)} = \frac{\gamma_k(\lambda) \cdot \mathcal{I}_k\big(\boldsymbol{p}(\lambda), \sigma_n^2/\mu\big)}{p_k(\lambda)} < \frac{\gamma_k(\lambda) \cdot \mathcal{I}_k\big(\boldsymbol{p}(\lambda), \sigma_n^2\big)}{p_k(\lambda)} . \tag{4.118}$$

From A3 it follows that inequality (4.114) is strict. Thus, $C\big(\boldsymbol{\gamma}(\lambda), \boldsymbol{\mathcal{I}}, P_{\max}\big) < 1$, which means that for any $\lambda \in (0,1)$, the point $\boldsymbol{q}(\lambda)$ is in the strict interior of the region, thus proving strict log-convexity of the SINR region.

Strict log-convexity implies Pareto optimality. It remains to show that $\mathcal{S}(P_{\max}) \in \mathcal{ST}_c$. The transformed set $\mathcal{L}og\big(\mathcal{S}(P_{\max})\big)$ is closed. This can be observed from definition (2.47). It is also upper-bounded because of the power constraint and the assumption of noise. Finally, the entire boundary is Pareto optimal, thus $\mathcal{S}(P_{\max}) \in \mathcal{ST}_c$ is fulfilled. □

Theorem 4.50 shows that the following problem of maximizing the sum of logarithmic SINR always has a single-valued solution, and this solution is the Nash bargaining solution.

$$\max_{\boldsymbol{\gamma} \in \mathcal{S}(P_{\max})} \sum_{k \in \mathcal{K}_u} \log \gamma_k . \tag{4.119}$$

Note, that $\log \gamma_k$ is a high-SNR approximation of the Shannon capacity $\log(1+\gamma_k)$. From the results of the previous Section 4.3 we know that the problem is convex after a change of variable. Hence, the NP-hard problem [123] of sum-rate maximization becomes convex as the SNR tends to infinity.

4.5 Individual Power Constraints – Pareto Optimality and Strict Convexity

In this section we discuss the SINR region of an interference-coupled multi-user system with individual power constraints. The assumption of individual power constraints does create some challenges. In particular, Pareto optimality and strict log-convexity depend on the way users are coupled by interference. This differs from the previous case of a sum power constraint. Under a sum-power constraint, the possible occurrence of interference-free users does not matter because the users are always coupled by sharing a common power budget. However, in order to analyze the behavior under *individual power constraints*, we need to take into account the interference coupling characterized by $\boldsymbol{D}_{\mathcal{I}}$. Therefore, large part of our analysis will focus on the effects of interference coupling. This requires a different mathematical approach involving combinatorial arguments.

4.5.1 Characterization of the Boundary for Individual Power Constraints

Consider log-convex interference functions and individual power limits $\boldsymbol{p}^{\max}$. Let $\boldsymbol{\gamma} > 0$ be any boundary point of the resulting region $\mathcal{S}(\boldsymbol{p}^{\max})$. The set of all power vectors achieving $\boldsymbol{\gamma}$ is

$$\mathcal{P}(\boldsymbol{\gamma}, \boldsymbol{p}^{\max}) = \{\mathbf{0} \leq \boldsymbol{p} \leq \boldsymbol{p}^{\max} : p_k \geq \gamma_k \mathcal{J}_k(\boldsymbol{p})\} . \tag{4.120}$$

For the following analysis, it is important to note that the set $\mathcal{P}(\boldsymbol{\gamma}, \boldsymbol{p}^{\max})$ can contain multiple elements. This is most easily explained by an example:

Example 4.51. Consider a 2-user Gaussian multiple access channel (MAC) with successive interference cancellation, normalized noise $\sigma_n^2 = 1$, and a given decoding order 1, 2. The SINR of the users are

$$\begin{aligned} \mathrm{SINR}_1(\boldsymbol{p}) &= \frac{p_1}{p_2 + 1} , \\ \mathrm{SINR}_2(\boldsymbol{p}) &= p_2 . \end{aligned}$$

Assuming power constraints $p_1 \leq p_1^{\max} = 1$ and $p_2 \leq p_2^{\max} = 1$, we obtain an SINR region as depicted in Fig. 4.6.

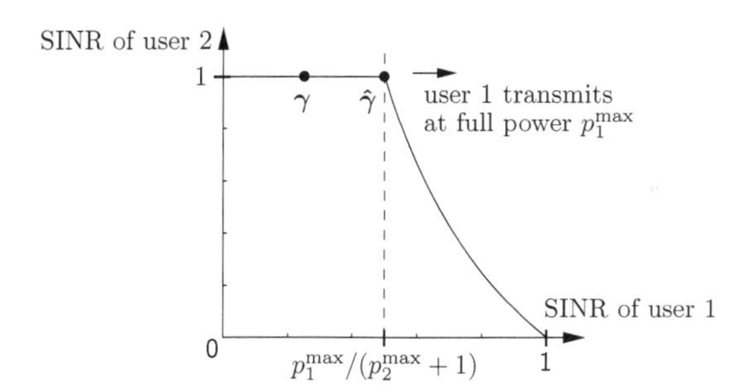

Fig. 4.6. Feasible SINR region for the 2-user MAC channel described in Example 4.51.

4.5.2 Properties of Boundary Points

Consider the boundary point $\boldsymbol{\gamma}$ depicted in Fig. 4.6. This point is achieved by $\boldsymbol{p}^* = [p_1^{\max}/2, p_2^{\max}]^T$, and therefore $\boldsymbol{p}^* \in \mathcal{P}(\boldsymbol{\gamma}, \boldsymbol{p}^{\max})$. This vector achieves $\boldsymbol{\gamma}$ with component-wise minimum power. However, $\boldsymbol{p}^*$ is not the only element of $\mathcal{P}(\boldsymbol{\gamma}, \boldsymbol{p}^{\max})$. Because of the assumed interference cancellation, we can increase the power (and thus the SINR) of User 1, without reducing the SINR at User 2. If both users transmit with maximum power $\boldsymbol{p}^{\max}$ then the corner point $\hat{\boldsymbol{\gamma}}$ is achieved. This power vector is also contained in $\mathcal{P}(\boldsymbol{\gamma}, \boldsymbol{p}^{\max})$ because $\hat{\boldsymbol{\gamma}} \geq \boldsymbol{\gamma}$, so the SINR targets $\boldsymbol{\gamma}$ are still fulfilled.

The following fixed point iteration will play an important role in our analysis.

$$p_k^{(n+1)} = \gamma_k \mathcal{J}_k(\boldsymbol{p}^{(n)}), \quad \forall k \in \mathcal{K}_u \ , \quad \boldsymbol{p}^{(0)} \in \mathcal{P}(\boldsymbol{\gamma}, \boldsymbol{p}^{\max}) \ . \tag{4.121}$$

Lemma 4.52. *Let $\boldsymbol{\gamma} > 0$ be an arbitrary boundary point, then the limit $\boldsymbol{p}^* = \lim_{n\to\infty} \boldsymbol{p}^{(n)} > 0$ achieves $\boldsymbol{\gamma}$ with component-wise minimum power. That is, $\boldsymbol{p}^* \leq \boldsymbol{p}$ for all $\boldsymbol{p} \in \mathcal{P}(\boldsymbol{\gamma}, \boldsymbol{p}^{\max})$.*

Proof. This lemma follows from [1]. A proof for the interference framework A1, A2, A3 plus strict monotonicity (2.22) was given in [2]. □

The next lemma shows that the inequality constraint in (4.120) is always fulfilled with equality for at least one component, otherwise $\boldsymbol{\gamma}$ could not be a boundary point.

Lemma 4.53. *For any boundary point $\boldsymbol{\gamma} > 0$, consider an arbitrary $\boldsymbol{p} \in \mathcal{P}(\boldsymbol{\gamma}, \boldsymbol{p}^{\max})$. There always exists a $k \in \mathcal{K}_u$ such that $p_k = \gamma_k \mathcal{J}_k(\boldsymbol{p})$.*

Proof. The proof is by contradiction. Suppose $p_k > \gamma_k \mathcal{J}_k(\boldsymbol{p})$ for all $k \in \mathcal{K}_u$. Then

$$\begin{aligned} C(\boldsymbol{\gamma}, \boldsymbol{p}^{\max}) &= \inf_{0 < \tilde{\boldsymbol{p}} \leq \boldsymbol{p}^{\max}} \left(\max_{k \in \mathcal{K}_u} \frac{\gamma_k \mathcal{J}_k(\tilde{\boldsymbol{p}})}{\tilde{p}_k} \right) \\ &\leq \max_{k \in \mathcal{K}_u} \frac{\gamma_k \mathcal{J}_k(\boldsymbol{p})}{p_k} < 1 \ . \end{aligned} \tag{4.122}$$

This is a contradiction because for any boundary point $\boldsymbol{\gamma}$ we have $C(\boldsymbol{\gamma}, \boldsymbol{p}^{\max}) = 1$. □

Sometimes we can find a power vector such that one or more components of $\boldsymbol{\gamma}$ are surpassed, as illustrated in Fig. 4.6. The corresponding indices are collected in an index set $\overline{\mathcal{K}_u}$ defined as follows.

Definition 4.54. *For any boundary point $\boldsymbol{\gamma} > 0$, let $\overline{\mathcal{K}_u}$ be the set of all $k \in \mathcal{K}_u$ such that there exists a $\boldsymbol{p}^{(k)} \in \mathcal{P}(\boldsymbol{\gamma}, \boldsymbol{p}^{\max})$ with $p_k^{(k)} > \gamma_k \mathcal{J}_k(\boldsymbol{p}^{(k)})$.*

For the point $\boldsymbol{\gamma}$, in Fig. 4.6, this is the first user, whose power can be increased without decreasing the performance of User 2. We are only interested in the case where $\overline{\mathcal{K}_u}$ is non-empty. Otherwise the fixed point is the unique solution, which is trivial. Also, we know from Lemma 4.53 that $\overline{\mathcal{K}_u} \neq \mathcal{K}_u$.

The next theorem shows that there always exists a vector $\hat{\boldsymbol{p}}$ for which strict inequality holds for all $k \in \overline{\mathcal{K}_u}$ simultaneously.

Theorem 4.55. *Let $\mathcal{I}_1, \ldots, \mathcal{I}_{K_u}$ be log-convex interference functions. Assume that $\boldsymbol{\gamma}$ is an arbitrary boundary point such that $\overline{\mathcal{K}_u}$ is non-empty. Then there exists a vector $\hat{\boldsymbol{p}} \in \mathcal{P}(\boldsymbol{\gamma}, \boldsymbol{p}^{\max})$ such that*

$$\hat{p}_k > \gamma_k \mathcal{J}_k(\hat{\boldsymbol{p}}), \quad \forall k \in \overline{\mathcal{K}_u} \ , \tag{4.123}$$

and for all $\boldsymbol{p} \in \mathcal{P}(\boldsymbol{\gamma}, \boldsymbol{p}^{\max})$ *we have*

$$p_k = \gamma_k \mathcal{J}_k(\boldsymbol{p}), \quad \forall k \in \mathcal{K}_u \backslash \overline{\mathcal{K}_u} \,. \tag{4.124}$$

Proof. Inequality (4.124) is a consequence of Definition 4.54. In order to show (4.123), consider arbitrary $k_1, k_2 \in \overline{\mathcal{K}_u}$, with $k_1 \neq k_2$, and vectors $\boldsymbol{p}^{(k_1)}, \boldsymbol{p}^{(k_2)}$ as in Definition 4.54. We define a vector $\boldsymbol{p}(\lambda)$ with components

$$p_l(\lambda) = (p_l^{(k_1)})^{1-\lambda} \cdot (p_l^{(k_2)})^{\lambda} \,, \quad l \in \mathcal{K}_u \,.$$

Log-convexity implies (1.5). Since $\boldsymbol{p}(\lambda) \leq \boldsymbol{p}^{\max}$, we have

$$\frac{\gamma_l \mathcal{J}_l(\boldsymbol{p}(\lambda))}{p_l(\lambda)} \leq \Big(\frac{\gamma_l \mathcal{J}_l(\boldsymbol{p}^{(k_1)})}{p_l^{(k_1)}}\Big)^{1-\lambda} \cdot \Big(\frac{\gamma_l \mathcal{J}_l(\boldsymbol{p}^{(k_2)})}{p_l^{(k_2)}}\Big)^{\lambda} \leq 1. \tag{4.125}$$

The last inequality holds because $\boldsymbol{p}^{(k_1)} \in \mathcal{P}(\boldsymbol{\gamma}, \boldsymbol{p}^{\max})$ implies that $p_l^{(k_1)} \geq \gamma_l \mathcal{J}_l(\boldsymbol{p}^{(k_1)})$, and the same holds for $\boldsymbol{p}^{(k_1)}$. It can be observed from (4.125) that $\boldsymbol{p}(\lambda) \in \mathcal{P}(\boldsymbol{\gamma}, \boldsymbol{p}^{\max})$ for $0 < \lambda < 1$. For indices $l = k_1$ or $l = k_2$, at least one factor on the right hand side of (4.125) is strictly less than one, and therefore

$$\frac{\gamma_l \mathcal{J}_l(\boldsymbol{p}(\lambda))}{p_l(\lambda)} < 1 \,, \quad l = k_1, k_2 \,.$$

In the same way, we can combine $\boldsymbol{p}(1/2)$ with another vector $\boldsymbol{p}^{(k_3)}$ with $k_3 \in \overline{\mathcal{K}_u}$. This leads to a new vector $\boldsymbol{p}'(\lambda)$ which fulfills

$$\frac{\gamma_l \mathcal{J}_l(\boldsymbol{p}'(\lambda))}{p'_l(\lambda)} < 1 \,, \quad l = k_1, k_2, k_3 \,.$$

Repeating this procedure for all $k \in \overline{\mathcal{K}_u}$, we obtain the desired vector $\hat{\boldsymbol{p}} \in \mathcal{P}(\boldsymbol{\gamma}, \boldsymbol{p}^{\max})$ fulfilling (4.123). □

The following corollary is an immediate consequence of Theorem 4.55.

Corollary 4.56. *Let* $\boldsymbol{p}^*$ *be the fixed point defined in Lemma 4.52. All other quantities are defined as in Theorem 4.53. We have* $\hat{\boldsymbol{p}} \geq \boldsymbol{p}^*$ *(Lemma 4.52) and thus for all* $k \in \overline{\mathcal{K}_u}$

$$\hat{p}_k > \gamma_k \mathcal{J}_k(\hat{\boldsymbol{p}}) \geq \gamma_k \mathcal{J}_k(\boldsymbol{p}^*) = p_k^* \,. \tag{4.126}$$

That is, the inequality $\hat{\boldsymbol{p}} \geq \boldsymbol{p}^*$ is strict for all components from $\overline{\mathcal{K}_u}$. In the following we will refer to $\overline{\mathcal{K}_u}$ as "oversized users".

The following theorem shows that the oversized users $\overline{\mathcal{K}_u}$ have no impact on the interference experienced by the other users $\mathcal{K}_u \backslash \overline{\mathcal{K}_u}$. That is, the interference is the same as if we would use the minimum-power vector $\boldsymbol{p}^*$. Also, the powers of users $\mathcal{K}_u \backslash \overline{\mathcal{K}_u}$ cannot be oversized.

Theorem 4.57. *Let $\mathcal{I}_1, \ldots, \mathcal{I}_{K_u}$ and γ be defined as in Theorem 4.55. Consider an arbitrary $\hat{\boldsymbol{p}} \in \mathcal{P}(\gamma, \boldsymbol{p}^{\max})$. For all $k \in \mathcal{K}_u \backslash \overline{\mathcal{K}_u}$, we have $\mathcal{J}_k(\boldsymbol{p}^*) = \mathcal{J}_k(\hat{\boldsymbol{p}})$ and $p_k^* = \hat{p}_k$.*

Proof. We are interested in the non-trivial case $\hat{\boldsymbol{p}} \neq \boldsymbol{p}^*$. Defining $\boldsymbol{\Gamma} = \text{diag}(\gamma)$, the fixed point iteration (4.121) can be written as $\boldsymbol{p}^{(n+1)} = \boldsymbol{\Gamma}\boldsymbol{\mathcal{J}}(\boldsymbol{p}^{(n)})$. By choosing the initialization $\boldsymbol{p}^{(0)} = \hat{\boldsymbol{p}}$, we obtain a monotone decreasing sequence [1]. Since $\boldsymbol{p}^{(0)} \geq \boldsymbol{p}^{(1)}$, we have

$$\boldsymbol{p}^{(1)} = \boldsymbol{\Gamma}\boldsymbol{\mathcal{J}}(\boldsymbol{p}^{(0)}) \geq \boldsymbol{\Gamma}\boldsymbol{\mathcal{J}}(\boldsymbol{p}^{(1)}) . \tag{4.127}$$

Thus $\boldsymbol{p}^{(1)} \in \mathcal{P}(\gamma, \boldsymbol{p}^{\max})$. For any $k \in \mathcal{K}_u \backslash \overline{\mathcal{K}_u}$ we have

$$p_k^{(1)} = \gamma_k \mathcal{J}_k(\boldsymbol{p}^{(0)}) = p_k^{(0)} = \hat{p}_k . \tag{4.128}$$

Likewise, we have for $\boldsymbol{p}^{(2)}$, $k \in \mathcal{K}_u \backslash \overline{\mathcal{K}_u}$,

$$p_k^{(2)} = \gamma_k \mathcal{J}_k(\boldsymbol{p}^{(1)}) = p_k^{(1)} = \hat{p}_k . \tag{4.129}$$

By induction, we have for all $n \in \mathbb{N}$

$$p_k^{(n)} = \hat{p}_k \quad \text{for all } k \in \mathcal{K}_u \backslash \overline{\mathcal{K}_u} .$$

With $\lim_{n\to\infty} \boldsymbol{p}^{(n)} = \boldsymbol{p}^*$ we have

$$p_k^* = \hat{p}_k \quad \text{for all } k \in \mathcal{K}_u \backslash \overline{\mathcal{K}_u} ,$$

thus proving the second statement. From the definition of $\overline{\mathcal{K}_u}$, we have

$$\gamma_k \mathcal{J}_k(\hat{\boldsymbol{p}}) = \hat{p}_k = p_k^* = \gamma_k \mathcal{J}_k(\boldsymbol{p}^*) , \quad \text{for all } k \in \mathcal{K}_u \backslash \overline{\mathcal{K}_u}$$

which concludes the proof. □

Corollary 4.58. *Consider an arbitrary $\hat{\boldsymbol{p}} \in \mathcal{P}(\gamma, \boldsymbol{p}^{\max})$. Then for all $\boldsymbol{p}$ with $\boldsymbol{p}^* \leq \boldsymbol{p} \leq \hat{\boldsymbol{p}}$ we have*

$$\mathcal{J}_k(\boldsymbol{p}^*) = \mathcal{J}_k(\boldsymbol{p}) = \mathcal{J}_k(\hat{\boldsymbol{p}}) \quad \textit{for all } k \in \mathcal{K}_u \backslash \overline{\mathcal{K}_u} . \tag{4.130}$$

Proof. This follows from Theorem 4.57 and the monotonicity axiom A3. □

Note that $\boldsymbol{p}^*$ and $\hat{\boldsymbol{p}}$ are both contained in $\mathcal{P}(\gamma, \boldsymbol{p}^{\max})$, so the resulting SINR values are contained in the feasible SINR region. However, we cannot infer from Corollary 4.58 that the same holds for $\boldsymbol{p}$.

For p_k from $\overline{\mathcal{K}_u}$ we can reduce this component without affecting the interference power. This holds for $\boldsymbol{p}^* \leq \boldsymbol{p} \leq \hat{\boldsymbol{p}}$ but not necessarily for vectors outside this area. This is because we cannot rely on strict monotonicity, as later in this section. We know that for $k \in \overline{\mathcal{K}_u}$ the interference functions for such $\boldsymbol{p}$ do not depend on the indices for which $p_k^* < \hat{p}_k$.

This is related to the discussion on Pareto optimality later in Section 4.5.3. Because of the individual power constraints, we cannot exploit strict monotonicity with respect to noise in order to guarantee a Pareto optimal boundary.

4.5.3 Analysis of the Pareto Boundary

Thus far we have focused on the interference coupling aspects. Now, we will analyze QoS sets resulting from these interference models. In this paper, *QoS* can stand for some arbitrary performance measure, which depends on the SINR by a strictly monotone and continuous function ϕ defined on $\mathbb{R}_+$. The QoS of user k is $u_k(\boldsymbol{p}) = \phi_k\big(\text{SINR}_k(\boldsymbol{p})\big)$, as discussed in Section 2.6.1.

Let γ_k be the inverse function of ϕ_k, then $\gamma_k(u_k)$ is the minimum SINR level needed by the kth user to satisfy the QoS target u_k. Let $\boldsymbol{u}$ be a vector of QoS values from some QoS region $\mathcal{U}$, then the associated SINR vector is

$$\boldsymbol{\gamma}(\boldsymbol{u}) = [\gamma_1(u_1), \ldots, \gamma_{K_u}(u_{K_u})]^T \,. \tag{4.131}$$

QoS values $\boldsymbol{u}$ are feasible if and only if $C\big(\boldsymbol{\gamma}(\boldsymbol{u}), \boldsymbol{\mathcal{I}}, \boldsymbol{p}^{\max})\big) \leq 1$. The QoS feasible set is the sub-level set (2.38). We are now interested in the boundary of $\mathcal{U}$, which is characterized by $C(\boldsymbol{\gamma}(\boldsymbol{u}), \boldsymbol{\mathcal{I}}, \boldsymbol{p}^{\max}) = 1$. Recall the definition of Pareto optimality from Definition 4.4.

Lemma 4.59. *A boundary point $\boldsymbol{u} \in \partial\mathcal{U}$ is Pareto optimal if and only if $\boldsymbol{\gamma}(\boldsymbol{u}) \in \partial\mathcal{S}(\boldsymbol{\mathcal{I}}, \boldsymbol{p}^{\max})$ is Pareto optimal.*

Proof. This is a direct consequence of the strictly monotone mapping (2.36). Pareto optimal points in $\mathcal{S}$ are mapped to Pareto optimal points in $\mathcal{U}$ and vice versa. Non-Pareto boundary segments in $\mathcal{S}$ are parallel to the coordinate axes. Those segments are mapped to parallel segments in $\mathcal{U}$ and vice versa. □

From Lemma 4.59 it follows that we can analyze Pareto optimality of $\mathcal{U}$ by focusing on the underlying SINR set instead. As an example, we discuss the capacity region resulting from Example 4.51.

Example 4.60. Consider the capacity region of the 2-user MAC with individual power limits as specified in Example 4.51. For a given decoding order 1, 2, the capacities of the users are

$$\begin{aligned} \text{capacity user 1:} \quad & C_1(\boldsymbol{p}) = \log_2(1 + \frac{p_1}{p_2 + 1}) \,, \\ \text{capacity user 2:} \quad & C_2(\boldsymbol{p}) = \log_2(1 + p_2) \,. \end{aligned}$$

Assuming that no time-sharing or rate-splitting can be performed, we obtain the capacity region depicted in Fig. 4.7. Note, that a part of the boundary is not Pareto optimal.

With Theorem 4.55 we show the following result.

Theorem 4.61. *Consider an arbitrary boundary point $\boldsymbol{\gamma} \in \partial\mathcal{S}(\boldsymbol{p}^{\max})$, with a fixed point $\boldsymbol{p}^* = \boldsymbol{p}^*(\boldsymbol{\gamma})$ as defined in Lemma 4.52. Then $\boldsymbol{\gamma}$ is Pareto optimal if and only if*

$$\mathcal{P}(\boldsymbol{\gamma}, \boldsymbol{p}^{\max}) = \{\boldsymbol{p}^*\} \,. \tag{4.132}$$

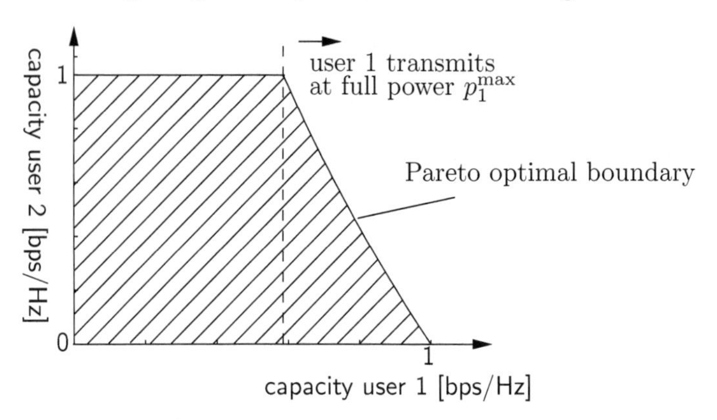

Fig. 4.7. Capacity region of a 2-user MAC with fixed decoding order 1, 2 and individual power limits.

Proof. We will prove the contrapositive statement. Suppose that $\mathcal{P}(\boldsymbol{\gamma}, \boldsymbol{p}^{\max})$ consists of multiple vectors. Then there is a non-empty set $\overline{\mathcal{K}_u}$ and a vector $\hat{\boldsymbol{p}}$ as in Theorem 4.55. For any $k \in \overline{\mathcal{K}_u}$ we have

$$\hat{\gamma}_k = \frac{\hat{p}_k}{\mathcal{J}_k(\hat{\boldsymbol{p}})} > \gamma_k \; .$$

Thus $\hat{\boldsymbol{\gamma}} \gneqq \boldsymbol{\gamma}$, and therefore $\boldsymbol{\gamma}$ is not Pareto optimal.

Conversely, assume that a boundary point $\boldsymbol{\gamma}$ is not Pareto optimal, then there exists a $\tilde{\boldsymbol{\gamma}}$ with $\tilde{\boldsymbol{\gamma}} \gneqq \boldsymbol{\gamma}$. This point is achieved by the power vector $\tilde{\boldsymbol{p}}$ fulfilling $\tilde{\boldsymbol{p}} = \mathrm{diag}(\tilde{\boldsymbol{\gamma}})\boldsymbol{\mathcal{J}}(\tilde{\boldsymbol{p}})$. We have $\tilde{\boldsymbol{p}} \neq \boldsymbol{p}^*$ and $\tilde{\boldsymbol{p}} \in \mathcal{P}(\tilde{\boldsymbol{\gamma}}, \boldsymbol{p}^{\max})$. We now show that any element of $\mathcal{P}(\tilde{\boldsymbol{\gamma}}, \boldsymbol{p}^{\max})$ is also contained in $\mathcal{P}(\boldsymbol{\gamma}, \boldsymbol{p}^{\max})$. To this end, consider an arbitrary $\boldsymbol{p} \in \mathcal{P}(\tilde{\boldsymbol{\gamma}}, \boldsymbol{p}^{\max})$. We have

$$p_k \geq \tilde{\gamma}_k \mathcal{J}_k(\boldsymbol{p}) \geq \gamma_k \mathcal{J}_k(\boldsymbol{p}) \; .$$

Thus $\boldsymbol{p} \in \mathcal{P}(\boldsymbol{\gamma}, \boldsymbol{p}^{\max})$, i.e., $\mathcal{P}(\tilde{\boldsymbol{\gamma}}, \boldsymbol{p}^{\max}) \subseteq \mathcal{P}(\boldsymbol{\gamma}, \boldsymbol{p}^{\max})$. Therefore we have determined two vectors $\tilde{\boldsymbol{p}} \neq \boldsymbol{p}^*$ that are both contained in $\mathcal{P}(\boldsymbol{\gamma}, \boldsymbol{p}^{\max})$. □

Next, we show how Pareto optimality is connected with the structure of the dependency matrix. To this end, consider again an arbitrary boundary point $\boldsymbol{\gamma} > 0$ and $\hat{\boldsymbol{p}}$ as defined in Theorem 4.55. The following Lemma 4.62 states that if $\mathcal{P}(\boldsymbol{\gamma}, \boldsymbol{p}^{\max})$ has multiple elements, then $\boldsymbol{D}_{\mathcal{I}}(\hat{\boldsymbol{p}})$ cannot be irreducible (see Definition A.1 in Section 2.2.1). Lemma 4.62 will be needed later for the proof of Theorem 4.66.

Lemma 4.62. *Consider an arbitrary boundary point* $\boldsymbol{\gamma} \in \partial\mathcal{S}(\boldsymbol{p}^{\max})$, *with a fixed point* $\boldsymbol{p}^* = \boldsymbol{p}^*(\boldsymbol{\gamma})$ *as defined in Lemma 4.52. If* $\mathcal{P}(\boldsymbol{\gamma}, \boldsymbol{p}^{\max}) \neq \{\boldsymbol{p}^*\}$, *then for any* $\hat{\boldsymbol{p}}$ *defined in Theorem 4.55, the local dependency matrix* $\boldsymbol{D}_{\mathcal{I}}(\hat{\boldsymbol{p}})$ *is reducible.*

Proof. The proof is by contradiction. Suppose that $\boldsymbol{D}_{\mathcal{I}}(\hat{\boldsymbol{p}})$ is irreducible. Assumption $\mathcal{P}(\boldsymbol{\gamma}, \boldsymbol{p}^{\max}) \neq \{\boldsymbol{p}^*\}$ implies the existence of an oversized user. Consequently, there are two complementary sets $\overline{\mathcal{K}_u}$ and $\mathcal{K}_u \backslash \overline{\mathcal{K}_u}$. Because of irreducibility there exists a connecting path between both sets. That is, there exist indices $k_1 \in \mathcal{K}_u \backslash \overline{\mathcal{K}_u}$ and $k_2 \in \overline{\mathcal{K}_u}$ such that

$$[\boldsymbol{D}_{\mathcal{I}}(\hat{\boldsymbol{p}})]_{k_1 k_2} > 0 . \tag{4.133}$$

We can reduce the power $\hat{p}_{k_2}$ of the oversized user without violating the feasibility condition. That is, there exists a $\delta > 0$ and a $\hat{p}_{k_2}^{(\delta)} = \hat{p}_{k_2} - \delta$ such that

$$\hat{p}_{k_2} > \hat{p}_{k_2}^{(\delta)} > \gamma_{k_2} \mathcal{J}_{k_2}(\hat{\boldsymbol{p}}) . \tag{4.134}$$

By keeping all the other components $l \neq k_2$ fixed, we obtain a new vector $\hat{\boldsymbol{p}}^{(\delta)} \lneqq \hat{\boldsymbol{p}}$. Because of monotonicity A3 we have $\mathcal{J}_{k_2}(\hat{\boldsymbol{p}}^{(\delta)}) \leq \mathcal{J}_{k_2}(\hat{\boldsymbol{p}})$, and with (4.134) we know that $\hat{\boldsymbol{p}}^{(\delta)} \in \mathcal{P}(\boldsymbol{\gamma}, \boldsymbol{p}^{\max})$.

From (4.133) we know that by reducing the power of user k_2 we reduce the interference of user k_1. Thus

$$\mathcal{J}_{k_1}(\hat{\boldsymbol{p}}^{(\delta)}) < \mathcal{J}_{k_1}(\hat{\boldsymbol{p}}), \quad k_1 \in \mathcal{K}_u \backslash \overline{\mathcal{K}_u} .$$

This contradicts (4.124) from Theorem 4.55, thus concluding the proof. □

4.5.4 Concept of Strongly Coupled Users

We will now introduce the new concept of *strongly coupled* users. This will prove useful in further characterizing the Pareto optimal boundary. It will turn out (Theorem 4.66) that this is an equivalent way of expressing Pareto optimality.

Definition 4.63. *A K_u-user system is said to be* strongly coupled *with power limits $\mathbf{p}^{\max}$, if for any point $\boldsymbol{\gamma}$, for which there is a $\mathbf{p} \in \mathcal{P}(\boldsymbol{\gamma}, \mathbf{p}^{\max})$ with*

$$\boldsymbol{\Gamma} \mathcal{J}(\mathbf{p}) \lneqq \mathbf{p} , \tag{4.135}$$

there exists a vector $\hat{\mathbf{p}} < \mathbf{p}$ such that

$$\boldsymbol{\Gamma} \mathcal{J}(\hat{\mathbf{p}}) < \hat{\mathbf{p}} . \tag{4.136}$$

The condition (4.136) reflects a practically relevant property: If it is possible to fulfill SINR requirements $\gamma_1, \ldots, \gamma_{K_u}$, and one user gets more than required, then all users are strongly coupled if and only if all users benefit from reducing the oversized user's power. This is an important aspect of "fairness" because it provides a mechanism for trading off resources between users.

Theorem 4.64. *If interference functions $\mathcal{I}_1, \ldots, \mathcal{I}_{K_u}$ with power limits $\mathbf{p}^{\max}$ are strongly coupled, then every boundary point $\boldsymbol{\gamma}$ is Pareto optimal.*

Proof. Assume that $\mathcal{I}_1, \dots, \mathcal{I}_{K_u}$ are strongly coupled. The proof is by contradiction. Suppose that there is a boundary point $\boldsymbol{\gamma}$ that is *not* Pareto optimal. Then there is a $\hat{\boldsymbol{\gamma}} \geq \boldsymbol{\gamma}$ such that we can find a k_0 with $\hat{\gamma}_{k_0} > \gamma_{k_0}$. Consider the indicator function C, as defined by (4.122). Because $C(\hat{\boldsymbol{\gamma}}, \boldsymbol{\mathcal{I}}, \boldsymbol{p}^{\max}) = 1$, the vector $\hat{\boldsymbol{\gamma}}$ is associated with a $\hat{\boldsymbol{p}} = \hat{\boldsymbol{p}}(\boldsymbol{\gamma})$ such that $\hat{\boldsymbol{p}} = \hat{\boldsymbol{\Gamma}} \boldsymbol{\mathcal{J}}(\hat{\boldsymbol{p}})$. Therefore,

$$\hat{p}_{k_0} = \hat{\gamma}_{k_0} \mathcal{J}_{k_0}(\hat{\boldsymbol{p}}) > \gamma_{k_0} \mathcal{J}_{k_0}(\hat{\boldsymbol{p}}) \,. \tag{4.137}$$

Because the interference functions are strongly coupled by assumption, there is a $\boldsymbol{p} < \hat{\boldsymbol{p}} \leq \boldsymbol{p}^{\max}$ such that

$$p_k > \gamma_k \mathcal{J}_k(\boldsymbol{p}) \quad k \in \mathcal{K}_u \,. \tag{4.138}$$

This would imply

$$C(\boldsymbol{\gamma}, \boldsymbol{\mathcal{I}}, \boldsymbol{p}^{\max}) \leq \max_{k \in \mathcal{K}_u} \frac{\gamma_k \mathcal{J}_k(\boldsymbol{p})}{p_k} < 1 \,, \tag{4.139}$$

which is a contradiction. Thus, every boundary point is Pareto optimal. □

An obvious question is: does the converse of Theorem 4.64 hold? That is, does a Pareto optimal boundary imply a strongly coupled system? This will be shown in Section 2.3 under the additional assumption of strict monotonicity. Without strict monotonicity we only have the following result:

Theorem 4.65. *Assume that every boundary point is Pareto optimal. Consider an arbitrary point $\boldsymbol{\gamma} > 0$. If there exists a $\boldsymbol{p} \leq \boldsymbol{p}^{\max}$ such that*

$$\frac{p_k}{\mathcal{J}_k(\boldsymbol{p})} \geq \gamma_k \quad \forall k \in \mathcal{K}_u \tag{4.140}$$

with strict inequality for at least one component, then there exists a $\tilde{\boldsymbol{p}} < \boldsymbol{p}^{\max}$ with

$$\frac{\tilde{p}_k}{\mathcal{J}_k(\tilde{\boldsymbol{p}})} > \gamma_k \quad \forall k \in \mathcal{K}_u \tag{4.141}$$

Proof. From assumption (4.140) it follows that $\boldsymbol{\gamma}$ is contained in the feasible region. However, it cannot be a boundary point because of the assumed strict inequality for one component. This would contradict the assumption of Pareto optimality. Thus, $\boldsymbol{\gamma}$ must be contained in the interior of the region, for which $C(\boldsymbol{\gamma}, \boldsymbol{p}^{\max}) < 1$. So there exists a vector $\tilde{\boldsymbol{p}} \leq \boldsymbol{p}^{\max}$ that fulfills the fixed point equation

$$\tilde{p}_k = \lambda \cdot \gamma_k \mathcal{J}_k(\tilde{\boldsymbol{p}}) > \gamma_k \mathcal{J}_k(\tilde{\boldsymbol{p}}), \quad k \in \mathcal{K}_u \,, \tag{4.142}$$

where $1 < \lambda = 1/C(\boldsymbol{\gamma}, \boldsymbol{p}^{\max})$. □

Note that Theorem 4.65 is not the converse of Theorem 4.64. The result only holds for interior points, not for the boundary. In the following Section 2.3 we will make the additional assumption of strict monotonicity. Under this additional condition the converse will be shown. Also, the connection with the dependency matrix $\boldsymbol{D}_{\mathcal{I}}$ will be explained.

4.5.5 Strict Monotonicity

Under the additional assumption of strict monotonicity (cf. Subsection 2.3) we can prove the converse of Theorem 4.64. In addition, this provides a link between the dependency matrix $\boldsymbol{D}_{\mathcal{I}}$ and Pareto optimality. This is summarized by the next theorem.

Theorem 4.66. *Consider a K_u-user system with individual power limits $\mathbf{p}^{\max}$ and interference functions $\mathcal{I}_1, \dots \mathcal{I}_{K_u}$ that are strictly monotone on their respective dependency sets. Then the following statements are equivalent.*

- *The system is strongly coupled (Definition 4.63)*
- *The dependency matrix $\boldsymbol{D}_{\mathcal{I}}$ is irreducible.*
- *Every boundary point is Pareto optimal.*

Proof. We first show that the dependency matrix $\boldsymbol{D}_{\mathcal{I}}$ is irreducible if and only if every boundary point $\boldsymbol{\gamma}$ is Pareto optimal.

The first part is by contradiction. Suppose that $\boldsymbol{D}_{\mathcal{I}}$ is irreducible but some boundary point $\boldsymbol{\gamma}$ is *not* Pareto optimal (see Definition 4.4). Then $\mathcal{P}(\boldsymbol{\gamma}, \boldsymbol{p}^{\max})$ has multiple elements and there is a vector $\hat{\boldsymbol{p}}$ as defined in Theorem 4.55. From Lemma 4.62 it follows that $\boldsymbol{D}_{\mathcal{I}}(\hat{\boldsymbol{p}})$ is reducible. However, this is a contradiction because irreducibility of $\boldsymbol{D}_{\mathcal{I}}$ implies irreducibility of $\boldsymbol{D}_{\mathcal{I}}(\hat{\boldsymbol{p}})$. This is shown as follows. The proof is again by contradiction. We need to show that $[\boldsymbol{D}_{\mathcal{I}}]_{kl} > 0$ implies $[\boldsymbol{D}_{\mathcal{I}}(\hat{\boldsymbol{p}})]_{kl} > 0$ for any $k, l \in \mathcal{K}_u$. Suppose that $[\boldsymbol{D}_{\mathcal{I}}(\hat{\boldsymbol{p}})]_{kl} = 0$, then we know from (2.9) that $f(\delta, \hat{\boldsymbol{p}}) = \mathcal{J}_k(\hat{\boldsymbol{p}} - \delta \boldsymbol{e}_l)$ is constant for all $\delta > 0$. This means that $\mathcal{I}_k$ does not depend on the lth component, which contradicts the assumption of strict monotonicity, thus proving that $\boldsymbol{D}_{\mathcal{I}}(\hat{\boldsymbol{p}})$ is irreducible. Thus, we have shown that an irreducible dependency matrix $\boldsymbol{D}_{\mathcal{I}}$ implies a Pareto optimal boundary.

Conversely, we need to show that if an arbitrary boundary point $\boldsymbol{\gamma}$ is Pareto optimal then $\boldsymbol{D}_{\mathcal{I}}$ is irreducible. The proof is by contradiction. Suppose that $\boldsymbol{D}_{\mathcal{I}}$ is reducible. Without loss of generality we can assume that $\boldsymbol{D}_{\mathcal{I}}$ has Frobenius normal form [57], with irreducible matrices $\boldsymbol{D}_1, \dots, \boldsymbol{D}_N$ along the main diagonal. Such a canonical form can always be achieved by a symmetric permutation of rows and columns of $\boldsymbol{D}_{\mathcal{I}}$. Suppose that the first (isolated) block has a dimension $k_1 \times k_1$. This means that the first k_1 interference functions do *not* depend on the components $p_{k_1+1}, \dots, p_K$. Thus, the vector $\boldsymbol{p}_1^* = [p_1^{\max}, \dots, p_{k_1}^{\max}]^T \in \mathbb{R}_{++}^{k_1}$ leads to SINR values $\gamma_k^* = p_k^{\max} / \mathcal{J}_k(\boldsymbol{p}_1^*)$ for $1 \leq k \leq k_1$. We introduce the set $\mathcal{M}_+ = \{\boldsymbol{p} \in \mathbb{R}_+^{K_u} : \boldsymbol{p} = [(\boldsymbol{p}_1^*)^T, p_{k_1+1}, \dots, p_{K_u}]^T\}$. For arbitrary $\boldsymbol{\gamma}^{(2)} = [\gamma_{k_1+1}, \dots, \gamma_{K_u}]^T > 0$ we define

$$C^{(2)}(\boldsymbol{\gamma}^{(2)}, \boldsymbol{p}^{\max}) = \inf_{\substack{\boldsymbol{p} \in \mathcal{M}_+ \\ 0 < \boldsymbol{p} \leq \boldsymbol{p}^{\max}}} \max_{k_1+1 \leq k \leq K} \frac{\gamma_k \mathcal{J}_k(\boldsymbol{p})}{p_k} . \tag{4.143}$$

Consider $\hat{\boldsymbol{\gamma}}^{(2)} > 0$ such that $C^{(2)}(\hat{\boldsymbol{\gamma}}^{(2)}, \boldsymbol{p}^{\max}) = 1$. Because of the noise and power constraints, we know from [1] that this point is feasible, i.e., there exists a $\boldsymbol{p}^{(2)} \in \mathbb{R}_+^{K-k_1}$ with $p_k^{(2)} \leq p_k^{\max}$ such that

$$\hat{\gamma}_k^{(2)} \mathcal{I}_k \left(\begin{bmatrix} \boldsymbol{p}_1^* \\ \boldsymbol{p}^{(2)} \\ \sigma_n^2 \end{bmatrix} \right) = p_k^{(2)} , \quad k_1 + 1 \le k \le K .$$

Here, $\boldsymbol{p}_1^*$ acts like additional noise. The complete K_u-dimensional vector of SINR values is $\tilde{\boldsymbol{\gamma}} = [\gamma_1^*, \ldots, \gamma_{k_1}^*, \hat{\gamma}_{k_1+1}^{(2)}, \ldots, \hat{\gamma}_K^{(2)}]^T$, and the vector that achieves this point is $\check{\boldsymbol{p}} = [(\boldsymbol{p}_1^*)^T, (\boldsymbol{p}^{(2)})^T]^T$. Using the definition (2.55) we obtain the inequality

$$\begin{aligned} C(\tilde{\boldsymbol{\gamma}}, \boldsymbol{\mathcal{I}}, \boldsymbol{p}^{\max}) &\ge \inf_{0 < \boldsymbol{p} \le \boldsymbol{p}^{\max}} \max_{1 \le k \le k_1} \frac{\gamma_k^* \mathcal{J}_k(\boldsymbol{p})}{p_k} \\ &= \frac{\gamma_k^* \mathcal{J}_k(\boldsymbol{p}_1^*)}{p_k^{\max}} = 1 . \end{aligned} \tag{4.144}$$

This can only be fulfilled with equality since we also have

$$C(\tilde{\boldsymbol{\gamma}}, \boldsymbol{\mathcal{I}}, \boldsymbol{p}^{\max}) \le \max_{1 \le k \le K} \frac{\tilde{\gamma}_k \mathcal{J}_k(\check{\boldsymbol{p}})}{\check{p}_k} = 1 .$$

Thus, $C(\tilde{\boldsymbol{\gamma}}, \boldsymbol{\mathcal{I}}, \boldsymbol{p}^{\max}) = 1$.

Next, consider an arbitrary λ with $0 < \lambda < 1$, and

$$\tilde{\boldsymbol{\gamma}}(\lambda) = [\gamma_1^*, \ldots, \gamma_{k_1}^*, \lambda\tilde{\gamma}_{k_1+1}, \ldots, \lambda\tilde{\gamma}_K]^T .$$

We have $C(\tilde{\boldsymbol{\gamma}}(\lambda), \boldsymbol{\mathcal{I}}, \boldsymbol{p}^{\max}) \le C(\tilde{\boldsymbol{\gamma}}, \boldsymbol{\mathcal{I}}, \boldsymbol{p}^{\max}) = 1$. Similar to (4.144) it is shown that $C(\tilde{\boldsymbol{\gamma}}(\lambda), \boldsymbol{\mathcal{I}}, \boldsymbol{p}^{\max}) \ge 1$. Thus, $C(\tilde{\boldsymbol{\gamma}}(\lambda), \boldsymbol{\mathcal{I}}, \boldsymbol{p}^{\max}) = 1$ holds, which means that both $\tilde{\boldsymbol{\gamma}}$ and $\tilde{\boldsymbol{\gamma}}(\lambda)$ are boundary points. That is, we can minimize components of $\tilde{\boldsymbol{\gamma}}$ without leaving the boundary, which contradicts the assumption of Pareto optimality.

Next, we show that the system is strongly coupled if and only if the dependency matrix $\boldsymbol{D}_{\mathcal{I}}$ is irreducible. Assume that $\boldsymbol{D}_{\mathcal{I}}$ is irreducible. Consider an arbitrary boundary point $\boldsymbol{\gamma}$ such that there is a $\boldsymbol{p} \in \mathcal{P}(\boldsymbol{\gamma}, \boldsymbol{p}^{\max})$ with $\boldsymbol{\Gamma}\boldsymbol{\mathcal{J}}(\boldsymbol{p}) \lneqq \boldsymbol{p}$. This inequality is strict for at least on component k_0, so we can decrease p_{k_0} without violating the inequality. Because of the assumed strict monotonicity, decreasing p_{k_0} decreases the interference of the users in the dependency set $\mathsf{L}(k_0)$. These users can in turn reduce their powers without without violating the above inequality. Irreducibility means that, in the graph of $\boldsymbol{D}_{\mathcal{I}}$ there is a path from any point to any other point. Thus, by successively decreasing the components of $\boldsymbol{p}$, we obtain a vector $\tilde{\boldsymbol{p}}$ with

$$\frac{\tilde{p}_k}{\mathcal{J}_k(\tilde{\boldsymbol{p}})} > \gamma_k , \quad \text{for all } k \in \mathcal{K}_u . \tag{4.145}$$

That is, the users are strongly coupled.

Having shown that irreducibility implies a strongly coupled system, it remains to show that if $\boldsymbol{D}_{\mathcal{I}}$ is reducible then the system cannot be strongly coupled. To this end, we use the SINR vectors $\tilde{\boldsymbol{\gamma}}$ and $\tilde{\boldsymbol{\gamma}}(\lambda)$ that were already introduced earlier in the proof. The point $\tilde{\boldsymbol{\gamma}}$ is achieved by $\check{\boldsymbol{p}}$, thus

$$\frac{\check{p}_k}{\mathcal{J}_k(\check{\boldsymbol{p}})} = \tilde{\gamma}_k > \lambda\tilde{\gamma}_k = \tilde{\gamma}_k(\lambda), \quad \text{for all } k_1 + 1 \le k \le K ,$$

$$\frac{\check{p}_k}{\mathcal{J}_k(\check{\boldsymbol{p}})} = \tilde{\gamma}_k(\lambda), \quad \text{for all } 1 \le k \le k_1 .$$

However, there is no vector $\boldsymbol{p}' \le \check{\boldsymbol{p}}$ with

$$\frac{p'_k}{\mathcal{J}_k(\boldsymbol{p}')} > \gamma_k(\lambda), \quad 1 \le k \le K .$$

Because then $\boldsymbol{\gamma}(\lambda)$ could not be a boundary point. □

Theorem 4.66 will be needed in the next section.

Further applications for strict monotonicity can be found in [124], where it was shown that strict monotonicity plays a central role in the proof of non-manipulability of certain resource allocation strategies.

4.5.6 Strict Log-Convexity

Next, we study under which conditions the SINR region $\mathcal{S}(\boldsymbol{p}^{\max})$, is strictly log-convex and contained in $\mathcal{ST}_c$. A necessary and sufficient condition is derived. To this end, we need the following result.

Lemma 4.67. *Let $\mathcal{I}_1, \dots \mathcal{I}_{K_u}$ be strictly log-convex interference functions, and each user affects the interference function of at least one other user, i.e., each column of $\boldsymbol{D}_{\mathcal{I}}$ has at least one non-zero entry off the main diagonal. Then for arbitrary $\hat{\mathbf{p}} \neq \check{\mathbf{p}}$ there exists at least one k_0 such that*

$$\mathcal{J}_{k_0}(\boldsymbol{p}(\lambda)) < (\mathcal{J}_{k_0}(\hat{\boldsymbol{p}}))^{1-\lambda} \cdot (\mathcal{J}_{k_0}(\check{\boldsymbol{p}}))^{\lambda} \quad \forall \lambda \in (0,1) . \tag{4.146}$$

Proof. If (4.146) is fulfilled for one λ_0 then it is fulfilled for all $\lambda \in (0,1)$. This follows from the strict log-convexity of the interference functions. The proof is by contradiction. Suppose that there is a $\lambda_0 \in (0,1)$ such that for all $k \in \mathcal{K}_u$

$$\mathcal{J}_k(\boldsymbol{p}(\lambda_0)) = (\mathcal{J}_k(\hat{\boldsymbol{p}}))^{1-\lambda_0} \cdot (\mathcal{J}_k(\check{\boldsymbol{p}}))^{\lambda_0} .$$

This can only be fulfilled if for all $k \in \mathcal{K}_u$ there exists a $c_k > 0$ such that

$$\hat{p}_l = c_k \check{p}_l \quad \text{for all } l \in \mathsf{L}_k$$

By assumption, each user depends on at least one other user, thus for each index l, there exists a $\tilde{k}$ such that $l \in \mathsf{L}_{\tilde{k}}$. Thus, we have equality for all components, which leads to the contradiction $\hat{\boldsymbol{p}} = \check{\boldsymbol{p}}$. □

With Lemma 4.67 we can derive a necessary and sufficient condition for strict log-convexity.

Theorem 4.68. *Let $\mathcal{I}_1, \dots \mathcal{I}_{K_u}$ be strictly log-convex interference functions. The transformed SINR region $\mathcal{L}og(\mathcal{S}(\boldsymbol{p}^{\max})))$ is strictly convex if and only if $\boldsymbol{D}_{\mathcal{I}}$ is irreducible.*

Proof. $\mathcal{I}_1, \dots \mathcal{I}_{K_u}$ are strictly log-convex and thus they are also strictly monotone on their respective dependency set (see Lemma 2.9). If the region is strictly convex, then the entire boundary is Pareto optimal. Theorem 4.66 implies that $\boldsymbol{D}_{\mathcal{I}}$ is irreducible.

It remains to show that irreducibility implies a strictly convex region. Consider arbitrary boundary points $\hat{\boldsymbol{\gamma}}$ and $\check{\boldsymbol{\gamma}}$ with corresponding power vectors $\hat{\boldsymbol{p}}$ and $\check{\boldsymbol{p}}$. As in the proof of Theorem 4.50, we use $\boldsymbol{\gamma}(\lambda)$ and $\boldsymbol{p}(\lambda)$. We have $\boldsymbol{p}(\lambda) = \hat{\boldsymbol{p}}^{1-\lambda} \cdot \check{\boldsymbol{p}}^{\lambda} \leq (\boldsymbol{p}^{\max})^{1-\lambda} \cdot (\boldsymbol{p}^{\max})^{\lambda} = \boldsymbol{p}^{\max}$. In [5, Appendix B] it was shown that $\gamma_k(\lambda) \leq p_k(\lambda)/\mathcal{J}_k(\boldsymbol{p}(\lambda))$. We now exploit that $\boldsymbol{D}_{\mathcal{I}}$ is irreducible, thus each column of $\boldsymbol{D}_{\mathcal{I}}$ has at least one non-zero entry outside the main diagonal. From Lemma 4.67 we know that for any $\hat{\boldsymbol{p}} \neq \check{\boldsymbol{p}}$ there is at least one component k_0 such that (4.146) is fulfilled. It follows that there exists a k_0 such that

$$\gamma_{k_0}(\lambda) < \frac{p_{k_0}(\lambda)}{\mathcal{J}_{k_0}(\boldsymbol{p}(\lambda))} . \tag{4.147}$$

From Theorem 4.66 we know that the system is strongly coupled, so there exists a $\tilde{\boldsymbol{p}}(\lambda) < \boldsymbol{p}(\lambda) \leq \boldsymbol{p}^{\max}$ such that for all k

$$\gamma_k(\lambda) < \frac{\tilde{p}_k(\lambda)}{\mathcal{J}_k(\tilde{\boldsymbol{p}}(\lambda))} , \quad k \in \mathcal{K}_u . \tag{4.148}$$

With the definition (2.54) we have

$$C(\boldsymbol{\gamma}(\lambda), \boldsymbol{\mathcal{I}}, \boldsymbol{p}^{\max}) \leq \max_{k \in \mathcal{K}_u} \frac{\gamma_k(\lambda)\mathcal{J}_k(\tilde{\boldsymbol{p}}(\lambda))}{\tilde{p}_k(\lambda)} < 1 .$$

Thus, $\boldsymbol{\gamma}(\lambda)$ is in the interior of the region. □

Next, consider the function

$$F_{\boldsymbol{w}}(s) = \sum_{k \in \mathcal{K}_u} w_k \log \frac{\mathcal{J}_k(\mathrm{e}^{\boldsymbol{s}})}{\mathrm{e}^{s_k}} \tag{4.149}$$

where we have used the change of variable $\boldsymbol{s} = \log \boldsymbol{p}$. The weights w_k account for possible user priorities. Note that minimizing (4.149) is equivalent to maximizing $\sum_k \log(p_k/\mathcal{J}_k(\boldsymbol{p}))$. For equal weights, this is a special case of the optimization problem (4.3), where the utility is the SINR.

Convexity of $F_{\boldsymbol{w}}(\boldsymbol{s})$ was already shown in [5]. However, in order to show that the SINR region is contained in $\mathcal{ST}_c$, we need strict convexity.

Theorem 4.69. *Let $\mathcal{I}_1, \dots \mathcal{I}_{K_u}$ be defined as in Lemma 4.67. Then $F_{\boldsymbol{w}}(\boldsymbol{s})$ is strictly convex for all $\boldsymbol{w} > 0$. That is, for all $\hat{\boldsymbol{s}} \neq \check{\boldsymbol{s}}$, we have*

$$F_{\boldsymbol{w}}(\boldsymbol{s}(\lambda)) < (1-\lambda)F_{\boldsymbol{w}}(\hat{\boldsymbol{s}}) + \lambda F_{\boldsymbol{w}}(\check{\boldsymbol{s}}) \quad \forall \lambda \in (0,1), \tag{4.150}$$

where $\boldsymbol{s}(\lambda) = \log \boldsymbol{p}(\lambda)$ is defined as by (2.13).

Proof. Assume an arbitrary $\boldsymbol{w} > 0$. For any $\lambda \in (0,1)$ there exists a k_0 such that (4.146) is fulfilled. Thus,

$$
\begin{aligned}
F_{\boldsymbol{w}}(\boldsymbol{s}(\lambda)) &= w_{k_0} \log \frac{\mathcal{J}_{k_0}(\mathrm{e}^{\boldsymbol{s}(\lambda)})}{\mathrm{e}^{s_{k_0}(\lambda)}} + \sum_{k \in \mathcal{K}_u \backslash k_0} w_k \log \frac{\mathcal{J}_k(\mathrm{e}^{\boldsymbol{s}(\lambda)})}{\mathrm{e}^{s_k(\lambda)}} \\
&< w_{k_0}(1-\lambda) \log \frac{\mathcal{J}_{k_0}(\mathrm{e}^{\hat{\boldsymbol{s}}})}{\mathrm{e}^{\hat{s}_{k_0}}} + w_{k_0} \lambda \log \frac{\mathcal{J}_{k_0}(\mathrm{e}^{\check{\boldsymbol{s}}})}{\mathrm{e}^{\check{s}_{k_0}}} + \\
&\quad + (1-\lambda) \sum_{k \in \mathcal{K}_u \backslash k_0} w_k \log \frac{\mathcal{J}_k(\mathrm{e}^{\hat{\boldsymbol{s}}})}{\mathrm{e}^{\hat{s}_k}} + \\
&\quad + \lambda \sum_{k \in \mathcal{K}_u \backslash k_0} w_k \log \frac{\mathcal{J}_k(\mathrm{e}^{\check{\boldsymbol{s}}})}{\mathrm{e}^{\check{s}_k}} \\
&= (1-\lambda) F_{\boldsymbol{w}}(\hat{\boldsymbol{s}}) + \lambda F_{\boldsymbol{w}}(\check{\boldsymbol{s}}) \, .
\end{aligned}
$$

Here we have exploited (4.146) and log-convexity of $\mathcal{I}_k$. □

The next corollary is an immediate consequence of Theorem 4.69.

Corollary 4.70. *The optimization problem*

$$
\min_{\boldsymbol{s} \leq \log \boldsymbol{p}^{\max}} F_{\boldsymbol{w}}(\boldsymbol{s}) \tag{4.151}
$$

has exactly one minimizer.

It was shown in Section 4.1.4 that for any set $\mathcal{U} \in \mathcal{ST}_c$ the properties of the classical Nash bargaining framework are preserved. The following theorem shows a sufficient condition for the SINR region $\mathcal{S}(\boldsymbol{p}^{\max})$ to be contained in $\mathcal{ST}_c$. The result builds on Theorem 4.69.

Note that sets from $\mathcal{ST}_c$ do not need to be strictly log-convex (see Figure 4.3). Thus, irreducibility of $\boldsymbol{D}_{\mathcal{I}}$, which was required in Theorem 4.68, is not necessary in this case.

Theorem 4.71. *Let $\mathcal{I}_1, \ldots \mathcal{I}_{K_u}$ be defined as in Lemma 4.67. Then the SINR region $\mathcal{S}(\boldsymbol{p}^{\max})$ is contained in $\mathcal{ST}_c$.*

Proof. The region is (relatively) closed and upper-bounded because of the power constraints. The image set $\mathcal{Q} = \mathcal{L}og\big(\mathcal{S}(\boldsymbol{p}^{\max})\big)$ is closed and upper-bounded. We need to show that for two arbitrary Pareto optimal boundary points $\hat{\boldsymbol{q}} \neq \check{\boldsymbol{q}}$, any point $\boldsymbol{q}(\lambda) = (1-\lambda)\hat{\boldsymbol{q}} + \lambda\check{\boldsymbol{q}}$, with $\lambda \in (0,1)$, is contained in the interior of the set. This is illustrated in Figure 4.3.

The proof is by contradiction. Suppose that there is a λ such that $\boldsymbol{q}(\lambda)$ is not in the interior. Since $\mathcal{Q}$ is convex comprehensive [5], this can only be fulfilled if

$$
C(\log \boldsymbol{q}(\lambda), \boldsymbol{\mathcal{I}}, \boldsymbol{p}^{\max}) = 1, \quad \forall \lambda \in (0,1) \, . \tag{4.152}
$$

Because of (4.152) there exists a vector $\hat{\boldsymbol{w}} > 0$ such that

$$\boldsymbol{q}(\lambda) \in \arg\max_{\boldsymbol{q} \in \mathcal{Q}} \sum_{k \in \mathcal{K}_u} \hat{w}_k q_k \,. \tag{4.153}$$

The set of maximizers of (4.153) is a convex set. For every maximizer $\boldsymbol{q}(\lambda)$ there is a corresponding vector $\boldsymbol{p}(\lambda) = \exp \boldsymbol{s}(\lambda)$ which fulfills the power constraints, and $\boldsymbol{s}(\lambda)$ is a solution of

$$\boldsymbol{s}(\lambda) \in \arg\max_{\mathbf{0} < \mathrm{e}^{\boldsymbol{s}} \leq \boldsymbol{p}^{\max}} F_{\boldsymbol{w}}(\boldsymbol{s}) \,. \tag{4.154}$$

That is, (4.154) has no unique optimum. This is a contradiction because $F_{\boldsymbol{w}}(\boldsymbol{s})$ is strictly convex. □

5

QoS-Constrained Power Minimization

In this chapter we discuss an algorithmic solution for the QoS-constrained power minimization problem, which was already introduced and discussed in Section 2.8. The following optimization framework is applicable to arbitrary systems of concave or convex standard interference functions $\mathcal{J}_k(\boldsymbol{p}) = \mathcal{I}_k(\underline{\boldsymbol{p}})$ (see Section 2.4 for a definition and discussion of standard interference functions).

We wish to achieve feasible SINR targets $\gamma_1, \ldots, \gamma_{K_u}$ with minimum total transmission power, i.e.,

$$\min_{\boldsymbol{p} \in \mathcal{P}} \sum_{l \in \mathcal{K}_u} p_l \quad \text{s.t.} \quad \min_{k \in \mathcal{K}_u} \frac{p_k}{\mathcal{J}_k(\boldsymbol{p})} \geq \gamma_k \ . \tag{5.1}$$

Since the interference functions are standard, the unique global optimum can be computed by the following fixed point iteration [1] with an arbitrary initialization $\boldsymbol{p}^{(0)} \geq 0$.

$$p_k^{(n+1)} = \gamma_k \mathcal{J}_k(\boldsymbol{p}^{(n)}), \quad \forall k \in \mathcal{K}_u \ . \tag{5.2}$$

This iteration has *geometric convergence* [7, 74]. Note, that no convexity is required for this method. Global convergence is ensured by certain contraction properties. An alternative convergence proof based on A1, A2, A3 plus strict monotonicity (2.22) was shown in [2]. A game-theoretic interpretation was given in [125].

Alternatively, we can use the results of Subsection 2.8, which show that problem (5.1) has an equivalent convex reformulation if the interference functions are convex, concave, or log-convex. Thus, standard convex optimization techniques can be applied. However, such a strategy can be inefficient if it fails to exploit the particular structure of the problem at hand.

In the following we discuss a technique that exploits the structure properties from Sections 3.3 and 3.4. Namely, every concave interference function can be expressed as a minimum of linear functions. This leads to an efficient algorithm with super-linear convergence . The result appeared in [7].

M. Schubert, H. Boche, *Interference Calculus*, Foundations in Signal Processing, Communications and Networking 7,

We begin by assuming that $\mathcal{J}_1, \ldots, \mathcal{J}_{K_u}$ are arbitrary concave standard interference functions (the convex case is similar and will be discussed in Section 5.4). From Theorem 3.63 we know that there exist upward-comprehensive closed convex sets $\mathcal{V}_1, \ldots, \mathcal{V}_{K_u} \subset \mathbb{R}_+^{K_u+1}$ such that

$$\mathcal{J}_k(\boldsymbol{p}) = \min_{\boldsymbol{v}' \in \mathcal{V}_k} \Big(\sum_{l \in \mathcal{K}_u} v'_l p_l + v'_{K_u+1} \Big) . \tag{5.3}$$

The first K_u components of $\boldsymbol{v}'$ determine the interference coupling between the users, while the last component can be interpreted as noise power. This is a typical structure for interference scenarios that involve adaptive receive or transmit strategies (see e.g. the examples from Section 1.4). We collect the interference coupling coefficients in a vector $\boldsymbol{v}$, and the noise is denoted by n.

$$\boldsymbol{v}' = \begin{bmatrix} \boldsymbol{v} \\ n \end{bmatrix} .$$

The coupling coefficients $\boldsymbol{v}$ and the noise n typically depend on certain system parameters. We introduce the parameterization $\boldsymbol{v} =: \boldsymbol{v}(z)$ and $n =: n(z)$, where the parameter z is from a compact set $\mathcal{Z}$. The functions $\boldsymbol{v}(z)$ and $n(z)$ are assumed to be continuous. If the goal is to minimize the interference, then the parameter z has an interpretation as a "receive strategy". In the context of robust designs (see Subsection 1.4.7) the parameter z can stand for *system uncertainties*. Introducing the notation

$$\boldsymbol{V}(z) = \begin{bmatrix} \boldsymbol{v}_1(z_1)^T \\ \vdots \\ \boldsymbol{v}_{K_u}(z_{K_u})^T \end{bmatrix} , \quad \boldsymbol{n}(z) = \begin{bmatrix} n_1(z_1) \\ \vdots \\ n_{K_u}(z_{K_u}) \end{bmatrix} ,$$

the interference function (5.3) can be rewritten as

$$\mathcal{J}_k(\boldsymbol{p}) = \min_{z_k \in \mathcal{Z}_k} [\boldsymbol{V}(z)\boldsymbol{p} + \boldsymbol{n}(z)]_k, \quad k \in \mathcal{K}_u . \tag{5.4}$$

In the following we will also use the vector notation $\boldsymbol{\mathcal{J}}(\boldsymbol{p}) = [\mathcal{J}_1(\boldsymbol{p}), \ldots, \mathcal{J}_{K_u}(\boldsymbol{p})]^T$.

A special case of (5.4) is the interference function (1.14) resulting from multi-user beamforming. Each beamformer is constrained to a compact set $\mathcal{Z}_k$, which is typically the unit sphere.[1] Then $\boldsymbol{V}(z)$ models the interference cross-talk caused by non-orthogonal beams and $\boldsymbol{n}(z)$ is the effective noise caused by the noise enhancement effect. This model was assumed, e.g., in [26,33,64,83]. But z_k could as well stand for some discrete choice between several receivers, or base stations, as in [36,37].

In oder to ensure the existence of an optimal strategy z, we assume that $\boldsymbol{V}(z)$ and $\boldsymbol{n}(z)$ are continuous and $\mathcal{Z}$ is non-empty. The overall strategy z

[1] Arbitrary constraints can be imposed on the beamformers, like the shaping constraints in [27]. We only require that $\mathcal{Z}_k$ is compact, to ensure that the minimum exists.

consists of K_u individual strategies $z = (z_1, \ldots, z_{K_u})$, where $z_k \in \mathcal{Z}_k$ is the receive strategy of the kth user, chosen from a compact set $\mathcal{Z}_k$. It is assumed that the kth row of $\boldsymbol{V}(z)$ and the kth component of $\boldsymbol{n}(z)$ only depend on z_k.

5.1 Matrix-Based Iteration

For any power allocation $\boldsymbol{p}$, the minimizer of (5.4) maximizes the signal-to-interference-plus-noise ratio

$$\mathrm{SINR}_k(\boldsymbol{p}) = \frac{p_k}{\mathcal{J}_k(\boldsymbol{p})} . \tag{5.5}$$

The SINR is typically defined as the ratio between the useful power and the interference+noise power at the *receiver*. The useful power is the product of the transmission power p_k and an effective path gain, which can be a function of z_k. This effective path gain is implicitly contained in $\mathcal{J}_k(\boldsymbol{p})$. Consider SINR targets $\boldsymbol{\gamma} = [\gamma_1, \ldots, \gamma_{K_u}]^T > 0$, with

$$\boldsymbol{\Gamma} = \mathrm{diag}\{\boldsymbol{\gamma}\} .$$

The condition

$$\mathrm{SINR}_k(\boldsymbol{p}) \geq \gamma_k, \quad \forall k ,$$

is fulfilled by all power allocations from the set

$$\mathcal{P}(\boldsymbol{\gamma}) = \{\boldsymbol{p} > 0 : p_k \geq \gamma_k \mathcal{J}_k(\boldsymbol{p}), \forall k\} . \tag{5.6}$$

In the following, we assume that the set $\mathcal{P}(\boldsymbol{\gamma})$ is non-empty. Among all feasible allocations, we are interested in the unique solution $\boldsymbol{p}^* > 0$ which minimizes the total power $\|\boldsymbol{p}\|_1$. Thus, the power minimization problem (5.1) can be rewritten as

$$\min_{\boldsymbol{p}} \sum_{l=1}^{K_u} p_l \quad \text{s.t.} \quad \boldsymbol{p} \in \mathcal{P}(\boldsymbol{\gamma}) . \tag{5.7}$$

Lemma 5.1. *The vector $\boldsymbol{p}^* > 0$ is the unique optimizer of (5.7) if and only if*

$$p_k^* = \gamma_k \mathcal{J}_k(\boldsymbol{p}^*), \quad k = 1, 2, \ldots, K_u . \tag{5.8}$$

Proof. This was shown in [1]. □

For an arbitrary initialization $\boldsymbol{p}^{(0)} \in \mathcal{P}(\boldsymbol{\gamma})$, it was shown in [126] that problem (5.7) is solved by the matrix-based iteration

$$\boldsymbol{p}^{(n+1)} = \left(\boldsymbol{I} - \boldsymbol{\Gamma}\boldsymbol{V}(z^{(n)})\right)^{-1} \boldsymbol{\Gamma}\boldsymbol{n}(z^{(n)}) \tag{5.9}$$

$$\text{with} \quad z_k^{(n)} = \arg\min_{z_k \in \mathcal{Z}_k} [\boldsymbol{V}(z)\boldsymbol{p}^{(n)} + \boldsymbol{n}(z)]_k \quad \forall k . \tag{5.10}$$

A max-min balancing strategy for finding a feasible initialization $\boldsymbol{p}^{(0)} \in \mathcal{P}(\boldsymbol{\gamma})$ will be discussed in the next Chapter 6. For the analysis in this chapter it is sufficient to assume that such a vector exists.

5.1.1 Optimal Matrices and Receive Strategies

For given $\boldsymbol{p}$, the set of receive strategies that are optimal with respect to the interference minimization problem (5.4) is defined as

$$\mathcal{Z}(\boldsymbol{p}) = \{z : [\boldsymbol{V}(z)\boldsymbol{p} + \boldsymbol{n}(z)]_k = \mathcal{J}_k(\boldsymbol{p}),\ k \in \mathcal{K}_u\} .$$

The associated matrices are contained in the set

$$\mathcal{M}(\boldsymbol{p}) = \{\boldsymbol{V}(z) : z \in \mathcal{Z}(\boldsymbol{p})\} . \tag{5.11}$$

Thus, for each matrix $\boldsymbol{V} \in \mathcal{M}(\boldsymbol{p})$, there exists an optimal receive strategy z^* such that $\boldsymbol{V} = \boldsymbol{V}(z^*)$.

Next, consider the iteration (5.9). For any given $\boldsymbol{p}^{(n)}$, there is an optimal coupling matrix $\boldsymbol{V}(z^{(n)}) \in \mathcal{M}(\boldsymbol{p}^{(n)})$, with a parameter $z^{(n)} \in \mathcal{Z}(\boldsymbol{p}^{(n)})$ such that

$$\boldsymbol{\mathcal{J}}(\boldsymbol{p}^{(n)}) = \boldsymbol{V}(z^{(n)})\boldsymbol{p}^{(n)} + \boldsymbol{n}(z^{(n)}) . \tag{5.12}$$

Thus, the allocation $\boldsymbol{p}^{(n+1)}$ is connected with $\boldsymbol{p}^{(n)}$ via an intermediary optimization of the receive strategy. Since the set $\mathcal{M}(\boldsymbol{p}^{(n)})$ can contain more than one element, the matrix representation (5.12) need not be unique. Thus, the allocation $\boldsymbol{p}^{(n+1)}$ (if existent) might depend on the choice of the coupling matrix $\boldsymbol{V}(z^{(n)})$ out of the set $\mathcal{M}(\boldsymbol{p}^{(n)})$.

One possible criterion for choosing $\boldsymbol{V}(z^{(n)}) \in \mathcal{M}(\boldsymbol{p}^{(n)})$ is feasibility. An allocation $\boldsymbol{p}^{(n+1)} > 0$ exists if and only if $\boldsymbol{\Gamma V}(z^{(n)})$ has a spectral radius

$$\rho\big(\boldsymbol{\Gamma V}(z^{(n)})\big) < 1 , \tag{5.13}$$

in which case $\boldsymbol{I} - \boldsymbol{\Gamma V}(z^{(n)})$ is non-singular and invertible, thus a positive power vector (5.9) exists. This aspect will be studied in the next section.

5.1.2 Feasibility

Consider the following system of equations

$$\boldsymbol{p} = \boldsymbol{\Gamma V p} + \boldsymbol{n} , \tag{5.14}$$

where $\boldsymbol{V} \geq 0$ and $\boldsymbol{n} = [n_1, \dots, n_{K_u}]^T > 0$. The system (5.14) has a unique solution $\boldsymbol{p} > 0$ if and only if $\rho(\boldsymbol{\Gamma V}) < 1$. The impact of a possible sum-power limit $P_{\max}$ will be shown by the following lemma.

Lemma 5.2. *Suppose that the linear system of equations (5.14) has a solution $\boldsymbol{p} > 0$ with $\|\boldsymbol{p}\|_1 = P_{\max}$, then*

$$\rho(\boldsymbol{\Gamma V}) < 1 - \frac{1}{P_{\max}} \cdot \min_{1 \leq k \leq K_u} n_k . \tag{5.15}$$

Proof. Let $\boldsymbol{q} > 0$ be arbitrary. Using (5.14) and $\boldsymbol{q}^T\boldsymbol{n}/\boldsymbol{q}^T\boldsymbol{p} \geq \min_k n_k/p_k$ (see Lemma A.12 in the Appendix A.7), we have

$$1 = \frac{\boldsymbol{q}^T\boldsymbol{\Gamma V p}}{\boldsymbol{q}^T\boldsymbol{p}} + \frac{\boldsymbol{q}^T\boldsymbol{n}}{\boldsymbol{q}^T\boldsymbol{p}} \geq \frac{\boldsymbol{q}^T\boldsymbol{\Gamma V p}}{\boldsymbol{q}^T\boldsymbol{p}} + \min_{1\leq k\leq K_u} \frac{n_k}{p_k}$$
$$> \frac{\boldsymbol{q}^T\boldsymbol{\Gamma V p}}{\boldsymbol{q}^T\boldsymbol{p}} + \frac{1}{P_{\max}} \min_{1\leq k\leq K_u} n_k \,.$$

Thus,

$$\begin{aligned}
1 &> \sup_{\boldsymbol{q}>0}\Big(\frac{\boldsymbol{q}^T\boldsymbol{\Gamma V p}}{\boldsymbol{q}^T\boldsymbol{p}} + \frac{1}{P_{\max}} \min_{1\leq k\leq K_u} n_k\Big) \\
&\geq \sup_{\boldsymbol{q}>0}\Big(\inf_{\boldsymbol{p}>0:\|\boldsymbol{p}\|_1=P_{\max}} \frac{\boldsymbol{q}^T\boldsymbol{\Gamma V p}}{\boldsymbol{q}^T\boldsymbol{p}}\Big) + \frac{1}{P_{\max}} \min_{1\leq k\leq K_u} n_k \\
&\geq \sup_{\boldsymbol{q}>0}\Big(\inf_{\boldsymbol{p}>0} \frac{\boldsymbol{q}^T\boldsymbol{\Gamma V p}}{\boldsymbol{q}^T\boldsymbol{p}}\Big) + \frac{1}{P_{\max}} \min_{1\leq k\leq K_u} n_k \\
&= \rho(\boldsymbol{\Gamma V}) + \frac{1}{P_{\max}} \min_{1\leq k\leq K_u} n_k \,,
\end{aligned}$$

where the last step follows from the Collatz-Wielandt type characterization (1.24) of the spectral radius. □

Of course, (5.15) holds as well if (5.14) is replaced by $\boldsymbol{p} \geq \boldsymbol{\Gamma V p} + \boldsymbol{n}$. We will now use Lemma 5.2 to analyze step (5.9).

Lemma 5.3. *Let $\boldsymbol{p} \in \mathcal{P}(\gamma)$ be arbitrary, then $\rho(\boldsymbol{\Gamma V}) < 1$ holds for all $\boldsymbol{V} \in \mathcal{M}(\boldsymbol{p})$.*

Proof. Since $\boldsymbol{p} \in \mathcal{P}(\gamma)$, we have $\|\boldsymbol{p}\|_1 < +\infty$ and

$$p_k \geq \gamma_k \mathcal{J}_k(\boldsymbol{p}) = \gamma_k[\boldsymbol{V}(z^*)\boldsymbol{p}]_k + \gamma_k n_k(z^*), \quad 1 \leq k \leq K_u \,,$$

where $\boldsymbol{V}(z^*) \in \mathcal{M}(\boldsymbol{p})$ is arbitrary. With Lemma 5.2 and $\boldsymbol{n}(z^*) > 0$ we obtain the desired result. □

It can be observed from Lemma 5.3 that the first step of the algorithm can be carried out for a feasible initialization $\boldsymbol{p}^{(0)}$.

Lemma 5.4. *Let $\boldsymbol{p}^{(0)} \in \mathcal{P}(\gamma)$ and $\boldsymbol{V}(z^{(0)}) \in \mathcal{M}(\boldsymbol{p}^{(0)})$, then*

$$\boldsymbol{p}^{(1)} = \big(\boldsymbol{I} - \boldsymbol{\Gamma V}(z^{(0)})\big)^{-1}\boldsymbol{\Gamma n}(z^{(0)})$$

always fulfills $\boldsymbol{p}^{(1)} \in \mathcal{P}(\gamma)$.

Proof. We have

$$\begin{aligned} p_k^{(1)} &= [\boldsymbol{\Gamma}\boldsymbol{V}(z^{(0)})\boldsymbol{p}^{(1)}]_k + \gamma_k n_k(z^{(0)}) \\ &\geq \min_{z_k \in \mathcal{Z}_k} \Big([\boldsymbol{\Gamma}\boldsymbol{V}(z)\boldsymbol{p}^{(1)}]_k + \gamma_k n_k(z) \Big) \\ &= \gamma_k \mathcal{J}_k(\boldsymbol{p}^{(1)}), \quad \forall k \end{aligned} \tag{5.16}$$

thus $\boldsymbol{p}^{(1)} \in \mathcal{P}(\gamma)$. □

In a similar way, feasibility can be shown for every step of the iteration (5.9). If $\boldsymbol{p}^{(0)} \in \mathcal{P}(\gamma)$, then every $\boldsymbol{p}^{(n)}$ belongs to the set $\mathcal{P}(\gamma)$. Notice that the actual sequence can still depend on the choice of $\boldsymbol{p}^{(0)}$, and also on the choice of matrices out of the set $\mathcal{M}(\boldsymbol{p}^{(n)})$ in every step. Further properties will be shown in the next section.

5.1.3 Monotonicity

The following lemma shows that the power vector obtained by step (5.9) is component-wise minimal.

Lemma 5.5. *Suppose that there exists a z' such that $\rho\big(\boldsymbol{\Gamma}\boldsymbol{V}(z')\big) < 1$, then*

$$\boldsymbol{p}' = \big(\boldsymbol{I} - \boldsymbol{\Gamma}\boldsymbol{V}(z')\big)^{-1}\boldsymbol{\Gamma}\boldsymbol{n}(z') \leq \boldsymbol{p} \tag{5.17}$$

for all $\boldsymbol{p} > 0$ which fulfill $\boldsymbol{p} \geq \boldsymbol{\Gamma}\boldsymbol{V}(z')\boldsymbol{p} + \boldsymbol{\Gamma}\boldsymbol{n}(z')$.

Proof. The vector $\boldsymbol{p}'$ is the fixed point which fulfills $\boldsymbol{p}' = \boldsymbol{\Gamma}\boldsymbol{V}(z')\boldsymbol{p}' + \boldsymbol{\Gamma}\boldsymbol{n}(z')$. Thus it has component-wise minimal powers among all feasible allocations [1]. □

Lemma 5.5 is now used to show component-wise monotonicity. This behavior is independent of the choice of matrix in each iteration step.

Lemma 5.6. *Let $\boldsymbol{p}^{(0)} \in \mathcal{P}(\gamma)$ be an arbitrary feasible initialization of (5.9), then for all n we have $\boldsymbol{p}^{(n+1)} \leq \boldsymbol{p}^{(n)}$.*

Proof. The initialization $\boldsymbol{p}^{(0)}$ is associated with some matrix $\boldsymbol{V}(z^{(0)}) \in \mathcal{M}(\boldsymbol{p}^{(0)})$. We have $\rho\big(\boldsymbol{\Gamma}\boldsymbol{V}(z^{(0)})\big) < 1$, thus iteration step (5.9) provides a power vector $\boldsymbol{p}^{(1)} > 0$, with

$$\boldsymbol{p}^{(1)} = \boldsymbol{\Gamma}\boldsymbol{V}(z^{(0)})\boldsymbol{p}^{(1)} + \boldsymbol{\Gamma}\boldsymbol{n}(z^{(0)})\,. \tag{5.18}$$

We know from Lemma 5.5 that $\boldsymbol{p}^{(1)}$ is component-wise minimal among all vectors $\boldsymbol{p} > 0$ which fulfill $\boldsymbol{p} \geq \boldsymbol{\Gamma}\boldsymbol{V}(z^{(0)})\boldsymbol{p} + \boldsymbol{\Gamma}\boldsymbol{n}(z^{(0)})$. Since $\boldsymbol{p}^{(0)} \in \mathcal{P}(\gamma)$, we have

$$\boldsymbol{p}^{(0)} \geq \boldsymbol{\Gamma}\boldsymbol{\mathcal{J}}(\boldsymbol{p}^{(0)}) = \boldsymbol{\Gamma}\boldsymbol{V}(z^{(0)})\boldsymbol{p}^{(0)} + \boldsymbol{\Gamma}\boldsymbol{n}(z^{(0)})\,. \tag{5.19}$$

It can be concluded that $\boldsymbol{p}^{(1)} \leq \boldsymbol{p}^{(0)}$. From the results of Section 5.1.2 we know that $\boldsymbol{p}^{(1)} \in \mathcal{P}(\gamma)$. In analogy, we can show that $\boldsymbol{p}^{(n)} \in \mathcal{P}(\gamma)$ implies $\boldsymbol{p}^{(n+1)} \leq \boldsymbol{p}^{(n)}$ and $\boldsymbol{p}^{(n+1)} \in \mathcal{P}(\gamma)$. It can be concluded that the entire sequence is monotone decreasing. □

5.1.4 Lipschitz Continuity

Every concave standard interference function has a form (5.4). The next lemma shows that this structure implies local Lipschitz continuity.

Lemma 5.7. *The function $\mathcal{J}_k(\boldsymbol{p})$ is locally Lipschitz-continuous. That is, for all $\boldsymbol{p} > 0$ there exist $C > 0$ and $\delta > 0$ such that for all $\hat{\boldsymbol{p}} > 0$ and $\|\boldsymbol{p}-\hat{\boldsymbol{p}}\|_1 < \delta$, we always have*

$$|\mathcal{J}_k(\boldsymbol{p}) - \mathcal{J}_k(\hat{\boldsymbol{p}})| \leq C\|\boldsymbol{p} - \hat{\boldsymbol{p}}\|_1 .$$

Proof. Let $\boldsymbol{p}^{(1)}$, $\boldsymbol{p}^{(2)}$, be arbitrary, with parameters $z^{(1)} \in \mathcal{Z}(\boldsymbol{p}^{(1)})$ and $z^{(2)} \in \mathcal{Z}(\boldsymbol{p}^{(2)})$. We have

$$\begin{aligned}
\mathcal{J}_k(\boldsymbol{p}^{(2)}) - \mathcal{J}_k(\boldsymbol{p}^{(1)}) &= \sum_{l=1}^{K_u} p_l^{(2)} V_{kl}(z^{(2)}) + n_k(z^{(2)}) \\
&\quad - \sum_{l=1}^{K_u} p_l^{(1)} V_{kl}(z^{(1)}) - n_k(z^{(1)}) \\
&\leq \sum_{l=1}^{K_u} (p_l^{(2)} - p_l^{(1)}) \cdot V_{kl}(z^{(1)}) \qquad (5.20)\\
&\leq \sum_{l=1}^{K_u} |p_l^{(2)} - p_l^{(1)}| \cdot V_{kl}(z^{(1)}) \\
&\leq \|\boldsymbol{p}^{(2)} - \boldsymbol{p}^{(1)}\|_\infty \cdot \sum_{l=1}^{K_u} V_{kl}(z^{(1)}) . \qquad (5.21)
\end{aligned}$$

Inequality (5.20) follows from the fact that $z^{(2)}$ minimizes the interference for given $\boldsymbol{p}^{(2)}$. If we replace $z^{(1)}$ by $z^{(2)}$ instead, then we obtain

$$\begin{aligned}
\mathcal{J}_k(\boldsymbol{p}^{(2)}) - \mathcal{J}_k(\boldsymbol{p}^{(1)}) &\geq \sum_{l=1}^{K_u} (p_l^{(2)} - p_l^{(1)}) \cdot V_{kl}(z^{(2)}) \\
&\geq -\|\boldsymbol{p}^{(2)} - \boldsymbol{p}^{(1)}\|_\infty \sum_{l=1}^{K_u} V_{kl}(z^{(2)}) . \qquad (5.22)
\end{aligned}$$

With (5.21) and (5.22), we can conclude that there exists a

$$C_1 = \max_{r=1,2} \sum_{l=1}^{K_u} V_{kl}(z^{(r)})$$

such that

$$|\mathcal{J}_k(\boldsymbol{p}^{(2)}) - \mathcal{J}_k(\boldsymbol{p}^{(1)})| \leq C_1 \cdot \|\boldsymbol{p}^{(2)} - \boldsymbol{p}^{(1)}\|_\infty .$$

The same can be shown for arbitrary $\boldsymbol{p}$. Because $\mathcal{Z}$ is closed and bounded, and $V_{kl}(z)$ is continuous, there is an optimal strategy z for every $\boldsymbol{p}$, so a constant $C_1 > 0$ exists. The constant $C_1 > 0$ depends on δ and the index k, but K_u is finite so a maximum can always be found. Finally, all norms on finite-dimensional vector spaces are equivalent, which proves the result. □

5.1.5 Global Convergence and Comparison with the Fixed Point Iteration

The global convergence of the matrix iteration (5.9) was first shown in [126]. An alternative, and maybe more intuitive proof appeared in [7]. It is based on a comparison with the fixed point iteration (5.2), which can be written in vector notation as follows.

$$\bar{\boldsymbol{p}}^{(n+1)} = \boldsymbol{\Gamma}\boldsymbol{\mathcal{J}}(\bar{\boldsymbol{p}}^{(n)}), \quad \bar{\boldsymbol{p}}^{(0)} \in \mathcal{P}(\gamma)\,. \tag{5.23}$$

Next, we show how this iteration is related to the matrix iteration (5.9).

If we choose $\mathcal{J}_k(\boldsymbol{p})$ as the special matrix-based function (5.4), then the iteration (5.23) becomes

$$\begin{aligned} \bar{\boldsymbol{p}}^{(n+1)} &= \boldsymbol{\Gamma}[\boldsymbol{V}(z^{(n)})\bar{\boldsymbol{p}}^{(n)} + \boldsymbol{n}(z^{(n)})] \\ \text{with} \quad z_k^{(n)} &= \arg\min_{z_k \in \mathcal{Z}_k}[\boldsymbol{V}(z)\bar{\boldsymbol{p}}^{(n)} + \boldsymbol{n}(z)]_k \quad \forall k\,. \end{aligned} \tag{5.24}$$

The interference functions $\boldsymbol{\mathcal{J}}$ are standard, i.e., they are positive, scalable, and monotone, as discussed in Subsection 2.4.1. Thus, iteration (5.24) has the following properties [1]

- If $\bar{\boldsymbol{p}}^{(0)} \in \mathcal{P}(\gamma)$ then the sequence is component-wise monotone decreasing, i.e,
$$\bar{\boldsymbol{p}}^{(n+1)} \leq \bar{\boldsymbol{p}}^{(n)}, \quad \text{for all } n. \tag{5.25}$$
- The sequence $\bar{\boldsymbol{p}}^{(n)}$ converges to the unique optimizer of the power minimization problem (5.7).

The next theorem shows a step-wise comparison of the fixed point iteration with the matrix iteration (5.9). This proves global convergence of the matrix iteration.

Theorem 5.8. *Starting the iterations (5.9) and (5.23) with the same feasible initialization $\boldsymbol{p}^{(0)} \in \mathcal{P}(\gamma)$, we have*

$$\boldsymbol{p}^{(n)} \leq \bar{\boldsymbol{p}}^{(n)}, \quad \textit{for all } n. \tag{5.26}$$

Thus, the sequence $\boldsymbol{p}^{(n)}$, as defined by (5.9), converges to the fixed point $\boldsymbol{p}^*$ which is the unique optimizer of the power minimization problem (5.7).

Proof. Let $\boldsymbol{p}^{(0)} = \bar{\boldsymbol{p}}^{(0)} \in \mathcal{P}(\gamma)$ be an arbitrary feasible initialization. Feasibility implies

$$\bar{\boldsymbol{p}}^{(0)} \geq \boldsymbol{\Gamma}\boldsymbol{\mathcal{J}}(\boldsymbol{p}^{(0)}) = \bar{\boldsymbol{p}}^{(1)}\,.$$

Because of monotonicity (5.25) we have

$$\bar{\boldsymbol{p}}^{(n+1)} = \boldsymbol{\Gamma}\boldsymbol{\mathcal{J}}(\bar{\boldsymbol{p}}^{(n)}) \leq \bar{\boldsymbol{p}}^{(n)} .$$

All $\bar{\boldsymbol{p}}^{(n)}$ are feasible and belong to $\mathcal{P}(\boldsymbol{\gamma})$. The initialization is associated with some arbitrary matrix $\boldsymbol{V}(z^{(0)}) \in \mathcal{M}(\boldsymbol{p}^{(0)})$. We have

$$\begin{aligned}\bar{\boldsymbol{p}}^{(1)} = \boldsymbol{\Gamma}\boldsymbol{\mathcal{J}}(\boldsymbol{p}^{(0)}) &= \boldsymbol{\Gamma}\boldsymbol{V}(z^{(0)})\boldsymbol{p}^{(0)} + \boldsymbol{\Gamma}\boldsymbol{n}(z^{(0)}) \\ &\geq \boldsymbol{\Gamma}\boldsymbol{V}(z^{(0)})\bar{\boldsymbol{p}}^{(1)} + \boldsymbol{\Gamma}\boldsymbol{n}(z^{(0)}) ,\end{aligned} \tag{5.27}$$

where the inequality follows from (5.25). Since $\boldsymbol{p}^{(1)}$ solves (5.18), and with Lemma 5.5, we have

$$\boldsymbol{p}^{(1)} \leq \boldsymbol{\Gamma}\boldsymbol{V}(z^{(0)})\bar{\boldsymbol{p}}^{(1)} + \boldsymbol{\Gamma}\boldsymbol{n}(z^{(0)}) \leq \bar{\boldsymbol{p}}^{(1)} . \tag{5.28}$$

Note that this inequality does not depend on the choice of $\boldsymbol{V}(z^{(0)}) \in \mathcal{M}(\boldsymbol{p}^{(0)})$.

Next, consider an arbitrary step n, and $\boldsymbol{p}^{(n-1)} \leq \bar{\boldsymbol{p}}^{(n-1)}$. Because of monotonicity this implies

$$\boldsymbol{\Gamma}\boldsymbol{\mathcal{J}}(\boldsymbol{p}^{(n-1)}) \leq \boldsymbol{\Gamma}\boldsymbol{\mathcal{J}}(\bar{\boldsymbol{p}}^{(n-1)}) = \bar{\boldsymbol{p}}^{(n)} . \tag{5.29}$$

Applying the fixed point iteration to the vector $\boldsymbol{p}^{(n-1)}$, we obtain $\boldsymbol{q}^{(n)} = \boldsymbol{\Gamma}\boldsymbol{\mathcal{J}}(\boldsymbol{p}^{(n-1)})$. With the monotonicity (5.25) we have $\boldsymbol{q}^{(n)} \leq \boldsymbol{p}^{(n-1)}$. This implies that for an arbitrary $\boldsymbol{V}(z^{(n-1)}) \in \mathcal{M}(\boldsymbol{p}^{(n-1)})$, we have

$$\begin{aligned}\boldsymbol{q}^{(n)} &= \boldsymbol{\Gamma}\boldsymbol{V}(z^{(n-1)})\boldsymbol{p}^{(n-1)} + \boldsymbol{\Gamma}\boldsymbol{n}(z^{(n-1)}) \\ &\geq \boldsymbol{\Gamma}\boldsymbol{V}(z^{(n-1)})\boldsymbol{q}^{(n)} + \boldsymbol{\Gamma}\boldsymbol{n}(z^{(n-1)}) .\end{aligned} \tag{5.30}$$

The vector $\boldsymbol{p}^{(n)}$ satisfies

$$\boldsymbol{p}^{(n)} = \boldsymbol{\Gamma}\boldsymbol{V}(z^{(n-1)})\boldsymbol{p}^{(n)} + \boldsymbol{\Gamma}\boldsymbol{n}(z^{(n-1)}) .$$

Thus, with Lemma 5.5 and (5.30) we know that $\boldsymbol{p}^{(n)} \leq \boldsymbol{q}^{(n)}$. From (5.29) we have $\boldsymbol{q}^{(n)} \leq \bar{\boldsymbol{p}}^{(n)}$, thus

$$\boldsymbol{p}^{(n)} \leq \boldsymbol{q}^{(n)} \leq \bar{\boldsymbol{p}}^{(n)} .$$

Starting with (5.28), the result (5.26) can be shown for all n by complete induction.

Since $\bar{\boldsymbol{p}}^{(n)}$ converges to the optimizer $\boldsymbol{p}^*$, and

$$\boldsymbol{p}^* \leq \boldsymbol{p}^{(n)} \leq \bar{\boldsymbol{p}}^{(n)} ,$$

we can conclude that also $\boldsymbol{p}^{(n)}$ converges to $\boldsymbol{p}^*$. □

Note that the convergence shown by Theorem 5.8 does not depend on which matrix from $\mathcal{M}(\boldsymbol{p}^{(n)})$ is chosen in each iteration step. However, the convergence behavior might depend on this choice. This will be studied in the next section.

5.2 Super-Linear Convergence

One difficulty in studying the above iterations is that $\boldsymbol{p}^{(n+1)}$ is not directly linked to $\boldsymbol{p}^{(n)}$, but indirectly via $z^{(n)}$. In order to study the convergence, we wish to express $\boldsymbol{p}^{(n+1)}$ as a function of $\boldsymbol{p}^{(n)}$. To this end, we define an auxiliary function

$$\boldsymbol{d}(\boldsymbol{p}) = [d_1(\boldsymbol{p}), \dots, d_{K_u}(\boldsymbol{p})]^T = \boldsymbol{p} - \boldsymbol{\Gamma}\boldsymbol{\mathcal{J}}(\boldsymbol{p}) . \tag{5.31}$$

The function $\boldsymbol{d}(\boldsymbol{p})$ is jointly convex as the sum of two convex functions. We have $\boldsymbol{d}(\boldsymbol{p}) \geq 0$ for all $\boldsymbol{p} \in \mathcal{P}(\boldsymbol{\gamma})$. From Lemma 5.1 we know that the optimizer $\boldsymbol{p}^*$ of the power minimization problem is completely characterized by $\boldsymbol{d}(\boldsymbol{p}^*) = \boldsymbol{0}$. In this sense, the function $\boldsymbol{d}(\boldsymbol{p})$ can be seen as a measure for the "distance" between some power allocation $\boldsymbol{p} \in \mathcal{P}(\boldsymbol{\gamma})$ and the optimizer $\boldsymbol{p}^*$.

5.2.1 Continuously Differentiable Interference Functions

Only in this section we will assume that $\boldsymbol{\mathcal{J}}(\boldsymbol{p})$ is continuously differentiable for $\boldsymbol{p} > 0$. Thus, $\boldsymbol{d}(\boldsymbol{p})$ is continuously differentiable as well. This simplification helps to understand the underlying concept. Later, the general case will be considered.

Assume that for each $\boldsymbol{p}$ there exists exactly one optimizer $z(\boldsymbol{p})$, thus

$$\boldsymbol{\mathcal{J}}(\boldsymbol{p}) = \boldsymbol{V}(z(\boldsymbol{p})) \cdot \boldsymbol{p} + \boldsymbol{n}(z(\boldsymbol{p})) . \tag{5.32}$$

Then, $\boldsymbol{\mathcal{J}}$ is continuously differentiable. In this case, the set $\mathcal{M}(\boldsymbol{p})$ always consists of a single element. The Jacobi matrix of $\boldsymbol{\mathcal{J}}(\boldsymbol{p})$, which contains the partial derivatives, is given as follows (see Appendix A.4).

$$\nabla\boldsymbol{\mathcal{J}}(\boldsymbol{p}) = \boldsymbol{V}\big(z(\boldsymbol{p})\big) . \tag{5.33}$$

Consequently,

$$\nabla\boldsymbol{d}(\boldsymbol{p}) = \boldsymbol{I} - \boldsymbol{\Gamma}\boldsymbol{V}\big(z(\boldsymbol{p})\big) . \tag{5.34}$$

The kth component of the manifold

$$\begin{aligned} \boldsymbol{g}^{(n)}(\boldsymbol{p}) &= \nabla\boldsymbol{d}(\boldsymbol{p})\big|_{\boldsymbol{p}=\boldsymbol{p}^{(n)}} \cdot (\boldsymbol{p} - \boldsymbol{p}^{(n)}) + \boldsymbol{d}(\boldsymbol{p}^{(n)}) \\ &= \boldsymbol{p} - \boldsymbol{\Gamma}\boldsymbol{V}\big(z(\boldsymbol{p}^{(n)})\big)\boldsymbol{p} - \boldsymbol{\Gamma}\boldsymbol{n}\big(z(\boldsymbol{p}^{(n)})\big) \end{aligned} \tag{5.35}$$

is a tangential hyperplane to the convex function $d_k(\boldsymbol{p})$ at the point $d_k(\boldsymbol{p}^{(n)})$, as illustrated in Fig. 5.1.

Assume that there is an initialization $\boldsymbol{p}^{(n)} \in \mathcal{P}(\boldsymbol{\gamma})$, such that $\boldsymbol{d}(\boldsymbol{p}^{(n)}) \geq 0$. Then, Newton's method can be applied in order to find a new power vector $\boldsymbol{p}^{(n+1)}$ which is closer to the global optimizer $\boldsymbol{p}^*$. This new point is characterized by $\boldsymbol{g}^{(n)}(\boldsymbol{p}) = \boldsymbol{0}$. The update formula is

$$\boldsymbol{p}^{(n+1)} = \boldsymbol{p}^{(n)} - \big(\nabla\boldsymbol{d}(\boldsymbol{p}^{(n)})\big)^{-1}\boldsymbol{d}(\boldsymbol{p}^{(n)}) . \tag{5.36}$$

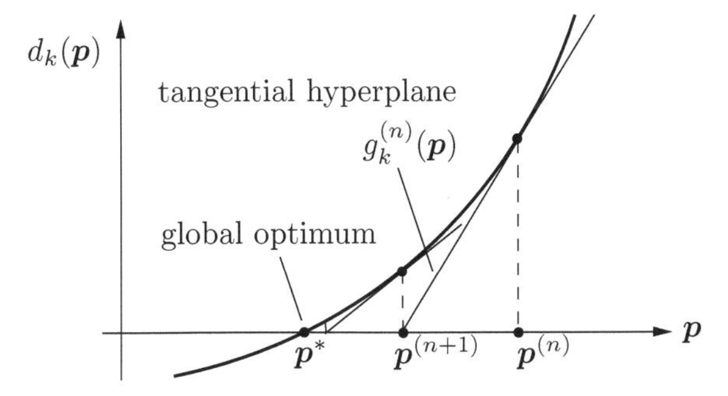

Fig. 5.1. Schematic illustration of the matrix iteration (5.9) under the assumption of continuously differentiable interference functions $d_k(\boldsymbol{p})$. The algorithm can be interpreted as a Newton iteration with quadratic convergence.

With (5.32) and (5.34) it can be verified that the Newton iteration (5.36) corresponds exactly to our algorithm (5.9), whose properties were analyzed in the previous sections. One result was that the inverse $\left(\nabla \boldsymbol{d}(\boldsymbol{p}^{(n)})\right)^{-1}$ is guaranteed to exist. It can therefore be concluded that *if* $\mathcal{J}(\boldsymbol{p})$ is continuously differentiable, then the iteration (5.9) can be interpreted as the classic Newton method. The algorithm finds the unique root of the equation $\boldsymbol{d}(\boldsymbol{p}) = \boldsymbol{0}$ with quadratic convergence speed.

5.2.2 Generalization to Non-Smooth Interference Functions

In this section, we will study the general case, where the functions $\mathcal{J}_k(\boldsymbol{p})$ are not guaranteed to be continuously differentiable. This is a consequence of model (5.4), which allows ambiguities in the choice of the receive strategy. In this case, the classic Newton iteration cannot be applied. We need some concepts from non-smooth analysis that are summarized in Appendix A.5.

Non-smooth versions of Newton's method exist. It was shown in [127] that a version based on Clarke's generalized Jacobian [128] does converge under certain conditions. We will now use the theoretical framework [127, 129] in order to show that the proposed iteration (5.9) always has super-linear convergence. To this end, it is important to exploit the special structure of the problem at hand: The function $\boldsymbol{d}(\boldsymbol{p})$ is convex, locally Lipschitz-continuous, and has certain monotonicity properties. Also, it should be exploited that the iteration steps (5.9) and (5.10) have a physical interpretation: Step (5.9) corresponds to power control, and (5.10) corresponds to the optimization of a receive strategy. It is therefore desirable to base the analysis only on matrices from the set $\mathcal{M}(\boldsymbol{p})$, which contains all coupling matrices resulting from optimal receive strategies for a given $\boldsymbol{p}$. The proposed iteration (5.9) can be rewritten as

$$\boldsymbol{p}^{(n+1)} = \boldsymbol{p}^{(n)} - (\boldsymbol{I} - \boldsymbol{\Gamma} \boldsymbol{V}_n)^{-1} \boldsymbol{d}(\boldsymbol{p}^{(n)}), \quad \boldsymbol{V}_n \in \mathcal{M}(\boldsymbol{p}^{(n)}) . \tag{5.37}$$

Super-linear convergence of this iteration will be shown in the remainder of this section. The following points should be emphasized:

- The results hold for arbitrary functions $\mathcal{J}_k(\boldsymbol{p})$, as defined by (5.4). No additional assumptions on smoothness are made.
- The matrices $(\boldsymbol{I} - \boldsymbol{\Gamma} \boldsymbol{V}_n)$ are chosen from a relatively small subset, as compared to Clarke's generalized Jacobian. This will be discussed later.
- The approach (5.37) still allows for a physical interpretation in terms of transmission powers and receive strategies.

Before stating the convergence theorem, a few properties need to be discussed.

5.2.3 Invertibility

For the characterization of the convergence behavior, it is important to control the norm of the matrices $(\boldsymbol{I} - \boldsymbol{\Gamma} \boldsymbol{V}_n)^{-1}$ for $\boldsymbol{V}_n \in \mathcal{M}(\boldsymbol{p}^{(n)})$. Iteration (5.37) requires that $\boldsymbol{I} - \boldsymbol{\Gamma} \boldsymbol{V}_n$ is invertible. This is shown by the following lemma:

Lemma 5.9. *Let $\boldsymbol{p}^{(0)} \in \mathcal{P}(\boldsymbol{\gamma})$ be an arbitrary initialization. Then there exists a constant $C_1 = C_1(\boldsymbol{p}^{(0)})$, with $0 < C_1 < 1$, such that for all $n \in \mathbb{N}$ and for all $\boldsymbol{V}_n \in \mathcal{M}(\boldsymbol{p}^{(n)})$ we have*

$$\rho(\boldsymbol{\Gamma} \boldsymbol{V}_n) \leq 1 - C_1 , \tag{5.38}$$

and the non-negative matrix $(\boldsymbol{I} - \boldsymbol{\Gamma} \boldsymbol{V}_n)^{-1}$ always fulfills

$$\rho\big((\boldsymbol{I} - \boldsymbol{\Gamma} \boldsymbol{V}_n)^{-1}\big) \leq \frac{1}{C_1} . \tag{5.39}$$

Proof. It was shown in Section 5.1.2 that $\rho(\boldsymbol{\Gamma} \boldsymbol{V}_n) < 1$ for arbitrary $\boldsymbol{V}_n \in \mathcal{M}(\boldsymbol{p}^{(n)})$. Thus, $\boldsymbol{I} - \boldsymbol{\Gamma} \boldsymbol{V}_n$ is non-singular and

$$\boldsymbol{p}^{(n+1)} = (\boldsymbol{I} - \boldsymbol{\Gamma} \boldsymbol{V}_n)^{-1} \boldsymbol{\Gamma} \boldsymbol{n}(\hat{z}) ,$$

where $\hat{z}$ is the parameter associated with $\boldsymbol{V}_n$, i.e., $\boldsymbol{V}(\hat{z}) = \boldsymbol{V}_n$, and $\boldsymbol{0} < \boldsymbol{p}^{(n+1)} \leq \boldsymbol{p}^{(n)}$. This follows from the assumption of a feasible initialization and the fact that all following steps are feasible as well (see Lemmas 5.3 and 5.4).

Defining $P^{(n)}_{\max} = \|\boldsymbol{p}^{(n)}\|_1$, and using Lemma 5.2, we have

$$\rho(\boldsymbol{\Gamma} \boldsymbol{V}_n) \leq 1 - \frac{1}{P^{(n+1)}_{\max}} \min_{1 \leq k \leq K} \gamma_k n_k(\hat{z}) .$$

Using $n'_k := \min_{z_k \in \mathcal{Z}_k} n_k(z_k) \leq n_k(\hat{z})$ and $P^{(n+1)}_{\max} \leq P^{(0)}_{\max}$, we obtain a positive constant

$$C_1(\boldsymbol{p}^{(0)}) = \frac{1}{P^{(0)}_{\max}} \min_k \gamma_k n'_k > 0$$

such that

$$\rho(\boldsymbol{\Gamma}\boldsymbol{V}_n) \leq 1 - C_1(\boldsymbol{p}^{(0)}) \ .$$

Consequently, $\rho(\boldsymbol{\Gamma}\boldsymbol{V}_n) < 1$ for all n, so $\boldsymbol{I} - \boldsymbol{\Gamma}\boldsymbol{V}_n$ is always non-singular. This follows from the convergence of the Neumann series

$$(\boldsymbol{I} - \boldsymbol{\Gamma}\boldsymbol{V}_n)^{-1} = \sum_{l=0}^{\infty} (\boldsymbol{\Gamma}\boldsymbol{V}_n)^l \ . \tag{5.40}$$

Since each summand is non-negative, also $(\boldsymbol{I} - \boldsymbol{\Gamma}\boldsymbol{V}_n)^{-1}$ is non-negative. Applying the infinite geometric series formula we get

$$\rho\big((\boldsymbol{I} - \boldsymbol{\Gamma}\boldsymbol{V}_n)^{-1}\big) = \frac{1}{1 - \rho(\boldsymbol{\Gamma}\boldsymbol{V}_n)} \leq \frac{1}{C_1(\boldsymbol{p}^{(0)})} \ ,$$

which leads to (5.39). □

5.2.4 Non-Smooth Versions of Newton's Method

Lemma 5.7 shows that $\boldsymbol{d}(\boldsymbol{p})$ is locally Lipschitz-continuous. This means that $\boldsymbol{d}$ is almost everywhere differentiable. In other words, the set of points, for which $\boldsymbol{d}$ is not differentiable, has measure zero. Let D_F be the set on which $\boldsymbol{d}$ is differentiable, and $\nabla \boldsymbol{d}(\boldsymbol{p})$ is the Jacobi matrix for $\boldsymbol{p} \in D_F$, as defined by (5.34). Since $\boldsymbol{d}(\boldsymbol{p})$ is locally Lipschitz-continuous and directionally differentiable, we know that it is also B-differentiable at the points of interest. The B-derivative $\partial_B \boldsymbol{d}(\boldsymbol{p})$ at the point $\boldsymbol{p}$ is defined as

$$\begin{aligned}\partial_B \boldsymbol{d}(\boldsymbol{p}) = \{\boldsymbol{A} \in \mathbb{R}^{K \times K} &: \text{there exists a sequence } \{\boldsymbol{p}_n\}_{k \in \mathbb{N}}, \\ &\text{with } \boldsymbol{p}_n \in D_F, \boldsymbol{p}_n \to \boldsymbol{p} \text{ and} \\ &\boldsymbol{A} = \lim_{k \to \infty} \nabla \boldsymbol{d}(\boldsymbol{p}_n) \ \} \ .\end{aligned}$$

Notice, that the set $\partial_B \boldsymbol{d}(\boldsymbol{p})$ can contain more than one element. As an example, consider the function $F(x) = |x|$, for which $\partial_B F(x)|_{x=0} = \{-1, 1\}$.

Clarke's generalized Jacobian $\partial \boldsymbol{d}(\boldsymbol{p})$ is defined as [128]

$$\partial \boldsymbol{d}(\boldsymbol{p}) = conv\big(\partial_B \boldsymbol{d}(\boldsymbol{p})\big) \tag{5.41}$$

which is the convex hull of the set given by the B-derivative. For the simple example $F(x) = |x|$, this is the interval $\partial F(x)|_{x=0} = [-1, 1]$.

A generalized Newton method based on Clarke's Jacobian was analyzed in [127].

$$\boldsymbol{p}^{(n+1)} = \boldsymbol{p}^{(n)} - \boldsymbol{V}_n^{-1} \boldsymbol{d}(\boldsymbol{p}^{(n)}) \ , \quad \boldsymbol{V}_n \in \partial \boldsymbol{d}(\boldsymbol{p}^{(n)}) \ . \tag{5.42}$$

It can be observed that $\boldsymbol{V}_n$ needs to be invertible in order for $\boldsymbol{p}^{(n+1)}$ to exist. However, the requirement that all elements of $\partial \boldsymbol{d}(\boldsymbol{p}^{(n)})$ must be invertible is quite strong and generally difficult to verify.

Another non-smooth version of Newton's method is the following iteration (see [127, 129] and the references therein)

$$\boldsymbol{p}^{(n+1)} = \boldsymbol{p}^{(n)} - \boldsymbol{V}_n^{-1}\boldsymbol{d}(\boldsymbol{p}^{(n)})\,, \quad \boldsymbol{V}_n \in \partial_B \boldsymbol{d}(\boldsymbol{p}^{(n)})\,. \tag{5.43}$$

The difference to (5.42) is that only matrices $\boldsymbol{V}_n$ from the B-derivative $\partial_B \boldsymbol{d}(\boldsymbol{p}^{(n)})$ are used. So only elements of this set need to be invertible. The local convergence behavior of the iteration (5.43) was studied in [127,129]. For the analysis it is required that $\boldsymbol{d}$ at a certain point $\hat{\boldsymbol{p}}$ is *strongly BD-regular* (B-derivative regular). This is fulfilled if all $\boldsymbol{V} \in \partial_B \boldsymbol{d}(\hat{\boldsymbol{p}})$ are nonsingular and thus invertible. Then, the iteration (5.43) has superlinear convergence [127, 129]. But the problem remains that $\partial_B \boldsymbol{d}(\hat{\boldsymbol{p}})$ can rarely be computed explicitly. There is no practical test for BD-regularity.

Fortunately, our function $\boldsymbol{d}(\boldsymbol{p})$ is always strongly BD-regular at the points of interest. In the next section it will be shown that this can be used in order to show super-linear convergence.

5.2.5 Superlinear Convergence

Under the given model (5.4), the following properties can be exploited:

- the sequence $\boldsymbol{p}^{(n)}$ is globally convergent, independent of the initialization $\boldsymbol{p}^{(0)} \in \mathcal{P}(\gamma)$.
- the algorithm is step-wise better than the fixed point iteration (see Theorem 5.8).
- monotony implies directions

$$\begin{aligned}
\boldsymbol{h}_1^{(n)} &= \boldsymbol{p}^{(n)} - \boldsymbol{p}^{(n+1)} \geq 0 \\
\boldsymbol{h}_2^{(n)} &= \boldsymbol{p}^{(n+1)} - \boldsymbol{p}^{(n)} \leq 0\,.
\end{aligned}$$

Thus the analysis of the convergence behavior can be restricted to these special cases.

For $\boldsymbol{p} \in \mathcal{P}(\gamma)$, we have $\mathcal{M}(\boldsymbol{p}) \subset \partial \boldsymbol{\mathcal{J}}(\boldsymbol{p})$, which is the generalized Jacobian of the interference function $\boldsymbol{\mathcal{J}}(\boldsymbol{p})$ at $\boldsymbol{p}$. The generalized Jacobian of $\boldsymbol{d}(\boldsymbol{p})$ at $\boldsymbol{p}$, as introduced in (5.41), can be rewritten as

$$\begin{aligned}
\partial \boldsymbol{d}(\boldsymbol{p}) &= \boldsymbol{I} - \boldsymbol{\Gamma}\, \partial \boldsymbol{\mathcal{J}}(\boldsymbol{p}) \\
&= \{\boldsymbol{V} \in \mathbb{R}^{K\times K} : \boldsymbol{V} = \boldsymbol{I} - \boldsymbol{\Gamma}\boldsymbol{A} \text{ with } \boldsymbol{A} \in \partial \boldsymbol{\mathcal{J}}(\boldsymbol{p})\}
\end{aligned} \tag{5.44}$$

That is, we only need to check the invertibility of the matrices $\boldsymbol{V} = \boldsymbol{I} - \boldsymbol{\Gamma}\boldsymbol{A}$, with $\boldsymbol{A} \in \mathcal{M}(\boldsymbol{p})$. At this point, we can apply Lemma 5.9, which shows that the norm of the inverse of $\boldsymbol{V}$ is always bounded. Thus, by exploiting the special structure of the given problem, we need no further restrictions on $\boldsymbol{\mathcal{J}}(\boldsymbol{p})$ and $\boldsymbol{d}(\boldsymbol{p})$. To conclude:

- Lemma 5.9 always ensures invertibility.
- The set $\mathcal{M}(\boldsymbol{p})$ and the corresponding inverses $(\boldsymbol{I}-\boldsymbol{\Gamma}\boldsymbol{A})^{-1}$, with $\boldsymbol{A} \in \mathcal{M}(\boldsymbol{p})$, can be described explicitly. The possible ambiguity of the receive strategy does not matter since the norm of the inverse is always bounded.

We are now in a position to show the super-linear convergence of the proposed iteration.

Theorem 5.10. *Assume an arbitrary initialization $\boldsymbol{p}^{(0)} \in \mathcal{P}(\gamma)$, then the matrix iteration (5.9) fulfills*

$$\lim_{n\to\infty} \frac{\|\boldsymbol{p}^{(n+1)} - \boldsymbol{p}^*\|}{\|\boldsymbol{p}^{(n)} - \boldsymbol{p}^*\|} = 0 \tag{5.45}$$

$$\lim_{n\to\infty} \frac{\|\boldsymbol{d}(\boldsymbol{p}^{(n+1)})\|}{\|\boldsymbol{d}(\boldsymbol{p}^{(n)})\|} = 0\,. \tag{5.46}$$

That is, the sequences $\boldsymbol{p}^{(n)}$ and $\boldsymbol{d}(\boldsymbol{p}^{(n)})$ have super-linear convergence.

Proof. The proof, which is shown here for completeness, uses the same technique as [127, 129]. We define $\boldsymbol{M}_n = \boldsymbol{I} - \boldsymbol{\Gamma V}(z^{(n)})$ and the direction $\boldsymbol{h}_n = \boldsymbol{p}^{(n)} - \boldsymbol{p}^*$. Using Definition A.5 from the appendix, and exploiting $\boldsymbol{d}(\boldsymbol{p}^*) = \boldsymbol{0}$, we have

$$\begin{aligned}
&\|\boldsymbol{p}^{(n+1)} - \boldsymbol{p}^*\| \\
&\quad = \|\boldsymbol{p}^{(n)} - \boldsymbol{p}^* - \boldsymbol{M}_n^{-1}\boldsymbol{d}(\boldsymbol{p}^{(n)})\| \\
&\quad = \|\boldsymbol{M}_n^{-1}\big(\boldsymbol{d}(\boldsymbol{p}^{(n)}) - \boldsymbol{d}(\boldsymbol{p}^*) - \boldsymbol{d}'(\boldsymbol{p}^*, \boldsymbol{h}_n)\big) + \\
&\qquad + \boldsymbol{M}_n^{-1}\big[-\boldsymbol{M}_n\boldsymbol{h}_n + \boldsymbol{d}'(\boldsymbol{p}^*, \boldsymbol{h}_n)\big]\| \\
&\quad \le \|\boldsymbol{M}_n^{-1}\| \cdot \|\boldsymbol{d}(\boldsymbol{p}^{(n)}) - \boldsymbol{d}(\boldsymbol{p}^*) - \boldsymbol{d}'(\boldsymbol{p}^*, \boldsymbol{h}_n)\| + \\
&\qquad + \|\boldsymbol{M}_n^{-1}\| \cdot \|\boldsymbol{M}_n\boldsymbol{h}_n - \boldsymbol{d}'(\boldsymbol{p}^*, \boldsymbol{h}_n)\|\,.
\end{aligned} \tag{5.47}$$

The norm is not specified because of the equivalence of norms on finite-dimensional vector spaces.

It remains to show that the upper bound (5.47) tends to zero as $\boldsymbol{h}_n \to 0$. We use some definitions from the appendix. Since $\boldsymbol{d}$ is locally Lipschitz-continuous and convex it is also B-differentiable. At the point $\boldsymbol{p}^*$ we have

$$\|\boldsymbol{d}(\boldsymbol{p}^{(n)}) - \boldsymbol{d}(\boldsymbol{p}^*) - \boldsymbol{d}'(\boldsymbol{p}^*, \boldsymbol{h}_n)\| = o(\|\boldsymbol{h}_n\|) \quad \text{as } \|\boldsymbol{h}_n\| \to 0\,.$$

The function $\boldsymbol{d}$ is also semi-smooth. Applying Lemma A.10 in the appendix, we have

$$\|\boldsymbol{M}_n\boldsymbol{h}_n - \boldsymbol{d}'(\boldsymbol{p}^*, \boldsymbol{h}_n)\| = o(\|\boldsymbol{h}_n\|)\,.$$

Consider an arbitrary $\epsilon > 0$. Because $\boldsymbol{d}$ is locally Lipschitz continuous (Lemma 5.7) and convex, we know from Lemma A.10 in the appendix that there exists a $\delta_1 = \delta_1(\epsilon, \boldsymbol{p}^*)$, such that

$$\|\boldsymbol{d}(\boldsymbol{p}) - \boldsymbol{d}'(\boldsymbol{p}^*, \boldsymbol{p} - \boldsymbol{p}^*)\| \le \epsilon\|\boldsymbol{p} - \boldsymbol{p}^*\|$$

for all points $\|\boldsymbol{p} - \boldsymbol{p}^*\| < \delta_1$. There exists a $n_0 = n_0(\epsilon)$ such that

$$\|\boldsymbol{p}^{(n+1)} - \boldsymbol{p}^*\| \le \epsilon\|\boldsymbol{p}^{(n)} - \boldsymbol{p}^*\| \tag{5.48}$$

for all $n \geq n_0(\epsilon)$. Thus,

$$0 \leq \limsup_{n\to\infty} \frac{\|\boldsymbol{p}^{(n+1)} - \boldsymbol{p}^*\|}{\|\boldsymbol{p}^{(n)} - \boldsymbol{p}^*\|} \leq \epsilon \,.$$

This implies (5.45).

Next, we prove (5.46). Let $n_1 \in \mathbb{N}$ be such that $\|\boldsymbol{p}^{(n)} - \boldsymbol{p}^*\| < \delta_1$ for all $n \geq n_1$. The number n_1 exists because of the convergence $\boldsymbol{p}^{(n)} \to \boldsymbol{p}^*$. Thus, we can use (5.48) to show

$$\begin{aligned} \|\boldsymbol{d}(\boldsymbol{p}^{(n+1)})\| &\leq \|\boldsymbol{d}'(\boldsymbol{p}^*, \boldsymbol{p}^{(n+1)} - \boldsymbol{p}^*)\| + \epsilon\|\boldsymbol{p}^{(n+1)} - \boldsymbol{p}^*\| \\ &\leq \big(L(\boldsymbol{p}^*) + \epsilon\big)\|\boldsymbol{p}^{(n+1)} - \boldsymbol{p}^*\| \\ &\leq \big(L(\boldsymbol{p}^*) + \epsilon\big)\cdot\epsilon\cdot\|\boldsymbol{p}^{(n)} - \boldsymbol{p}^*\| \end{aligned} \tag{5.49}$$

where $L(\boldsymbol{p}^*)$ is the Lipschitz constant of $\boldsymbol{d}$ at the point $\boldsymbol{p}^*$. The last inequality follows from (5.48). Since $\|\boldsymbol{M}_n^{-1}\| \leq C_2 < +\infty$, we have

$$\begin{aligned} \|\boldsymbol{p}^{(n+1)} - \boldsymbol{p}^{(n)}\| &= \|\boldsymbol{M}_n^{-1}\cdot\boldsymbol{d}(\boldsymbol{p}^{(n)})\| \\ &\leq \|\boldsymbol{M}_n^{-1}\|\cdot\|\boldsymbol{d}(\boldsymbol{p}^{(n)})\| \\ &\leq C_2\cdot\|\boldsymbol{d}(\boldsymbol{p}^{(n)})\| \,. \end{aligned} \tag{5.50}$$

Thus, for all $n \geq \max(n_0, n_1)$, we have

$$\begin{aligned} \|\boldsymbol{p}^{(n)} - \boldsymbol{p}^*\| &\leq \|\boldsymbol{p}^{(n+1)} - \boldsymbol{p}^{(n)}\| + \|\boldsymbol{p}^{(n+1)} - \boldsymbol{p}^*\| \\ &\leq C_2\|\boldsymbol{d}(\boldsymbol{p}^{(n)})\| + \epsilon\|\boldsymbol{p}^{(n)} - \boldsymbol{p}^*\| \,. \end{aligned} \tag{5.51}$$

Consequently,

$$\|\boldsymbol{p}^{(n)} - \boldsymbol{p}^*\| \leq \frac{C_2}{1-\epsilon}\cdot\|\boldsymbol{d}(\boldsymbol{p}^{(n)})\| \,. \tag{5.52}$$

Combining (5.49) and (5.52), we obtain for all $n \geq \max(n_0, n_1)$,

$$\|\boldsymbol{d}(\boldsymbol{p}^{(n+1)})\| \leq C_2\frac{L(\boldsymbol{p}^*)+\epsilon}{1-\epsilon}\cdot\epsilon\cdot\|\boldsymbol{d}(\boldsymbol{p}^{(n)})\| \,.$$

Therefore,

$$0 \leq \limsup_{n\to\infty}\frac{\|\boldsymbol{d}(\boldsymbol{p}^{(n+1)})\|}{\|\boldsymbol{d}(\boldsymbol{p}^{(n)})\|} \leq C_2\frac{L(\boldsymbol{p}^*)+\epsilon}{1-\epsilon}\cdot\epsilon \,. \tag{5.53}$$

Inequality (5.53) holds for arbitrary $\epsilon > 0$. For $\epsilon \to 0$, we obtain (5.46). □

5.2.6 Quadratic Convergence for Degree-2 Semi-Continuous Interference Functions

In the previous section, superlinear convergence has been shown for general interference functions of the form (5.4). The iteration has even quadratic convergence if additional properties are fulfilled [129].

Theorem 5.11. *Let $\mathcal{J}$ be semi-continuous of degree 2 at point $\boldsymbol{p}^*$, then there exists a constant C_1 such that*

$$\|\boldsymbol{p}^{(n+1)} - \boldsymbol{p}^*\| \leq C_1(\|\boldsymbol{p}^{(n)} - \boldsymbol{p}^*\|)^2 , \quad \text{for all } n \in \mathbb{N}.$$

Proof. The proof is similar to the one of Theorem 5.10. Here, we use the second result in Lemma A.10 in the appendix. □

Next, we show that the convergence accelerates near the optimum. Assume that $\boldsymbol{p}^{(0)} \in \mathcal{P}(\gamma)$ is an initialization. Then, $\boldsymbol{p}^{(n)}$ is a monotone sequence which converges to $\boldsymbol{p}^*$. There exists an $m \in \mathbb{N}$ such that

$$C_1\|\boldsymbol{p}^* - \boldsymbol{p}^{(m)}\| < 1 . \tag{5.54}$$

Beyond this point, the convergence behavior can be further specified. For all $l \geq 1$ we have

$$\begin{aligned} \|\boldsymbol{p}^* - \boldsymbol{p}^{(m+l)}\| &\leq C_1(\|\boldsymbol{p}^* - \boldsymbol{p}^{(m+l-1)}\|)^2 \\ &\leq C_1C_1^2(\|\boldsymbol{p}^* - \boldsymbol{p}^{(m+l-2)}\|)^4 \\ &\leq \prod_{k=0}^{l-1} C_1^{2^k} \cdot (\|\boldsymbol{p}^* - \boldsymbol{p}^{(m)}\|)^{2^l} . \end{aligned} \tag{5.55}$$

If $C_1 \leq 1$, then super-exponential convergence becomes evident from (5.55). If $C_1 > 1$, then

$$\prod_{k=0}^{l-1} C_1^{2^k} = C_1^{\sum_{k=0}^{l-1} 2^k} = C_1^{2^l-1} \leq C_1^{2^l} .$$

Thus

$$\|\boldsymbol{p}^* - \boldsymbol{p}^{(m+l)}\| \leq (C_1\|\boldsymbol{p}^* - \boldsymbol{p}^{(m)}\|)^{2^l} . \tag{5.56}$$

As soon as the iteration achieves the point where condition (5.54) is fulfilled, it has super-exponential convergence. This explains the rapid convergence observed from numerical simulations for the beamforming problem in [33]. Typically, only a few steps are required.

5.3 Convergence of the Fixed Point Iteration

The super-linear convergence of the matrix iteration (Theorem 5.10) can generally not be achieved by the fixed point iteration. The fixed point iteration was shown to have geometric convergence [74].

In order to illustrate the convergence bahavior of the fixed point iteration, consider the example of a simple linear interference function

$$\mathcal{J}_k(\boldsymbol{p}) = [\boldsymbol{V}\boldsymbol{p} + \boldsymbol{n}]_k, \quad k = 1, 2, \ldots, K , \tag{5.57}$$

with a fixed coupling matrix $\boldsymbol{V} \geq 0$. For this special model, the matrix iteration (5.9) does even converge in a single step. The optimizer $\boldsymbol{p}^*$ is simply found by solving the system of equations (5.8). Thus, it can be expected that also the fixed point iteration performs well for this model. However, the following analysis shows that only linear convergence is achieved.

5.3.1 Linear Convergence

Using the linear interference functions (5.57) and an initialization $\boldsymbol{p}^{(0)}$, step $n+1$ of the fixed point iteration (5.23) yields

$$\begin{aligned}\boldsymbol{p}^{(n+1)} &= (\boldsymbol{\Gamma V})^{n+1}\boldsymbol{p}^{(0)} + \sum_{l=0}^{n}(\boldsymbol{\Gamma V})^{l}\boldsymbol{\Gamma n} \\ &= (\boldsymbol{\Gamma V})^{n+1}\boldsymbol{p}^{(0)} + (\boldsymbol{I} - \boldsymbol{\Gamma V})^{-1}\boldsymbol{\Gamma n} \\ &\quad - (\boldsymbol{\Gamma V})^{n+1}(\boldsymbol{I} - \boldsymbol{\Gamma V})^{-1}\boldsymbol{\Gamma n}\,. \end{aligned} \tag{5.58}$$

Since $\boldsymbol{p}^* = (\boldsymbol{I} - \boldsymbol{\Gamma V})^{-1}\boldsymbol{\Gamma n}$, we have

$$\boldsymbol{p}^{(n+1)} - \boldsymbol{p}^* = (\boldsymbol{\Gamma V})^{n+1}\big(\boldsymbol{p}^{(0)} - \boldsymbol{p}^*\big)\,. \tag{5.59}$$

With $\boldsymbol{p}^{(n+1)} = \boldsymbol{\Gamma V p}^{(n)} + \boldsymbol{\Gamma n}$ and (5.59), we obtain

$$\begin{aligned}\boldsymbol{p}^{(n)} - \boldsymbol{p}^* &= (\boldsymbol{\Gamma V})^{-1}(\boldsymbol{p}^{(n+1)} - \boldsymbol{\Gamma V p}^* - \boldsymbol{\Gamma n}) \\ &= (\boldsymbol{\Gamma V})^{-1}(\boldsymbol{p}^{(n+1)} - \boldsymbol{p}^*) \\ &= (\boldsymbol{\Gamma V})^{n}\big(\boldsymbol{p}^{(0)} - \boldsymbol{p}^*\big)\,. \end{aligned} \tag{5.60}$$

It can be observed from (5.60) that

$$\boldsymbol{p}^{(n+1)} - \boldsymbol{p}^* = \boldsymbol{\Gamma V}(\boldsymbol{p}^{(n)} - \boldsymbol{p}^*)\,. \tag{5.61}$$

Relation (5.61) can be used to show the connection with the spectral radius. We have

$$\begin{aligned}\sup_{\boldsymbol{x}>0, \|\boldsymbol{x}\|_1 = 1} \frac{\boldsymbol{x}^T(\boldsymbol{p}^{(n+1)} - \boldsymbol{p}^*)}{\boldsymbol{x}^T(\boldsymbol{p}^{(n)} - \boldsymbol{p}^*)} &= \sup_{\boldsymbol{x}>0} \frac{\boldsymbol{x}^T\boldsymbol{\Gamma V}(\boldsymbol{p}^{(n)} - \boldsymbol{p}^*)}{\boldsymbol{x}^T(\boldsymbol{p}^{(n)} - \boldsymbol{p}^*)} \\ &\geq \sup_{\boldsymbol{x}>0}\inf_{\boldsymbol{y}>0} \frac{\boldsymbol{x}^T\boldsymbol{\Gamma V y}}{\boldsymbol{x}^T\boldsymbol{y}} = \rho(\boldsymbol{\Gamma V})\,. \end{aligned} \tag{5.62}$$

Since $\sup_{\boldsymbol{x}>0} \boldsymbol{x}^T\boldsymbol{a}/\boldsymbol{x}^T\boldsymbol{b} = \max_k [\boldsymbol{a}]_k/[\boldsymbol{b}]_k$ for $\boldsymbol{a}, \boldsymbol{b} > 0$ (see Lemma A.12 in the Appendix A.7), we have

$$\sup_{\boldsymbol{x}>0} \frac{\boldsymbol{x}^T(\boldsymbol{p}^{(n+1)} - \boldsymbol{p}^*)}{\boldsymbol{x}^T(\boldsymbol{p}^{(n)} - \boldsymbol{p}^*)} = \max_k \frac{p_k^{(n+1)} - p_k^*}{p_k^{(n)} - p_k^*} \geq \rho(\boldsymbol{\Gamma V})\,. \tag{5.63}$$

We have $\boldsymbol{p}^* \leq \boldsymbol{p}^{(n+1)} \leq \boldsymbol{p}^{(n)}$, thus

$$\boldsymbol{x}^T(\boldsymbol{p}^{(n+1)} - \boldsymbol{p}^*) = \|\boldsymbol{p}^{(n+1)} - \boldsymbol{p}^*\|_{\boldsymbol{x}}$$

is a norm, so

$$\sup_{\boldsymbol{x}>0} \frac{\|\boldsymbol{p}^{(n+1)} - \boldsymbol{p}^*\|_{\boldsymbol{x}}}{\|\boldsymbol{p}^{(n)} - \boldsymbol{p}^*\|_{\boldsymbol{x}}} \geq \rho(\boldsymbol{\Gamma V}) . \tag{5.64}$$

This "worst case" characterization shows that the relative mismatch is always bounded by the spectral radius $\rho(\boldsymbol{\Gamma V})$.

The convergence behavior can be further specified.

Theorem 5.12. *Consider the linear interference model (5.57). If each column of* $\boldsymbol{V}$ *contains at least one non-zero entry, then the fixed point iteration can only have linear convergence.*

Proof. With (5.61) we have

$$\begin{aligned}
\|\boldsymbol{p}^{(n+1)} - \boldsymbol{p}^*\|_1 &= \sum_{k=1}^{K_u} (p_k^{(n+1)} - p_k^*) \\
&= \sum_{l=1}^{K_u} \Big(\sum_{k=1}^{K_u} \gamma_k V_{kl} \Big) (p_l^{(n)} - p_l^*) \\
&= \sum_{l=1}^{K_u} c_l (p_l^{(n)} - p_l^*) \\
&\geq (\min_l c_l) \cdot \|\boldsymbol{p}^{(n)} - \boldsymbol{p}^*\|_1 ,
\end{aligned} \tag{5.65}$$

where $c_l = \Big(\sum_{k=1}^{K_u} \gamma_k V_{kl} \Big)$ is constant. From (5.65) we have

$$\frac{\|\boldsymbol{p}^{(n+1)} - \boldsymbol{p}^*\|_1}{\|\boldsymbol{p}^{(n)} - \boldsymbol{p}^*\|_1} \geq (\min_l c_l) > 0 . \tag{5.66}$$

This inequality is strict because of the assumed structure of $\boldsymbol{V}$, i.e., each user causes interference to at least one other user. Thus, even for $n \to \infty$, the ratio in (5.66) is always lower bounded by a positive constant, which shows linear convergence for the ℓ_1 norm. All norms on finite-dimensional vector spaces are equivalent, so the result extends to other norms as well. □

Theorem 5.12 shows for the linear interference model $\boldsymbol{Vp} + \boldsymbol{n}$, that the fixed point iteration cannot achieve the same superlinear convergence as in (5.45) (we exclude trivial cases, like $\boldsymbol{V} = \boldsymbol{0}$). The linear interference model is not a worst-case scenario, and the observed convergence behavior is typical for more complicated interference models as well. As an example, we will analyze an interference function with adaptive beamforming in the next section.

5.3.2 Geometrical Interpretation for the 2-User Beamforming Case

Consider the beamforming scenario discussed in Subsection 1.4.2. The average interference (normalized by the useful power) observed by the kth user is

$$\mathcal{J}_k(\boldsymbol{p}) = \min_{\|\boldsymbol{u}_k\|=1} \frac{\boldsymbol{u}_k^H \big(\sum_{l\neq k} p_l \boldsymbol{R}_l + \sigma_n^2 \boldsymbol{I}\big) \boldsymbol{u}_k}{\boldsymbol{u}_k^H \boldsymbol{R}_k \boldsymbol{u}_k} . \tag{5.67}$$

The function (5.67) is a concave standard interference function. The receive strategy is the beamforming vector $\boldsymbol{u}_k$. In this case, "receive strategy" means a choice between filter coefficients from a compact set (since $\|\boldsymbol{u}_k\| = 1$).

The beamforming model (5.67) was studied in [26, 33, 64, 83], where iterative algorithms were proposed. The strategies [64, 83] can be understood as special cases of the fixed point iteration (5.24), which is again a special case of (5.23). The algorithm [33] is a special case of the matrix-based iteration (5.9). The superlinear convergence of the matrix iteration explains the rapid convergence that was observed in [33].

Now, consider the special case $K = 2$. The covariance matrices $\boldsymbol{R}_k$ are assumed to have full rank. In this case, the users are mutually coupled by interference functions

$$\mathcal{J}_1(p_2) = \min_{\|\boldsymbol{u}\|=1} \frac{\boldsymbol{u}^H (p_2 \boldsymbol{R}_2 + \sigma_n^2 \boldsymbol{I}) \boldsymbol{u}}{\boldsymbol{u}^H \boldsymbol{R}_1 \boldsymbol{u}}$$

and

$$\mathcal{J}_2(p_1) = \min_{\|\boldsymbol{u}\|=1} \frac{\boldsymbol{u}^H (p_1 \boldsymbol{R}_1 + \sigma_n^2 \boldsymbol{I}) \boldsymbol{u}}{\boldsymbol{u}^H \boldsymbol{R}_2 \boldsymbol{u}} ,$$

It is known (see e.g. [33] and the references therein) that SINR targets γ_1 and γ_2 are jointly achievable iff $\rho < 1$, where

$$\rho = \inf_{\boldsymbol{p}>0} \max_k \frac{\gamma_k \mathcal{J}_k(\boldsymbol{p})}{p_k} = \sqrt{\gamma_1 \gamma_2 \cdot \frac{\lambda_{\min}}{\lambda_{\max}}} .$$

Here, $\lambda_{\max}$ and $\lambda_{\min}$ are the maximum and minimum eigenvalue of the matrix $\boldsymbol{R}_1^{-1} \boldsymbol{R}_2$. Thus, mutual interference depends both on the targets γ_k, and on the eigenvalue spread, which becomes larger if the channels $\boldsymbol{R}_1$ and $\boldsymbol{R}_2$ become more "distinctive". If both users use the same channel, i.e., $\boldsymbol{R}_1 = \boldsymbol{R}_2$, then $\rho = \sqrt{\gamma_1 \gamma_2}$. In this extreme case, the targets can only be supported if $\gamma_1 \gamma_2 < 1$. Otherwise, beamforming helps separating the users.

In order to illustrate the effects, consider two randomly chosen covariance matrices $\boldsymbol{R}_1$ and $\boldsymbol{R}_2$. By varying the targets γ_k, we can influence how close the scenario is to infeasibility. Choosing the spectral radius ρ between 0 and 1, we obtain different convergence behaviors for the fixed point iteration.

From Lemma 5.1 we know that (5.7) has a unique optimizer $\boldsymbol{p}^*$, which is characterized by the two following equations.

$$p_1^* = \gamma_1 \mathcal{J}_1(p_2^*) \tag{5.68}$$
$$p_2^* = \gamma_2 \mathcal{J}_2(p_1^*) \ . \tag{5.69}$$

In Fig. 5.2 and 5.3, the function $\gamma_1 \mathcal{J}_1(p_2)$ is plotted over the y-axis and $\gamma_2 \mathcal{J}_2(p_1)$ is plotted over the x-axis. Since the optimal p_1^* and p_2^* are simultaneously connected by (5.68) and (5.69), the optimum $\boldsymbol{p}^*$ is characterized by the unique intersection of both curves.

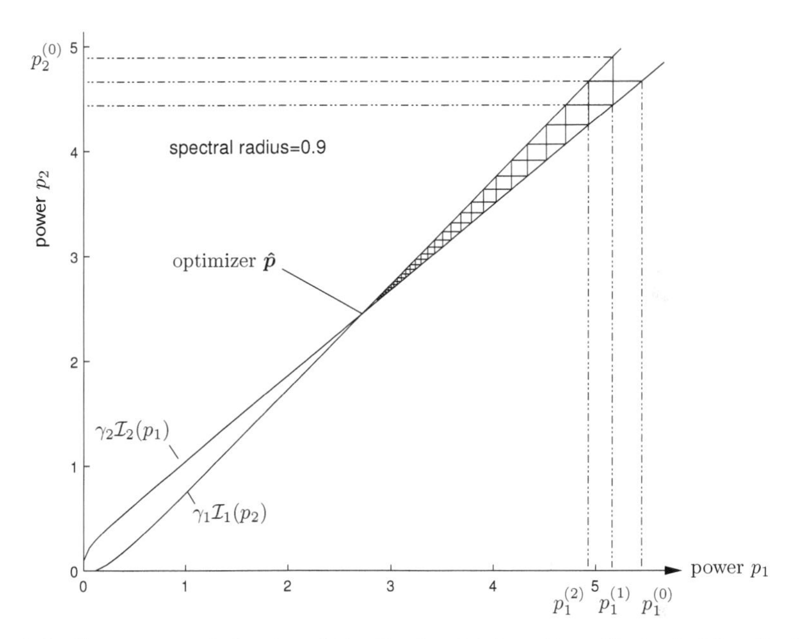

Fig. 5.2. Convergence behavior of the fixed point iteration, illustrated for the first user. The spectral radius is 0.9, i.e., the system is close to infeasibility.

It can be observed that the number of iterations depends on the opening angle between the curves $\gamma_1 \mathcal{J}_1(p_2)$ and $\gamma_2 \mathcal{J}_2(p_1)$, which are monotone increasing. For $p_1 \to \infty$ and $p_2 \to \infty$ we have

$$\gamma_1 \mathcal{J}_1(p_2) = \gamma_1 \lambda_{\min} \cdot p_2$$
$$\gamma_2 \mathcal{J}_2(p_1) = \gamma_2 \lambda_{\max} \cdot p_1 \ .$$

The lines intersect only if $\gamma_1 \gamma_2 \lambda_{\min} / \lambda_{\max} < 1$. Otherwise, no solution exists and the problem is infeasible. If $\gamma_1 \gamma_2 \lambda_{\min} / \lambda_{\max} \lessapprox 1$, then the lines intersect in an acute angle. Which means that many iterations are required in order to achieve the optimum (see Fig. 5.2). The number of iterations is not bounded and can tend to infinity. The angle becomes large if $\gamma_1 \gamma_2 \ll \lambda_{\max} / \lambda_{\min}$. In this case, only a few iterations are required (see Fig. 5.3). This illustrates how the convergence behavior of the fixed point iteration is connected with the spectral

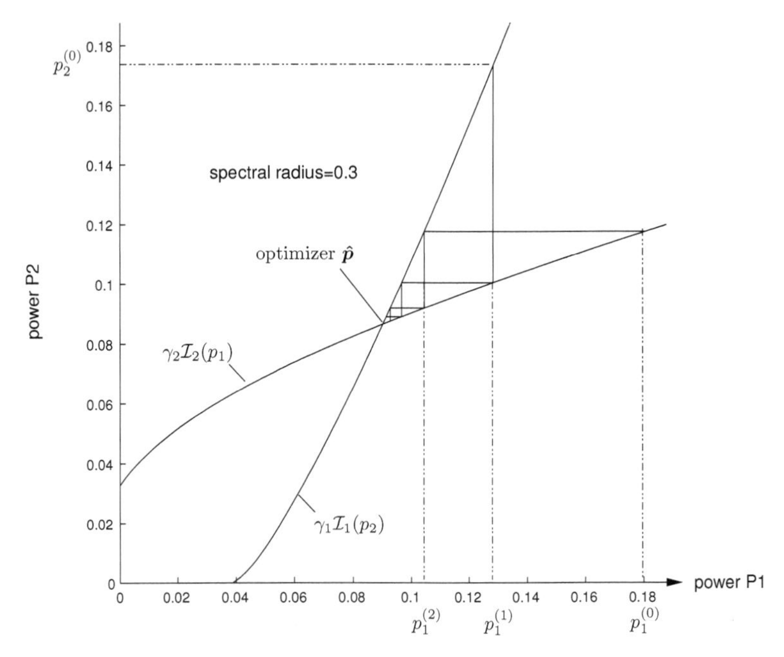

Fig. 5.3. Convergence behavior of the fixed point iteration, illustrated for the first user. The spectral radius is 0.3, i.e., the system is stable and far from infeasibility.

radius. This behavior was already observed from numerical simulations in the literature (see e.g. [126]).

In Section 5.3 it was shown for the linear interference function (5.57) that the fixed point iteration can generally not achieve super-linear convergence. Using the above geometrical illustration for the 2-user case, we are now able to show that this behavior also holds for the case when the receive strategies are chosen adaptively.

The convergence behavior near the fixed point is illustrated in Fig. 5.4 for $K = 2$. The curves $\mathcal{J}_1(p_2)$ and $\mathcal{J}_2(p_1)$ can be bounded by lines. The angle between the lines determines the convergence of the iteration. If we would update the powers with respect to these lines, then we improve the convergence. So the convergence defined by the lines can be seen as an upper bound of the actual convergence of the fixed point iteration. It can be observed that even for this upper bound, it is generally not possible to achieve super-linear convergence. Thus, it can be concluded that the behavior shown for linear interference functions in Section 5.3 is also typical for interference functions with an adaptive receive strategy.

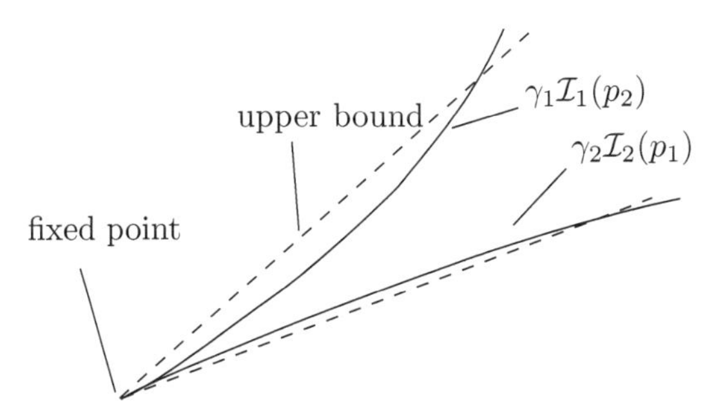

Fig. 5.4. The upper bound on the opening angle illustrates that the fixed point iteration does generally not achieve super-linear convergence

5.4 Worst-Case Interference and Robust Designs

In this section we will discuss the power minimization problem (5.1) under the assumption of *convex* standard interference functions $\mathcal{J}_1, \ldots, \mathcal{J}_{K_u}$. Similar to the concave case, the problem can be solved with super-linear convergence, as shown in [9].

From Theorem 3.62 we know that there exist convex compact downward-comprehensive sets $\mathcal{V}_1, \ldots, \mathcal{V}_{K_u} \subset \mathbb{R}_+^{K_u+1}$ such that

$$\mathcal{J}_k(\boldsymbol{p}) = \max_{\boldsymbol{v}' \in \mathcal{V}_k} \Big(\sum_{l \in \mathcal{K}_u} v'_l p_l + v'_{K_u+1} \Big) . \tag{5.70}$$

Following the same reasoning as in the beginning of Chapter 5, the interference functions have the following representation.

$$\mathcal{J}_k(\boldsymbol{p}) = \max_{z_k \in \mathcal{Z}_k} [\boldsymbol{V}(z)\boldsymbol{p} + \boldsymbol{n}(z)]_k, \quad k \in \mathcal{K}_u . \tag{5.71}$$

Here, $\boldsymbol{V}(z)$ is the interference coupling matrix and $\boldsymbol{n}(z)$ is the effective noise vector. Both depend on a parameter z, which is discussed in Subsection 1.4.7. The interference function models the worst-case interference of user k.

5.4.1 Matrix Iteration for Convex Interference Functions

The matrix iteration for the convex case is as follows.

$$\boldsymbol{p}^{(n+1)} = \big(\boldsymbol{I} - \boldsymbol{\Gamma}\boldsymbol{V}(z^{(n)})\big)^{-1}\boldsymbol{\Gamma}\boldsymbol{n}(z^{(n)}) \tag{5.72}$$

$$\text{with} \quad z_k^{(n)} = \arg\max_{z_k \in \mathcal{Z}_k} [\boldsymbol{V}(z)\boldsymbol{p}^{(n)} + \boldsymbol{n}(z)]_k \quad \forall k . \tag{5.73}$$

This algorithm has a similar structure as the iterative receiver optimization strategy from Section 5.1. However, there are also major differences. The iteration (5.10) aims at minimizing the interference and requires a feasible initialization, whereas the iteration (5.72) can always be initialized with $\boldsymbol{p}^{(0)} = \mathbf{0}$.

In the remainder of this Subsection, we prove that the iteration (5.72) converges to the unique global optimum of the power minimization problem (5.1).

Theorem 5.13. *Assume that the power minimization problem (5.1) has an optimizer $\mathbf{p}^*$, then the iteration (5.72), with an initialization $\mathbf{p}^{(0)} \leq \mathbf{p}^*$, is component-wise monotone increasing and upper-bounded by $\mathbf{p}^*$, i.e.,*

$$\boldsymbol{p}^{(n)} \leq \boldsymbol{p}^{(n+1)} \leq \boldsymbol{p}^*, \quad \forall n . \tag{5.74}$$

Proof. For an arbitrary step n, with parameter $z^{(n)}$, consider the function

$$f_n(\boldsymbol{p}) = \boldsymbol{\Gamma V}(z^{(n)})\boldsymbol{p} + \boldsymbol{\Gamma n}(z^{(n)}) .$$

Applying f_n recursively m times, we have

$$f_n^m(\boldsymbol{p}) = \underbrace{f_n(\dots f_n}_{m \text{ times}}(\boldsymbol{p})) .$$

Since $\boldsymbol{\Gamma V}(z^{(0)}) \geq 0$, the inequality $\boldsymbol{p}^{(0)} \leq \boldsymbol{p}^*$ implies

$$\begin{aligned} f_0^1(\boldsymbol{p}^{(0)}) &= \boldsymbol{\Gamma V}(z^{(0)})\boldsymbol{p}^{(0)} + \boldsymbol{\Gamma n}(z^{(0)}) \\ &\leq \boldsymbol{\Gamma V}(z^{(0)})\boldsymbol{p}^* + \boldsymbol{\Gamma n}(z^{(0)}) \\ &\leq \boldsymbol{\Gamma V}(z^*)\boldsymbol{p}^* + \boldsymbol{\Gamma n}(z^*) = \boldsymbol{p}^* , \end{aligned}$$

where the last inequality follows from (5.73). Likewise, it can be shown that $f_0^m(\boldsymbol{p}^{(0)}) \leq \boldsymbol{p}^*$ implies $f_0^{m+1}(\boldsymbol{p}^{(0)}) \leq \boldsymbol{p}^*$. Thus,

$$\lim_{m \to \infty} f_0^m(\boldsymbol{p}^{(0)}) = \boldsymbol{p}^{(1)} \leq \boldsymbol{p}^* .$$

Since $\boldsymbol{p}^{(0)} \leq \boldsymbol{p}^*$, the fixed point iteration (5.2) is monotone increasing, thus

$$\boldsymbol{p}^{(0)} \leq \boldsymbol{\Gamma} \mathcal{J}(\boldsymbol{p}^{(0)}) = \boldsymbol{\Gamma V}(z^{(0)})\boldsymbol{p}^{(0)} + \boldsymbol{\Gamma n}(z^{(0)}) .$$

Solving for $\boldsymbol{p}^{(0)}$ leads to

$$\boldsymbol{p}^{(0)} \leq \big(\boldsymbol{I} - \boldsymbol{\Gamma V}(z^{(0)})\big)^{-1} \boldsymbol{\Gamma n}(z^{(0)}) = \boldsymbol{p}^{(1)} .$$

The inverse exists because of the assumption $\rho^* < 1$. Thus,

$$\boldsymbol{p}^{(0)} \leq \boldsymbol{p}^{(1)} \leq \boldsymbol{p}^* . \tag{5.75}$$

Now, assume that $\boldsymbol{p}^{(n)} \leq \boldsymbol{p}^*$ for the nth step. In analogy to the above reasoning, it can be shown that

$$\lim_{m \to \infty} f_n^m(\boldsymbol{p}^{(n)}) = \boldsymbol{p}^{(n+1)} \leq \boldsymbol{p}^* . \tag{5.76}$$

Since $z^{(n)}$ maximizes the interference for given $\boldsymbol{p}^{(n)}$, we have

$$\begin{aligned}\boldsymbol{p}^{(n)} &= \boldsymbol{\Gamma V}(z^{(n-1)})\boldsymbol{p}^{(n)} + \boldsymbol{\Gamma n}(z^{(n-1)}) \\ &\leq \boldsymbol{\Gamma V}(z^{(n)})\boldsymbol{p}^{(n)} + \boldsymbol{\Gamma n}(z^{(n)}) \,. \end{aligned} \tag{5.77}$$

Solving this inequality for $\boldsymbol{p}^{(n)}$, we obtain

$$\boldsymbol{p}^{(n)} \leq \big(\boldsymbol{I} - \boldsymbol{\Gamma V}(z^{(n)})\big)^{-1}\boldsymbol{\Gamma n}(z^{(n)}) = \boldsymbol{p}^{(n+1)} \,. \tag{5.78}$$

Again, the inverse exists because of the assumption $\rho^* < 1$. Combining (5.76) and (5.78), we have

$$\boldsymbol{p}^{(n)} \leq \boldsymbol{p}^{(n+1)} \leq \boldsymbol{p}^* \,. \tag{5.79}$$

With (5.75) and (5.79), we have a monotone increasing sequence which is bounded by $\boldsymbol{p}^*$. □

Theorem 5.14. *Let $\mathbf{p}^{(n)}$ be the sequence obtained by the proposed iteration (5.72) and $\bar{\mathbf{p}}^{(n)}$ is obtained from the fixed point iteration (5.23). Starting with the same initialization $\mathbf{p}^{(0)} \leq \mathbf{p}^*$, we have*

$$\bar{\boldsymbol{p}}^{(n)} \leq \boldsymbol{p}^{(n)}, \quad \textit{for all } n. \tag{5.80}$$

That is, $\mathbf{p}^{(n)}$ is lower bounded by the fixed point iteration and both sequences converge to the unique optimizer of the power minimization problem (5.1).

Proof. Starting both iterations with the same $\boldsymbol{p}^{(0)} \leq \boldsymbol{p}^*$, we have

$$\begin{aligned}\bar{\boldsymbol{p}}^{(1)} &= \boldsymbol{\Gamma V}(z^{(0)})\boldsymbol{p}^{(0)} + \boldsymbol{\Gamma n}(z^{(0)}) \\ &\leq \boldsymbol{\Gamma V}(z^{(0)})\boldsymbol{p}^{(1)} + \boldsymbol{\Gamma n}(z^{(0)}) = \boldsymbol{p}^{(1)} \,, \end{aligned}$$

where the inequality follows from (5.74). Thus, $\bar{\boldsymbol{p}}^{(1)} \leq \boldsymbol{p}^{(1)}$. Because of monotonicity, $\bar{\boldsymbol{p}}^{(n)} \leq \boldsymbol{p}^{(n)}$ implies

$$\bar{\boldsymbol{p}}^{(n+1)} = \boldsymbol{\Gamma}\boldsymbol{\mathcal{J}}(\bar{\boldsymbol{p}}^{(n)}) \leq \boldsymbol{\Gamma}\boldsymbol{\mathcal{J}}(\boldsymbol{p}^{(n)}) \,. \tag{5.81}$$

Since $\boldsymbol{\Gamma V}(z) \geq 0$, monotonicity (5.74) leads to

$$\begin{aligned}\boldsymbol{\Gamma}\boldsymbol{\mathcal{J}}(\boldsymbol{p}^{(n)}) &= \boldsymbol{\Gamma V}(z^{(n)})\boldsymbol{p}^{(n)} + \boldsymbol{\Gamma n}(z^{(n)}) \\ &\leq \boldsymbol{\Gamma V}(z^{(n)})\boldsymbol{p}^{(n+1)} + \boldsymbol{\Gamma n}(z^{(n)}) = \boldsymbol{p}^{(n+1)}. \end{aligned} \tag{5.82}$$

By combining (5.81) and (5.82), it can be concluded that $\bar{\boldsymbol{p}}^{(n)} \leq \boldsymbol{p}^{(n)}$ implies $\bar{\boldsymbol{p}}^{(n+1)} \leq \boldsymbol{p}^{(n+1)}$. Thus, (5.80) follows by complete induction.

The sequence $\boldsymbol{p}^{(n)}$ is lower-bounded by $\bar{\boldsymbol{p}}^{(n)}$ and upper-bounded by the optimizer $\boldsymbol{p}^*$. Since $\bar{\boldsymbol{p}}^{(n)}$ converges to $\boldsymbol{p}^*$ [1], the proposed sequence $\boldsymbol{p}^{(n)}$ converges to $\boldsymbol{p}^*$ as well. □

5.4.2 Convergence Analysis

Having shown global convergence, we are now interested in the convergence behavior. One difficulty in studying convergence is that $\boldsymbol{p}^{(n+1)}$ is not directly linked to $\boldsymbol{p}^{(n)}$, but indirectly via $z^{(n)}$. Moreover, z need not be unique. That is, for any given $\boldsymbol{p}$, there is a set of optimal receive strategies

$$\mathcal{Z}(\boldsymbol{p}) = \{z : [\boldsymbol{V}(z)\boldsymbol{p} + \boldsymbol{n}(z)]_k = \mathcal{J}_k(\boldsymbol{p}), 1 \leq k \leq K\} .$$

The associated optimal coupling matrices are collected in the set

$$\mathcal{M}(\boldsymbol{p}) = \{\boldsymbol{V}(z) : z \in \mathcal{Z}(\boldsymbol{p})\} . \tag{5.83}$$

For each matrix $\boldsymbol{V} \in \mathcal{M}(\boldsymbol{p})$, there exists an optimal parameter z' such that $\boldsymbol{V} = \boldsymbol{V}(z')$.

In order to better understand the convergence behavior, it is helpful to consider the function

$$\boldsymbol{d}(\boldsymbol{p}) = \boldsymbol{p} - \boldsymbol{\Gamma}\boldsymbol{\mathcal{J}}(\boldsymbol{p}) . \tag{5.84}$$

The function $\boldsymbol{d}(\boldsymbol{p})$ is jointly concave as the sum of two concave monotone functions. We have $\boldsymbol{d}(\boldsymbol{p}) \leq 0$ for all $\boldsymbol{p} \leq \boldsymbol{p}^*$. From Lemma 5.1 we know that the optimizer $\boldsymbol{p}^*$ is completely characterized by $\boldsymbol{d}(\boldsymbol{p}^*) = 0$. Hence, the power minimization problem (5.1) can be interpreted as the search for the unique root of the concave function $\boldsymbol{d}(\boldsymbol{p})$. If $\boldsymbol{d}(\boldsymbol{p})$ is continuously differentiable, then Newton's Method can be applied, as illustrated in Fig. 5.5.

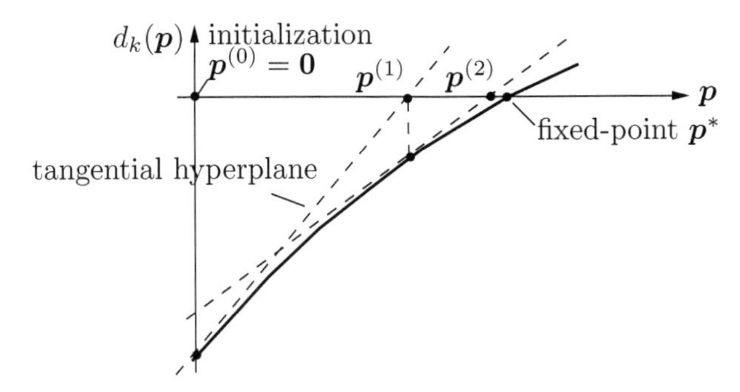

Fig. 5.5. *For convex interference functions, the optimum of the power minimization problem (5.1) is given as the unique root of the concave function* $\boldsymbol{d}(\boldsymbol{p})$. *Note that* $\boldsymbol{d}(\boldsymbol{p})$ *need not be smooth, thus a direct application of Newton's Method is not possible.*

In this case, the Jacobi matrix, which contains the partial derivatives, is given as

$$\nabla \boldsymbol{d}(\boldsymbol{p}) = \boldsymbol{I} - \boldsymbol{\Gamma}\boldsymbol{V}\big(z(\boldsymbol{p})\big) . \tag{5.85}$$

The representation (5.85) assumes that $z(\boldsymbol{p})$ is a non-ambiguous function of the power allocation $\boldsymbol{p}$. However, this need not be true if the set $\mathcal{Z}(\boldsymbol{p})$ contains more than one element. This possible ambiguity in the choice of z means that $\boldsymbol{d}(\boldsymbol{p})$ is generally non-smooth. Nevertheless, super-linear can be shown.

Let D_F be the set on which $\boldsymbol{d}$ is differentiable, and $\nabla \boldsymbol{d}(\boldsymbol{p})$ is the Jacobi matrix for $\boldsymbol{p} \in D_F$, as defined in (5.85). Since $\boldsymbol{d}(\boldsymbol{p})$ is locally Lipschitz-continuous and directionally differentiable (see Lemma 5.7), we know that it is also B-differentiable at the points of interest. The B-derivative $\partial_B \boldsymbol{d}(\boldsymbol{p})$ at the point $\boldsymbol{p}$ is defined as

$$\partial_B \boldsymbol{d}(\boldsymbol{p}) = \{\boldsymbol{A} \in \mathbb{R}^{K \times K} : \text{there exists a sequence } \{\boldsymbol{p}_k\}_{k \in \mathbb{N}}, \text{ with } \boldsymbol{p}_k \in D_F, \boldsymbol{p}_k \to \boldsymbol{p} \text{ and } \boldsymbol{A} = \lim_{k \to \infty} \nabla \boldsymbol{d}(\boldsymbol{p}_k)\} .$$

Notice, that the set $\partial_B \boldsymbol{d}(\boldsymbol{p})$ can contain more than one element. As an example, consider the function $F(x) = |x|$, for which $\partial_B F(0) = \{-1, 1\}$.

Clarke's generalized Jacobian $\partial \boldsymbol{d}(\boldsymbol{p})$ is defined as

$$\partial \boldsymbol{d}(\boldsymbol{p}) = conv\big(\partial_B \boldsymbol{d}(\boldsymbol{p})\big) , \tag{5.86}$$

which is the convex hull of the set given by the B-derivative. For the above example function $F(x)$ at the point $x = 0$, this is the interval $\partial F(0) = [-1, 1]$.

Based on this definition of Clarke's generalized Jacobian, a generalized Newton iteration can be defined. Applied to our function $\boldsymbol{d}(\boldsymbol{p})$, the generalized form can be written as

$$\boldsymbol{p}^{(n+1)} = \boldsymbol{p}^{(n)} - \boldsymbol{V}_n^{-1} \boldsymbol{d}(\boldsymbol{p}^{(n)}) , \quad \boldsymbol{V}_n \in \partial \boldsymbol{d}(\boldsymbol{p}^{(n)}) . \tag{5.87}$$

It can be observed that $\boldsymbol{V}_n$ needs to be invertible in order to compute $\boldsymbol{p}^{(n+1)}$. This requirement is quite strong and difficult to verify.

In [127, 129] it was proposed

$$\boldsymbol{p}^{(n+1)} = \boldsymbol{p}^{(n)} - \boldsymbol{V}_n^{-1} \boldsymbol{d}(\boldsymbol{p}^{(n)}) , \quad \boldsymbol{V}_n \in \partial_B \boldsymbol{d}(\boldsymbol{p}^{(n)}) . \tag{5.88}$$

The difference to (5.87) is that $\boldsymbol{V}_n$ is chosen from the set given by the B-derivative. So only elements of this set need to be invertible. The local convergence behavior of the iteration (5.88) was studied in [127, 129]. For the analysis it is required that $\boldsymbol{d}$ at a certain point $\hat{\boldsymbol{p}}$ is *strongly BD-regular* (B-derivative regular). This is fulfilled if all $\boldsymbol{V} \in \partial_B \boldsymbol{d}(\hat{\boldsymbol{p}})$ are nonsingular and thus invertible. If this is fulfilled, then the iteration (5.88) has super-linear convergence [127, 129]. But in general, $\partial_B \boldsymbol{d}(\hat{\boldsymbol{p}})$ can rarely be computed explicitly, and there is no practical test for BD-regularity.

Fortunately, we can exploit the special properties of the problem at hand.

- The sequence $\boldsymbol{p}^{(n)}$ is monotone and globally convergent, independent of the initialization $\boldsymbol{p}^{(0)} \leq \boldsymbol{p}^*$.
- The sequence $\boldsymbol{p}^{(n)}$ is lower-bounded by the fixed point iteration.

- Monotonicity implies directions

$$\begin{aligned} \boldsymbol{h}_1^{(n)} &= \boldsymbol{p}^{(n)} - \boldsymbol{p}^{(n+1)} \geq 0 \\ \boldsymbol{h}_2^{(n)} &= \boldsymbol{p}^{(n+1)} - \boldsymbol{p}^{(n)} \leq 0 \,. \end{aligned}$$

Thus the analysis of the convergence behavior can be restricted to these special conditions. We only need to check the invertibility of the matrices $\boldsymbol{V} = \boldsymbol{I} - \boldsymbol{\Gamma} \boldsymbol{A}$, with $\boldsymbol{A} \in \mathcal{M}(\boldsymbol{p})$. For given $\boldsymbol{V}_n \in \mathcal{M}(\boldsymbol{p}^{(n)})$, the derivative is $\boldsymbol{I} - \boldsymbol{\Gamma} \boldsymbol{V}_n$. We have the following iteration

$$\boldsymbol{p}^{(n+1)} = \boldsymbol{p}^{(n)} - (\boldsymbol{I} - \boldsymbol{\Gamma} \boldsymbol{V}_n)^{-1} \boldsymbol{d}(\boldsymbol{p}^{(n)}), \quad \boldsymbol{V}_n \in \mathcal{M}(\boldsymbol{p}^{(n)}) \,. \tag{5.89}$$

It can be verified that the iteration (5.89) is equivalent to the proposed iteration (5.72). Since we assume that the targets $\boldsymbol{\Gamma}$ are achievable, i.e., $\rho^* < 1$, the inverse $(\boldsymbol{I} - \boldsymbol{\Gamma} \boldsymbol{V}_n)^{-1}$ always exists. This can be shown by decomposing the inverse matrix in a convergent Neumann series with non-negative terms.

We are now in the position to characterize the convergence behavior of the proposed iteration.

Theorem 5.15. *Let $\mathbf{p}^{(0)} \leq \mathbf{p}^*$ be an arbitrary initialization, then the iteration (5.72) fulfills*

$$\lim_{n \to \infty} \frac{\|\boldsymbol{p}^{(n+1)} - \boldsymbol{p}^*\|_1}{\|\boldsymbol{p}^{(n)} - \boldsymbol{p}^*\|_1} = 0 \tag{5.90}$$

$$\lim_{n \to \infty} \frac{\|\boldsymbol{d}(\boldsymbol{p}^{(n+1)})\|_1}{\|\boldsymbol{d}(\boldsymbol{p}^{(n)})\|_1} = 0 \,. \tag{5.91}$$

That is, the sequences $\mathbf{p}^{(n)}$ and $\mathbf{d}(\mathbf{p}^{(n)})$ have super-linear convergence.

Proof. The proof is similar to the proof of Theorem 5.10, based on results from [127, 129]. □

An even better convergence behavior can be shown for certain semi-continuous functions.

Theorem 5.16. *Let $\mathcal{J}$ be semi-continuous of degree 2 at point $\mathbf{p}^*$, then there exists a constant C_1 such that*

$$\|\boldsymbol{p}^{(n+1)} - \boldsymbol{p}^*\|_1 \leq C_1 (\|\boldsymbol{p}^{(n)} - \boldsymbol{p}^*\|_1)^2$$

for all $n \in \mathbb{N}$.

Proof. The proof is similar to the proof of Theorem 5.11. □

6

Weighted SIR Balancing

In this chapter we discuss the SIR balancing problem

$$C(\gamma) = \inf_{\boldsymbol{p}>0} \Big(\max_{1 \leq k \leq K} \frac{\gamma_k \mathcal{I}_k(\boldsymbol{p})}{p_k} \Big) , \tag{6.1}$$

which was already introduced in Subsection 1.4.5. We assume general *concave* interference functions $\mathcal{I}_1, \ldots, \mathcal{I}_K$. An extension to convex interference functions is straightforward and will not be discussed here. Under an additional assumption on the interference coupling (as defined later), inf can be replaced by min. Two optimization strategies from [7] will be discussed in the following.

While the algorithm from Chapter 5 achieves a point in the interior of the QoS regions, the objective of (6.1) is to minimize the largest *inverse* signal-to-interference ratio (SIR), weighted by some target values. This problem can be reformulated as the maximization of the smallest weighted SIR (see Appendix A.2), thus we refer to (6.1) as the *max-min SIR balancing* problem. For equal weights, this strategy is also known as *max-min fairness*.

Problem (6.1) is fundamental for the analysis and optimization of interference-coupled systems. The function $C(\gamma)$ characterizes the multi-user SIR region (see Subsection 1.4.5). By analyzing the min-max problem, we extend certain results of the linear Perron-Frobenius theory to the broader class of concave and convex functions. The optimum $C(\gamma)$ provides an indicator for the feasibility of SIR targets γ. If the infimum is attained, then (6.1) yields a solution $\boldsymbol{p}^*$ that balances the weighted SIR values at a common level. Then, $\boldsymbol{p}^*$ is a valid initialization of the matrix iteration (5.9) from Chapter 5. Solving (6.1) is also of interest, e.g., for the power minimization algorithm [1], which diverges if the chosen targets γ are infeasible. Infeasible scenarios can be detected by computing $C(\gamma)$, thus it can be used as a measure for congestion (see e.g. [130]).

Since the interference functions are concave, we can apply Theorem 3.23, which shows the existence of upward-comprehensive closed convex sets $\mathcal{V}_1, \ldots, \mathcal{V}_K \subset \mathbb{R}^K_+$, such that

M. Schubert, H. Boche, *Interference Calculus*, Foundations in Signal Processing, Communications and Networking 7,

$$\mathcal{I}_k(\boldsymbol{p}) = \min_{\boldsymbol{v} \in \mathcal{V}_k} \Big(\sum_{l \in \mathcal{K}} v_l p_l \Big) .$$

As motivated at the beginning of Chapter 5, we introduce parameters $z = (z_1, \dots, z_K)$, which are referred to as "receive strategies", and a coupling matrix

$$\boldsymbol{V}(z) = \begin{bmatrix} \boldsymbol{v}_1(z_1)^T \\ \vdots \\ \boldsymbol{v}_K(z_K)^T \end{bmatrix} .$$

We assume that $\mathcal{Z}_k$ is the closed and bounded strategy space of user k, and $\mathcal{Z} = \mathcal{Z}_1 \times \cdots \times \mathcal{Z}_K$. The function $\boldsymbol{v}_k(z_k)$ is assumed to be continuous. These sets $\mathcal{Z}_k$ can also be discrete, e.g. when we are interested in the best choice between several receivers, as in [36,37]. For the special case of adaptive beamforming, the set $\mathcal{Z}_k$ is the complex unit sphere.

With this parametrization, the interference functions can be rewritten as follows.

$$\mathcal{I}_k(\boldsymbol{p}) = \min_{z_k \in \mathcal{Z}_k} \boldsymbol{p}^T \boldsymbol{v}_k(z_k), \quad k \in \mathcal{K} . \tag{6.2}$$

A globally convergent algorithm for solving the min-max problem (6.1) was derived in [8]. The results will be discussed in the following. To this end, we assume that $\boldsymbol{V}(z)$ is *irreducible* for every $z \in \mathcal{Z}$ (see Appendix A.1 for a definition of irreducibility). This assumption is indispensable for the model under consideration, as shown in [81]. The infimum (6.1) is only guaranteed to be attainable if $\boldsymbol{V}(z)$ is (block) irreducible. Roughly speaking, this means that all users are coupled by interference and the SIR is always well-defined. Without irreducibility, one interference term and its associated power component can tend to zero, thus causing numerical problems in the algorithm.

It was shown in [131], for interference functions of the form (6.2), that the max-min-SIR problem (6.1) can be reformulated as the *Perron root* minimization problem

$$C(\gamma) = \min_{z \in \mathcal{Z}} \rho\big(\boldsymbol{\Gamma} \boldsymbol{V}(z)\big) . \tag{6.3}$$

The spectral radius ρ is an indicator for feasbility (see Subsection 1.4.4). Since $\boldsymbol{\Gamma V}$ is non-negative irreducible, ρ is the maximum eigenvalue. It is referred to as the *Perron root.*

The matrix $\boldsymbol{V}$ is generally non-symmetric. Such eigenvalue optimization problems are considered as complicated [132]. But due to the special structure of the concave interference functions, the problem can be solved by a globally convergent algorithm. If $C(\gamma) \leq 1$, then there exists a receive strategy z and a power vector $\boldsymbol{p}^*$ which jointly achieve the targets γ.

6.1 The Max-Min Optimum

Max-min SIR balancing is a classical problem in power control theory, dating back to [45,48–50,52,54,61,133,134], where linear interference functions were investigated under the assumption of an irreducible coupling matrix. Based on this property, it was shown in [48] that the joint achievability of SIR values γ is completely characterized by the spectral radius $\rho(\boldsymbol{\Gamma V})$ (see Subsection 1.4.1 and the overview paper [46]). This line of work is closely connected with the Perron-Frobenius theory of non-negative irreducible matrices [56, 57]. The unique max-min optimizer (up to a scaling) is the right-hand principal eigenvector of the weighted coupling matrix $\boldsymbol{\Gamma V}$. The optimum $C(\gamma)$ is given by the Perron root $\rho(\boldsymbol{\Gamma V})$.

The linear case can be extended to adaptive receive or transmit strategies. An example is the problem of max-min SIR balancing for adaptive downlink beamforming [59,60,62,82,135]. Using a duality between uplink and downlink beamforming channels [61, 82], this problem can be solved by optimizing a an equivalent "virtual uplink channel" (see Subsection 1.4.6). This problem was studied in [8, 59, 60]. The beamforming algorithm from [8, 60] is in fact a special case of the PEV iteration that will be discussed later in Section 6.2.

In the remainder of this section we will discuss some general aspects of the min-max problem (6.1). We assume general concave interference functions of the form (6.2). This includes the beamforming scenario as a special case. Two algorithms for solving will be discussed later, in Sections (6.2) and (6.3).

6.1.1 Set of Optimal Receive Strategies

We say that a receive strategy $z = (z_1, \dots, z_K)$ is optimal if it solves (6.3). For a given γ, the set of optimal receive strategies is

$$\mathcal{Z}(\gamma) = \{z \in \mathcal{Z} : C(\gamma) = \rho\big(\boldsymbol{\Gamma V}(z)\big)\} \,. \tag{6.4}$$

The set $\mathcal{Z}(\gamma)$ is non-empty since $\mathcal{Z}$ is non-empty by assumption, so it is always possible to find an optimal strategy $\hat{z} \in \mathcal{Z}(\gamma)$. Since $\boldsymbol{V}(\hat{z})$ is irreducible by assumption, we know from the Perron-Frobenius theorem (see e.g. [48,56,57]), that there exists a principal eigenvector $\hat{\boldsymbol{p}} > 0$ such that

$$\boldsymbol{\Gamma V}(\hat{z})\hat{\boldsymbol{p}} = C(\gamma)\hat{\boldsymbol{p}} \,. \tag{6.5}$$

The next lemma shows that an optimal receive strategy minimizes the interference of each user. This fundamental property will be exploited later.

The principal eigenvector of a matrix $\boldsymbol{A}$, denoted by $\mathrm{pev}(\boldsymbol{A})$, is the eigenvector associated with the maximum eigenvalue. This vector can be arbitrarily scaled. In this book we agree on $\|\mathrm{pev}(\boldsymbol{A})\|_1 = 1$.

Lemma 6.1. *Assume that $\boldsymbol{V}(z)$ is irreducible for all $z \in \mathcal{Z}$, then for any $\hat{z} \in \mathcal{Z}(\gamma)$, with a principal eigenvector $\hat{\boldsymbol{p}} = \mathrm{pev}\big(\boldsymbol{\Gamma V}(\hat{z})\big)$, we have*

$$\min_{z_k \in \mathcal{Z}_k} \hat{\boldsymbol{p}}^T \boldsymbol{v}_k(z_k) = \hat{\boldsymbol{p}}^T \boldsymbol{v}_k(\hat{z}_k), \quad \forall k \in \mathcal{K} \,. \tag{6.6}$$

Proof. The proof is by contradiction. Suppose that there exists an index k_0 such that

$$\hat{\boldsymbol{p}}^T \boldsymbol{v}_{k_0}(\tilde{z}_{k_0}) = \min_{z_{k_0} \in \mathcal{Z}_{k_0}} \hat{\boldsymbol{p}}^T \boldsymbol{v}_{k_0}(z_{k_0}) < \hat{\boldsymbol{p}}^T \boldsymbol{v}_{k_0}(\hat{z}_{k_0}) . \tag{6.7}$$

That is,

$$\boldsymbol{\Gamma V}(\tilde{z})\hat{\boldsymbol{p}} \leq \boldsymbol{\Gamma V}(\hat{z})\hat{\boldsymbol{p}} = C(\gamma)\hat{\boldsymbol{p}} . \tag{6.8}$$

Since inequality (6.8) is strict for component k_0, it follows that $\hat{\boldsymbol{p}}$ is not the principal eigenvector of $\boldsymbol{\Gamma V}(\tilde{z})$. This eigenvector is unique, thus

$$\begin{aligned} \rho\big(\boldsymbol{\Gamma V}(\tilde{z})\big) &= \inf_{\boldsymbol{p}>0}\Big(\max_{k\in\mathcal{K}} \frac{[\boldsymbol{\Gamma V}(\tilde{z})\boldsymbol{p}]_k}{p_k}\Big) \\ &< \max_{k\in\mathcal{K}} \frac{[\boldsymbol{\Gamma V}(\tilde{z})\hat{\boldsymbol{p}}]_k}{\hat{p}_k} . \end{aligned} \tag{6.9}$$

Combining (6.8) and (6.9), we have

$$\rho\big(\boldsymbol{\Gamma V}(\tilde{z})\big) < \max_{k\in\mathcal{K}} \frac{[\boldsymbol{\Gamma V}(\hat{z})\hat{\boldsymbol{p}}]_k}{\hat{p}_k} = C(\gamma) . \tag{6.10}$$

This is a contradiction since $C(\gamma)$, as defined by (6.3), is the global minimum. Thus, (6.6) holds. □

Using the vector notation

$$\boldsymbol{\mathcal{I}}(\boldsymbol{p}) = [\mathcal{I}_1(\boldsymbol{p}), \ldots, \mathcal{I}_K(\boldsymbol{p})]^T \tag{6.11}$$

where $\mathcal{I}_k(\boldsymbol{p})$ is defined by (6.2), Lemma 6.1 implies

$$\boldsymbol{\Gamma \mathcal{I}}(\hat{\boldsymbol{p}}) = \boldsymbol{\Gamma V}(\hat{z})\hat{\boldsymbol{p}} = C(\gamma)\hat{\boldsymbol{p}} . \tag{6.12}$$

It can be observed that the eigenvector $\hat{\boldsymbol{p}} > 0$ is a fixed point of the function $\boldsymbol{\Gamma \mathcal{I}}(\hat{\boldsymbol{p}})/C(\gamma)$. This implies [2] that $\hat{\boldsymbol{p}}$ is an optimizer of the SIR balancing problem (6.1). Conversely, every $\hat{\boldsymbol{p}}$ solving (6.1) fulfills (6.12), with an optimal receive strategy $\hat{z}$.

In the following section it will be shown that such an optimal power allocation always exists. Even though there can be multiple optimal receive strategies $z \in \mathcal{Z}(\gamma)$, every strategy is associated with the same power vector. This behavior is due to the assumption of irreducibility, which is fundamental.

6.1.2 Existence and Uniqueness of the Optimal Power Allocation

For a given power allocation $\boldsymbol{p}$, the set

$$\mathcal{M}(\boldsymbol{p}) = \{\boldsymbol{V} : \boldsymbol{\mathcal{I}}(\boldsymbol{p}) = \boldsymbol{V}\boldsymbol{p}\} \tag{6.13}$$

contains all coupling matrices resulting from optimum receive strategies. Note, that different coupling matrices can lead to the same interference value $\boldsymbol{\mathcal{I}}(\boldsymbol{p})$, i.e., the representation is not unique.

In [2] it was shown for general interference functions characterized by axioms A1, A2, A3, that there always exists a vector $\boldsymbol{p}^* \geq \mathbf{0}$, $\boldsymbol{p}^* \neq \mathbf{0}$, such that

$$\boldsymbol{\Gamma}\boldsymbol{\mathcal{I}}(\boldsymbol{p}^*) = C(\boldsymbol{\gamma})\boldsymbol{p}^* . \tag{6.14}$$

That is, the vector $\boldsymbol{p}^*$ balances all values SIR_k/γ_k at a common level $C(\boldsymbol{\gamma})$. For the special case of interference functions (6.2), based on irreducible matrices, we even know that there is a positive fixed point $\boldsymbol{p}^* > 0$ fulfilling (6.14). Uniqueness of the fixed point, as stated by the following lemma, will be important for the convergence proofs in Sections 6.2 and 6.3. The result was shown in [81].

Lemma 6.2. *If $\boldsymbol{V}(z)$ is irreducible for all z, then problem (6.1) has a unique optimizer $\mathbf{p}^* > 0$ (unique up to an arbitrary scaling). Only $\mathbf{p}^*$ fulfills*

$$\boldsymbol{\Gamma}\boldsymbol{\mathcal{I}}(\boldsymbol{p}^*) = C(\boldsymbol{\gamma})\boldsymbol{p}^* . \tag{6.15}$$

All matrices $\boldsymbol{V}(z)$ with

$$\rho\big(\boldsymbol{\Gamma}\boldsymbol{V}(z)\big) = C(\boldsymbol{\gamma})$$

are contained in the set $\mathcal{M}(\boldsymbol{p}^*)$, thus all optimal matrices $\boldsymbol{\Gamma}\boldsymbol{V}(z)$, with $z \in \mathcal{Z}(\boldsymbol{\gamma})$, have the same principal right eigenvector $\boldsymbol{p}^*$.

6.1.3 Equivalence of Max-Min and Min-Max Balancing

We now compare the optimum $C(\boldsymbol{\gamma})$ of the min-max balancing problem (6.1), with the max-min optimum

$$c(\boldsymbol{\gamma}) = \sup_{\boldsymbol{p}>0}\Big(\min_{k\in\mathcal{K}} \frac{\gamma_k \mathcal{I}_k(\boldsymbol{p})}{p_k}\Big) . \tag{6.16}$$

That is, instead of minimizing the largest component, the objective is now to maximize the smallest component. In general, we have [131]

$$c(\boldsymbol{\gamma}) \leq C(\boldsymbol{\gamma}) . \tag{6.17}$$

Note, that (6.17) is not a simple consequence of Fan's minimax inequality since we do not only interchange the optimization order, but also the domain. Inequality (6.17) holds because of the special structure of the interference functions.

The next lemma shows equivalence of min-max and max-min balancing for the given interference model.

Lemma 6.3. *Let $\mathcal{I}_1, \ldots, \mathcal{I}_K$ be defined as in (6.2), based on irreducible coupling matrices, then*

$$\sup_{\boldsymbol{p}>0}\Big(\min_{k\in\mathcal{K}} \frac{\gamma_k \mathcal{I}_k(\boldsymbol{p})}{p_k}\Big) = \inf_{\boldsymbol{p}>0}\Big(\max_{k\in\mathcal{K}} \frac{\gamma_k \mathcal{I}_k(\boldsymbol{p})}{p_k}\Big) . \tag{6.18}$$

Proof. According to Lemma 6.2, there exists a $\tilde{\boldsymbol{p}} > 0$ such that

$$c(\gamma) = \sup_{\boldsymbol{p}>0}\Big(\min_{k\in\mathcal{K}} \frac{\gamma_k \mathcal{I}_k(\boldsymbol{p})}{p_k}\Big)$$
$$\geq \min_{1\leq k\leq K} \frac{\gamma_k \mathcal{I}_k(\tilde{\boldsymbol{p}})}{\tilde{p}_k} = C(\gamma) . \tag{6.19}$$

With (6.17), this must be satisfied with equality. □

Lemma 6.3 shows that both optimization problems are equivalent if all users in the system are "coupled by interference".

Equality (6.18) also holds for a fixed z, i.e.,

$$\sup_{\boldsymbol{p}>0}\Big(\min_{k\in\mathcal{K}} \frac{[\boldsymbol{\Gamma V}(z)\boldsymbol{p}]_k}{p_k}\Big) = \inf_{\boldsymbol{p}>0}\Big(\max_{k\in\mathcal{K}} \frac{[\boldsymbol{\Gamma V}(z)\boldsymbol{p}]_k}{p_k}\Big) . \tag{6.20}$$

Now, we can exploit (see e.g. [82])

$$\begin{matrix}(\inf)\\ \sup\\ {\boldsymbol{x}>0}\end{matrix} \frac{\boldsymbol{x}^T\boldsymbol{b}}{\boldsymbol{x}^T\boldsymbol{c}} = \begin{matrix}(\min)\\ \max\\ {1\leq k\leq K}\end{matrix} \frac{b_k}{c_k}, \quad \text{for any } \boldsymbol{b}, \boldsymbol{c} > 0 . \tag{6.21}$$

Applying (6.21) to (6.20), we obtain the following equivalent characterizations of the Perron root.

$$\rho\big(\boldsymbol{\Gamma V}(z)\big) = \inf_{\boldsymbol{p}>0}\Big(\sup_{\boldsymbol{x}>0} \frac{\boldsymbol{x}^T\boldsymbol{\Gamma V}(z)\boldsymbol{p}}{\boldsymbol{x}^T\boldsymbol{p}}\Big)$$
$$= \sup_{\boldsymbol{p}>0}\Big(\inf_{\boldsymbol{x}>0} \frac{\boldsymbol{x}^T\boldsymbol{\Gamma V}(z)\boldsymbol{p}}{\boldsymbol{x}^T\boldsymbol{p}}\Big) . \tag{6.22}$$

This representation will play an important role for the convergence proof in the next section.

6.2 Principal Eigenvector (PEV) Iteration

In this section we present an efficient iterative solution for the SIR balancing problem (6.1), under the assumption of interference functions (6.2) with irreducible coupling matrices. As discussed earlier, any system of coupled concave interference functions can be expressed in this way. The iteration is referred to as the *principal eigenvector (PEV) iteration.* It converges to a unique global optimizer $\boldsymbol{p}^*$. From Lemma 6.1 it is clear that $\boldsymbol{p}^*$ is associated with a z^*, which solves the Perron root minimization problem (6.3).

Let the superscript $(\cdot)^{(n)}$ denote the nth iteration step. Starting with an arbitrary initialization $\boldsymbol{p}^{(0)} > 0$, the PEV iteration is

$$\boldsymbol{p}^{(n+1)} = \mathrm{pev}(\boldsymbol{\Gamma V}(z^{(n)})) \quad \text{(principal eigenvector)} \tag{6.23}$$
$$\text{where} \quad z_k^{(n)} = \arg\min_{z_k\in\mathcal{Z}_k} (\boldsymbol{p}^{(n)})^T \boldsymbol{v}_k(z_k), \quad \text{for all } k \in \mathcal{K} .$$

This iteration was already outlined in [81], however without convergence analysis. One difficulty with showing convergence lies in the possible ambiguity of $\boldsymbol{V}(z^{(n)})$. For a given $\boldsymbol{p}^{(n)}$, there can be different optimal $z^{(n)}$ leading to different matrices $\boldsymbol{V}(z^{(n)})$. This means that the interference functions are generally not continuously differentiable with respect to $\boldsymbol{p}$. Despite this complicated dependency, global convergence can be shown by exploiting the special structure of the underlying interference model. Regardless of the chosen initialization and the actual choice of the matrices $\boldsymbol{V}(z^{(n)}) \in \mathcal{M}(\boldsymbol{p}^{(n)})$, the sequence (6.23) always converges to the unique optimizer $\boldsymbol{p}^*$. This optimizer is associated with a unique minimum Perron root $\rho^* = C(\gamma)$, as defined by (6.3). Our results are summarized by the following theorem.

Theorem 6.4. *Let $\boldsymbol{V}(z)$ be irreducible for all $z \in \mathcal{Z}$. For an arbitrary initialization $\boldsymbol{p}^{(0)} > 0$, we have*

$$\rho\big(\boldsymbol{\Gamma}\boldsymbol{V}(z^{(n+1)})\big) \leq \rho\big(\boldsymbol{\Gamma}\boldsymbol{V}(z^{(n)})\big) \quad \textit{for all } n \in \mathbb{N}\,, \tag{6.24}$$

$$\lim_{n\to\infty} \rho\big(\boldsymbol{\Gamma}\boldsymbol{V}(z^{(n)})\big) = \rho^*\,, \tag{6.25}$$

$$\lim_{n\to\infty} \boldsymbol{p}^{(n)} = \boldsymbol{p}^* \quad \textit{(component-wise convergence)}\,. \tag{6.26}$$

Proof. For any initialization $\boldsymbol{p}^{(0)} > 0$, $\boldsymbol{p}^{(0)} \neq \boldsymbol{p}^*$, the algorithm yields sequences $\rho^{(n)} := \rho\big(\boldsymbol{\Gamma}\boldsymbol{V}(z^{(n)})\big)$ and $\boldsymbol{p}^{(n)}$.

We begin by showing monotonicity (6.24). The matrix $\boldsymbol{\Gamma}\boldsymbol{V}(z^{(0)})$ has a maximal eigenvalue $\rho^{(0)}$ associated with an eigenvector $\boldsymbol{p}^{(1)}$, so

$$\begin{aligned}
\rho^{(0)} &= \max_{k\in\mathcal{K}} \frac{[\boldsymbol{\Gamma}\boldsymbol{V}(z^{(0)})\boldsymbol{p}^{(1)}]_k}{p_k^{(1)}} \\
&\geq \max_{k\in\mathcal{K}} \frac{[\boldsymbol{\Gamma}\boldsymbol{V}(z^{(1)})\boldsymbol{p}^{(1)}]_k}{p_k^{(1)}} \\
&\geq \inf_{\boldsymbol{p}>0} \max_{k\in\mathcal{K}} \frac{[\boldsymbol{\Gamma}\boldsymbol{V}(z^{(1)})\boldsymbol{p}]_k}{p_k} = \rho^{(1)}\,.
\end{aligned}$$

The first inequality follows from the optimality of the receive strategies, and the second inequality follows from the Collatz-Wielandt-type representation (1.24) of the Perron root. Likewise, we have $\rho^{(n)} \geq \rho^{(n+1)}$ for any $n \in \mathbb{N}$, thus (6.24) follows by induction.

In order to show convergence, consider the set

$$\mathcal{P}_C = \{\boldsymbol{p} > 0 : \|\boldsymbol{p}\|_1 = 1,\ C\boldsymbol{p} \geq \boldsymbol{\Gamma}\boldsymbol{\mathcal{I}}(\boldsymbol{p})\}, \quad \text{for } C \geq \rho^*\,.$$

Any $\boldsymbol{p} \in \mathcal{P}_C$ fulfills $C \cdot \mathrm{SIR}_k(\boldsymbol{p}) \geq \gamma_k$ for all $k \in \mathcal{K}$. That is, by decreasing C, the required SIR level becomes larger. For $C_1 < C_2$, we have $\mathcal{P}_{C_1} \subset \mathcal{P}_{C_2}$. By assumption of irreducibility, there is a unique optimizer (cf. Lemma 6.2), so $\mathcal{P}_{\rho^*} = \{\boldsymbol{p}^*\}$ with $\rho^*\boldsymbol{p}^* = \boldsymbol{\Gamma}\boldsymbol{\mathcal{I}}(\boldsymbol{p}^*)$.

For an arbitrary $C > \rho^*$ and $\hat{\boldsymbol{p}} \in \mathcal{P}_C$, $\hat{\boldsymbol{p}} \neq \boldsymbol{p}^*$, we have

$$C\hat{\boldsymbol{p}} \geq \boldsymbol{\Gamma}\boldsymbol{\mathcal{I}}(\hat{\boldsymbol{p}}) . \tag{6.27}$$

This inequality must be strict for at least one component since $\hat{\boldsymbol{p}} \neq \boldsymbol{p}^*$. This is a consequence of Lemma 6.2.

Let $\hat{z}$ be an optimum receive strategy associated with $\hat{\boldsymbol{p}}$. The resulting coupling matrix is $\boldsymbol{V}(\hat{z}) \in \mathcal{M}(\hat{\boldsymbol{p}})$. With

$$\boldsymbol{\Gamma}\boldsymbol{\mathcal{I}}(\hat{\boldsymbol{p}}) = \boldsymbol{\Gamma}\boldsymbol{V}(\hat{z})\hat{\boldsymbol{p}} ,$$

inequality (6.27) can be rewritten as

$$C \geq \frac{[\boldsymbol{\Gamma}\boldsymbol{V}(\hat{z})\hat{\boldsymbol{p}}]_k}{\hat{p}_k}, \quad \forall k \in \mathcal{K} .$$

Exploiting that $\boldsymbol{V}(\hat{z})$ is irreducible by assumption, and using (1.24) and Lemma A.12 from Appendix A.7, we have

$$\begin{aligned} C &\geq \max_{k \in \mathcal{K}} \frac{[\boldsymbol{\Gamma}\boldsymbol{V}(\hat{z})\hat{\boldsymbol{p}}]_k}{\hat{p}_k} = \sup_{\boldsymbol{x}>0} \frac{\boldsymbol{x}^T\boldsymbol{\Gamma}\boldsymbol{V}(\hat{z})\hat{\boldsymbol{p}}}{\boldsymbol{x}^T\hat{\boldsymbol{p}}} \\ &> \inf_{\boldsymbol{p}>0}\Big(\sup_{\boldsymbol{x}>0} \frac{\boldsymbol{x}^T\boldsymbol{\Gamma}\boldsymbol{V}(\hat{z})\boldsymbol{p}}{\boldsymbol{x}^T\boldsymbol{p}}\Big) = \rho\big(\boldsymbol{\Gamma}\boldsymbol{V}(\hat{z})\big) \end{aligned} \tag{6.28}$$

This inequality is strict because $\hat{\boldsymbol{p}}$ is not the principal eigenvector of $\boldsymbol{\Gamma}\boldsymbol{V}(\hat{z})$, otherwise $\hat{\boldsymbol{p}}$ would be the fixed point fulfilling (6.15), which is ruled out by the assumption $\hat{\boldsymbol{p}} \neq \boldsymbol{p}^*$. Thus, for $\hat{\boldsymbol{p}} \in \mathcal{P}_C$ with $\hat{\boldsymbol{p}} \neq \boldsymbol{p}^*$, we have

$$C > \rho\big(\boldsymbol{\Gamma}\boldsymbol{V}(\hat{z})\big) \tag{6.29}$$

for any choice of $\boldsymbol{V}(\hat{z}) \in \mathcal{M}(\hat{\boldsymbol{p}})$. This implies

$$C > \sup_{\boldsymbol{V} \in \mathcal{M}(\hat{\boldsymbol{p}})} \rho\big(\boldsymbol{\Gamma}\boldsymbol{V}\big) . \tag{6.30}$$

The set $\mathcal{M}(\hat{\boldsymbol{p}})$ is closed and bounded. Also, $\rho\big(\boldsymbol{\Gamma}\boldsymbol{V}\big)$ is a continuous function of the matrix components $[\boldsymbol{V}]_{kl}$, so the supremum (6.30) is attained.

In order to prove convergence, we need to tightly control the maximum value. So we introduce the set

$$\tilde{\mathcal{P}}_C = \{\boldsymbol{p} \in \mathcal{P}_C : \max_{k \in \mathcal{K}} \frac{\gamma_k \mathcal{I}_k(\boldsymbol{p})}{p_k} = C\} . \tag{6.31}$$

The set $\tilde{\mathcal{P}}_C$ is a closed subset of $\mathcal{P}_C$. In order to show this, let $\boldsymbol{p}^{(n)} \in \tilde{\mathcal{P}}_C$ be an arbitrary Cauchy sequence with limit $\boldsymbol{p}$, then

$$C\, p_k = \lim_{n\to\infty} C\, p_k^{(n)} \geq \lim_{n\to\infty} \gamma_k \mathcal{I}_k(\boldsymbol{p}^{(n)}) = \gamma_k \mathcal{I}_k(\boldsymbol{p}) , \quad \forall k .$$

Here we have used Lemma 2.15 from Section 2.5, where continuity of $\mathcal{I}_k(\boldsymbol{p})$ was shown. Thus $\boldsymbol{p} \in \mathcal{P}_C$, which implies $\tilde{\mathcal{P}}_C \subseteq \mathcal{P}_C$. In order to show that $\tilde{\mathcal{P}}_C$ is closed, consider

$$\frac{\gamma_k \mathcal{I}_k(\boldsymbol{p})}{p_k} = \frac{\lim_{n\to\infty} \gamma_k \mathcal{I}_k(\boldsymbol{p}^{(n)})}{\lim_{n\to\infty} p_k^{(n)}} = \lim_{n\to\infty} \frac{\gamma_k \mathcal{I}_k(\boldsymbol{p}^{(n)})}{p_k^{(n)}} . \tag{6.32}$$

The second equality holds because of Lemma A.11 in Appendix A.6. Because $\boldsymbol{p}^{(n)} \in \tilde{\mathcal{P}}_C$, we have

$$\max_{k\in\mathcal{K}} \frac{\gamma_k \mathcal{I}_k(\boldsymbol{p})}{p_k} = \max_{k\in\mathcal{K}} \left(\lim_{n\to\infty} \frac{\gamma_k \mathcal{I}_k(\boldsymbol{p}^{(n)})}{p_k^{(n)}} \right) = C .$$

It follows that the limit $\boldsymbol{p}$ is contained in $\tilde{\mathcal{P}}_C$. Hence, $\tilde{\mathcal{P}}_C$ is a closed subset of $\mathcal{P}_C$.

Next, we show that (6.25) holds, i.e., $\lim_{n\to\infty} \rho^{(n)} = \rho^*$. For any $\hat{\boldsymbol{p}} \in \tilde{\mathcal{P}}_C$ there is a receive strategy $\hat{z}$, as defined earlier. Consider the function

$$f_1(C) = \min_{\hat{\boldsymbol{p}}\in\tilde{\mathcal{P}}_C} \big(C - \rho\big(\boldsymbol{\Gamma V}(\hat{z})\big)\ \big), \quad \text{with } \boldsymbol{V}(\hat{z}) \in \mathcal{M}(\hat{\boldsymbol{p}}) .$$

It can be shown that the minimum is attained since $\rho\big(\boldsymbol{\Gamma V}(z)\big)$ is continuous and $\tilde{\mathcal{P}}_C$ is a closed and bounded set. The optimum ρ^* is characterized by $f_1(\rho^*) = 0$. For $C > \rho^*$ we have $f_1(C) > 0$. For each n we have $\boldsymbol{p}^{(n+1)} \in \tilde{\mathcal{P}}_{\rho^{(n)}}$.

The proof is by contradiction. Suppose that $\lim_{n\to\infty} \rho^{(n)} > \rho^*$. For an arbitrary $\epsilon > 0$, and $\rho^*(\epsilon) = \rho^* + \epsilon$, there exists an $n_0 = n_0(\epsilon)$ such that $\boldsymbol{p}^{(n)} \in \tilde{\mathcal{P}}_{\rho^*(\epsilon)}$ for all $n \geq n_0(\epsilon)$. From the definition of f_1 we obtain

$$f_1(\rho^*(\epsilon)) \leq \rho^*(\epsilon) - \rho\big(\boldsymbol{\Gamma V}(z^{(n)})\big) \quad \text{for all } n \geq n_0(\epsilon).$$

Since $\rho\big(\boldsymbol{\Gamma V}(z^{(n)})\big) \geq \rho^*$ for all $n \geq n_0(\epsilon)$, we have

$$f_1(\rho^*(\epsilon)) \leq \rho^*(\epsilon) - \rho^* = \epsilon .$$

Thus

$$0 = f_1(\rho^*) < \lim_{n\to\infty} f_1(\rho^{(n)}) \leq f_1(\rho^* + \epsilon) = f_1(\rho^*(\epsilon)) \leq \epsilon$$

for all $\epsilon > 0$. Letting $\epsilon \to 0$ we obtain the contradiction

$$0 = f_1(\rho^*) < \liminf_{\epsilon\to 0} f_1(\rho^* + \epsilon) \leq 0 ,$$

thus proving $\lim_{n\to\infty} \rho^{(n)} = \rho^*$.

It remains to show component-wise convergence (6.26) of the sequence $\boldsymbol{p}^{(n)}$. We have

$$\rho^{(n)}\boldsymbol{p}^{(n+1)} = \boldsymbol{\Gamma V}(z^{(n)})\boldsymbol{p}^{(n+1)} \geq \boldsymbol{\Gamma \mathcal{I}}(\boldsymbol{p}^{(n+1)}) . \tag{6.33}$$

The sequence $\boldsymbol{p}^{(n)}$ has a subsequence $\boldsymbol{p}^{(n_l)}$, $l \in \mathbb{N}$ with a limit

$$\lim_{l\to\infty} \boldsymbol{p}^{(n_l)} = \tilde{\boldsymbol{p}}$$

where $\tilde{\boldsymbol{p}} > 0$ and $\|\tilde{\boldsymbol{p}}\|_1 = 1$. Convergence $\lim_{l\to\infty} \rho^{(n_l)} = \rho^*$ has already been shown, thus

$$\rho^* \tilde{\boldsymbol{p}} = \lim_{l\to\infty} \rho^{(n_l)} \boldsymbol{p}^{(n_l)} \geq \lim_{l\to\infty} \boldsymbol{\Gamma}\boldsymbol{\mathcal{I}}(\boldsymbol{p}^{(n_l)}) = \boldsymbol{\Gamma}\boldsymbol{\mathcal{I}}(\tilde{\boldsymbol{p}}) \,.$$

Consequently, $\tilde{\boldsymbol{p}} \in \mathcal{P}_1$. Because of the uniqueness of the optimizer $\boldsymbol{p}^*$, it can be concluded that $\tilde{\boldsymbol{p}} = \boldsymbol{p}^*$, thus $\boldsymbol{p}^{(n)}$ converges to $\boldsymbol{p}^*$. □

The proof shows direct and component-wise convergence of $\boldsymbol{p}^{(n)}$ to the optimum $\boldsymbol{p}^*$. From Lemma 6.2 and the assumption of irreducibility, we know that there is a unique limit point, despite the possible ambiguity of the parameter $z^{(n)}$.

6.3 Fixed Point Iteration

Next, consider the fixed point iteration (5.2), which can be written as follows.

$$\boldsymbol{p}^{(n+1)} = \boldsymbol{\Gamma}\boldsymbol{\mathcal{I}}(\boldsymbol{p}^{(n)}) \,, \quad \boldsymbol{q}^{(0)} \geq 0 \quad \text{(arbitrary)}. \tag{6.34}$$

The fixed point iteration was proposed for solving the problem of QoS-constrained power minimization (5.1). Now, an interesting question is whether the same algorithm can be applied to the min-max problem (6.1), with the specific interference model (6.2). A different interference model was used in [1], were the assumption of noise played an important role. Namely, scalability $\alpha\mathcal{I}_k(\boldsymbol{p}) > \mathcal{I}_k(\alpha\boldsymbol{p})$, for all $\alpha > 1$, was required. This property is not fulfilled for the interference function used here. So it is not clear whether any of the results in [1] can be transfered to the problem at hand.

For the special interference function (6.2), the iteration (6.34) can be rewritten as

$$\boldsymbol{p}^{(n+1)} = \boldsymbol{\Gamma}\boldsymbol{V}(z^{(n)})\boldsymbol{p}^{(n)} \tag{6.35}$$

$$\text{where} \quad z_k^{(n)} = \arg\min_{z_k \in \mathcal{Z}_k} (\boldsymbol{p}^{(n)})^T \boldsymbol{v}_k(z_k), \quad k = 1, 2, \ldots, K \,.$$

We begin by assuming that the SIR target vector $\boldsymbol{\gamma}$ lies on the boundary of the SIR region (1.23), so $C(\boldsymbol{\gamma}) = 1$. Under this assumption, the next theorem shows that the sequence $\boldsymbol{p}^{(n)}$ always converges to the min-max optimizer $\boldsymbol{p}^*$.

Theorem 6.5. *Assume that $\boldsymbol{V}(z)$ is irreducible for all $z \in \mathcal{Z}$, and $\|\boldsymbol{p}^{(n)}\|_1 = 1$ for all $n \in \mathbb{N}$. Let $\boldsymbol{p}^{(0)} > 0$ be an arbitrary initialization, and $\boldsymbol{p}^* > 0$ the unique min-max optimizer characterized by $\boldsymbol{p}^* C(\boldsymbol{\gamma}) = \boldsymbol{\Gamma}\boldsymbol{\mathcal{I}}(\boldsymbol{p}^*)$, where $C(\boldsymbol{\gamma}) = 1$, then*

$$\lim_{n\to\infty} \boldsymbol{p}^{(n)} = \boldsymbol{p}^* \quad \textit{(component-wise convergence)} \,. \tag{6.36}$$

Proof. There exist constants $\mu_1, \mu_2 \in \mathbb{R}_{++}$ such that

$$\mu_1 \boldsymbol{p}^* \leq \boldsymbol{p}^{(0)} \leq \mu_2 \boldsymbol{p}^* .$$

Because of the monotonicity axiom A3 this implies

$$\mu_1 \boldsymbol{p}^* = \boldsymbol{\Gamma \mathcal{I}}(\mu_1 \boldsymbol{p}^*) \leq \boldsymbol{\Gamma \mathcal{I}}(\boldsymbol{p}^{(0)}) \leq \boldsymbol{\Gamma \mathcal{I}}(\mu_2 \boldsymbol{p}^*) = \mu_2 \boldsymbol{p}^* .$$

Since $\boldsymbol{\Gamma \mathcal{I}}(\boldsymbol{p}^{(0)}) = \boldsymbol{p}^{(1)}$, we have $\mu_1 \boldsymbol{p}^* \leq \boldsymbol{p}^{(1)} \leq \mu_2 \boldsymbol{p}^*$. In the same way, we show

$$\mu_1 \boldsymbol{p}^* \leq \boldsymbol{p}^{(n)} \leq \mu_2 \boldsymbol{p}^*, \quad \text{for all } n \in \mathbb{N} .$$

Now, consider sequences

$$\boldsymbol{p}^{(n)} = \sup_{l \geq n} \boldsymbol{p}^{(l)} \tag{6.37}$$

$$\boldsymbol{p}^{(n)} = \inf_{l \geq n} \boldsymbol{p}^{(l)} . \tag{6.38}$$

We have $\mu_1 \boldsymbol{p}^* \leq \boldsymbol{p}^{(n)} \leq \boldsymbol{p}^{(n)} \leq \mu_2 \boldsymbol{p}^*$, and because of the definitions (6.37) and (6.38),

$$\boldsymbol{p}^{(n)} \leq \boldsymbol{p}^{(n+1)} \leq \boldsymbol{p}^{(n+1)} \leq \boldsymbol{p}^{(n)} .$$

Thus, there exist vectors $\boldsymbol{p}^*$, $\boldsymbol{p}^*$, such that

$$\lim_{n\to\infty} \boldsymbol{p}^{(n)} = \boldsymbol{p}^* \leq \boldsymbol{p}^* = \lim_{n\to\infty} \boldsymbol{p}^{(n)} .$$

Because of the continuity of the interference functions (see Section 2.5), we have

$$\boldsymbol{\Gamma \mathcal{I}}(\boldsymbol{p}^*) = \lim_{n\to\infty} \boldsymbol{\Gamma \mathcal{I}}(\boldsymbol{p}^{(n)}) . \tag{6.39}$$

For a fixed $n \in \mathbb{N}$, and all $l \geq n$, we have $\boldsymbol{p}^{(n)} \leq \boldsymbol{p}^{(l)}$, so $\boldsymbol{\Gamma \mathcal{I}}(\boldsymbol{p}^{(n)}) \leq \boldsymbol{\Gamma \mathcal{I}}(\boldsymbol{p}^{(l)})$. Thus,

$$\begin{aligned} \boldsymbol{\Gamma \mathcal{I}}(\boldsymbol{p}^{(n)}) &\leq \inf_{l \geq n} \boldsymbol{\Gamma \mathcal{I}}(\boldsymbol{p}^{(l)}) \\ &= \inf_{l \geq n} \boldsymbol{p}^{(l+1)} = \boldsymbol{p}^{(n+1)} \end{aligned} \tag{6.40}$$

With (6.39), this implies

$$\boldsymbol{\Gamma \mathcal{I}}(\boldsymbol{p}^*) = \lim_{n\to\infty} \boldsymbol{\Gamma \mathcal{I}}(\boldsymbol{p}^{(n)}) \leq \lim_{n\to\infty} \boldsymbol{p}^{(n+1)} = \boldsymbol{p}^* . \tag{6.41}$$

There exists an irreducible matrix $\boldsymbol{V}(\underline{z}^*) \in \mathcal{M}(\boldsymbol{p}^*)$ such that

$$\boldsymbol{\Gamma \mathcal{I}}(\boldsymbol{p}^*) = \boldsymbol{\Gamma V}(\underline{z}^*) \boldsymbol{p}^* \leq \boldsymbol{p}^* . \tag{6.42}$$

It will now be shown that inequality (6.42) can only be satisfied with equality. Since $\boldsymbol{p}^* > 0$, we can rewrite (6.42) as

$$\frac{[\boldsymbol{\Gamma V}(\underline{z}^*)\boldsymbol{p}^*]_k}{p_k^*} \leq 1, \quad \forall k \in \mathcal{K} \,. \tag{6.43}$$

Suppose that component k_0 of inequality (6.42) is strict, then

$$\begin{aligned} \rho\big(\boldsymbol{\Gamma V}(\underline{z}^*)\big) &= \inf_{\boldsymbol{p}>0} \max_{1\leq k \leq K} \frac{\gamma_k\big[\boldsymbol{V}(\underline{z}^*)\boldsymbol{p}\big]_k}{p_k} \\ &< \frac{\gamma_k\big[\boldsymbol{V}(\underline{z}^*)\boldsymbol{p}^*\big]_{k_0}}{p_{k_0}^*} \leq 1 \end{aligned}$$

This leads to the contradiction $1 = \min_{z\in\mathcal{Z}} \rho\big(\boldsymbol{\Gamma V}(z)\big) < 1$, thus

$$\boldsymbol{\Gamma V}(\underline{z}^*)\boldsymbol{p}^* = \boldsymbol{p}^* \,. \tag{6.44}$$

Since $\boldsymbol{V}(\underline{z}^*)$ is irreducible by assumption, and $C(\boldsymbol{\gamma}) = 1$, the vector $\boldsymbol{p}^*$ is the principal eigenvector of $\boldsymbol{\Gamma V}(\underline{z}^*)$. Also, the receive strategy $\underline{z}^*$ is optimal as discussed in Section 6.1.1.

Next, consider the global optimizer $\boldsymbol{p}^*$ with a receive strategy $z^* \in \mathcal{Z}(\boldsymbol{\gamma})$. We have

$$\boldsymbol{\Gamma V}(z^*)\boldsymbol{p}^* = \boldsymbol{p}^* \,. \tag{6.45}$$

It remains to show $\boldsymbol{p}^* = \boldsymbol{p}^*$. Comparing (6.44) and (6.45), it can be observed that both vectors achieve the same boundary point $\boldsymbol{\gamma}$ with equality. Since all $\boldsymbol{V}(z)$ are irreducible by assumption, we can use Lemma 6.2, which states that even if the coupling matrices $\boldsymbol{V}(\underline{z}^*)$ and $\boldsymbol{V}(z^*)$ are different, the associated eigenvectors are the same, thus $\boldsymbol{p}^* = \boldsymbol{p}^*$.

In a similar way, it can be shown that $\boldsymbol{p}^* = \boldsymbol{p}^*$, thus $\lim_{n\to\infty} \boldsymbol{p}^{(n)} = \boldsymbol{p}^*$, which concludes the proof. □

Theorem 6.5 shows that the min-max problem is solved by the fixed point iteration (6.34), provided that the point $\boldsymbol{\gamma}$ lies on the boundary of the SIR region $\mathcal{S}$. This can easily be extended to arbitrary targets $\boldsymbol{\gamma}'$ by introducing a normalization $\boldsymbol{\gamma} = \boldsymbol{\gamma}'/C(\boldsymbol{\gamma}')$. However, this approach is not practical because beforehand knowledge of the global optimum $C(\boldsymbol{\gamma}')$ would be required.

The result was extended in [136], where the following iteration was proposed.

$$\boldsymbol{p}^{(n+1)} = \frac{1}{\|\boldsymbol{p}^{(n)}\|} \boldsymbol{\Gamma}\boldsymbol{\mathcal{I}}(\boldsymbol{p}^{(n)}) \,. \tag{6.46}$$

In every iteration step, the function is weighted by an arbitrary monotone norm $\|\cdot\|$. We have $\alpha\boldsymbol{\mathcal{I}}(\boldsymbol{p}^{(n)}) = \boldsymbol{\mathcal{I}}(\alpha\boldsymbol{p}^{(n)})$ (Axiom A2), which means that $\boldsymbol{p}^{(n)}$ is scaled to norm one.

Since we have a ssumed a fully coupled system of concave interference functions with irreducible coupling matrices, it follows from the results [136] that the modified iteration (6.46) converges to the unique global optimum of the min-max problem (6.1).

6.4 Convergence Behavior of the PEV Iteration

The proposed PEV iteration (6.23) converges very fast for the beamforming model (1.14), which is a special case of the assumed concave interference function. It was observed [8] that the convergence speed is relatively independent of the required accuracy of the iteration. Typically, only a few iteration steps are required, even when the required accuracy is increased by orders of magnitudes. Also, it was observed that the convergence is not much influeneced by the choice of initialization. However, the actual convergence rate is unknown, and a formal analysis seems to be difficult.

One difficulty is the non-uniqueness of the optimal receive strategies, as mentioned in Section 6.2. As a consequence, the interference functions are not continuously differentiable. In Chapter 5 a similar problem occurred in the context of constrained power minimization. In this case, the convergence behavior was successfully analyzed by applying results from non-smooth analysis. By exploiting semi-smoothness of the interference functions, superlinear convergence of the matrix iteration was shown [7]. Also, the convergence of the fixed point iteration was successfully analyzed [7, 74, 130]. However, the same approach cannot be used for the max-min SIR balancing problem, which seems to be more difficult to handle.

The crucial observation that enabled us to understand the convergence behavior of the matrix iteration in Chapter 5 was, that the power minimization problem can be reformulated as the search for the unique root of the convex semi-smooth function $\boldsymbol{d}(\boldsymbol{p})$, as defined by (5.31). Unfortunately, it seems not possible to apply the same approach to the PEV iteration (6.23). Assume that γ is a boundary point, i.e., $C(\gamma) = 1$. Then the optimizer $\boldsymbol{p}^*$ of the max-min SIR balancing problem (6.1) is also characterized as a fixed point. It is obtained as the unique root of the function

$$\boldsymbol{d}(\boldsymbol{p}) = \boldsymbol{p} - \boldsymbol{\Gamma}\boldsymbol{\mathcal{I}}(\boldsymbol{p}) \,. \tag{6.47}$$

However, this is where the similarity to the power minimization problem seems to end. The SIR balancing problem has a different mathematical structure than the power minimization problem. Namely, there is only a single point $\hat{\boldsymbol{p}} > 0$ such that $\boldsymbol{d}(\hat{\boldsymbol{p}}) \geq 0$ is fulfilled, and this point is the optimum, i.e., $\hat{\boldsymbol{p}} = \boldsymbol{p}^*$ with $\boldsymbol{d}(\hat{\boldsymbol{p}}) = 0$. Also, each non-optimal power allocation $\boldsymbol{p} > 0$, $\boldsymbol{p} \neq \boldsymbol{p}^*$ is associated with a matrix $\boldsymbol{V}(z(\boldsymbol{p})) \in \mathcal{M}(\boldsymbol{p})$, which has a Perron root $\rho\big(\boldsymbol{\Gamma}\boldsymbol{V}(z(\boldsymbol{p}))\big) > 1$, This is because $\min_{z\in\mathcal{Z}} \rho\big(\boldsymbol{\Gamma}\boldsymbol{V}(z)\big) = 1$ is the optimum. This means, that even if $\big(\boldsymbol{I} - \boldsymbol{\Gamma}\boldsymbol{V}(z(\boldsymbol{p}))\big)$ is non-singular, its inverse contains negative components. Thus, a key property of the matrix iteration (5.9), is not fulfilled for the SIR balancing problem.

A

Appendix

A.1 Irreducibility

Definition A.1 (irreducibility). *Any $K \times K$ non-negative matrix $\boldsymbol{D}$ is irreducible if and only if its directed graph $\mathcal{G}(\boldsymbol{D})$ is strongly connected. The graph $\mathcal{G}(\boldsymbol{D})$ consists of K nodes. A pair of nodes (N_i, N_j) is connected by a directed edge if $[\boldsymbol{D}]_{ij} > 0$. A graph is called* strongly connected *if for each pair of nodes (N_i, N_j) there is a sequence of directed edges leading from N_i to N_j. Matrices which are not irreducible are said to be* reducible.

This is illustrated in Fig. A.1 and Fig. A.2.

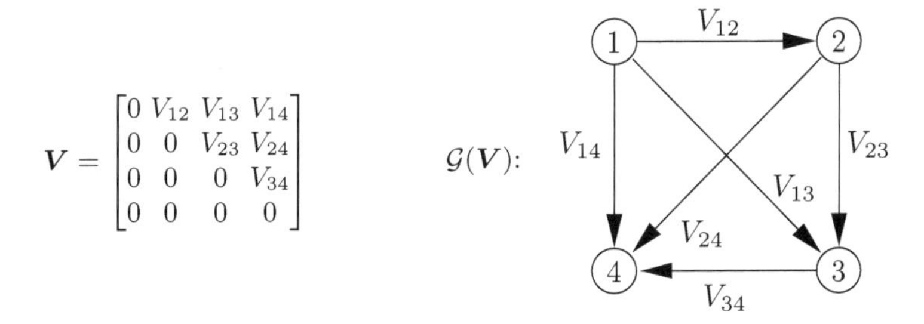

Fig. A.1. $\boldsymbol{V}$ is reducible $\Leftrightarrow$ the directed graph $\mathcal{G}(\boldsymbol{V})$ is not fully connected

A square matrix is *reducible* if there is a simultaneous permutation of rows and columns (=renumbering of users) such that

$$\begin{bmatrix} A & 0 \\ C & B \end{bmatrix}$$

where A and B are square matrices which are not necessarily the same size.

$$V = \begin{bmatrix} 0 & V_{12} & V_{13} & V_{14} \\ 0 & 0 & V_{23} & V_{24} \\ 0 & 0 & 0 & V_{34} \\ V_{41} & 0 & 0 & 0 \end{bmatrix}$$

$\mathcal{G}(V)$:

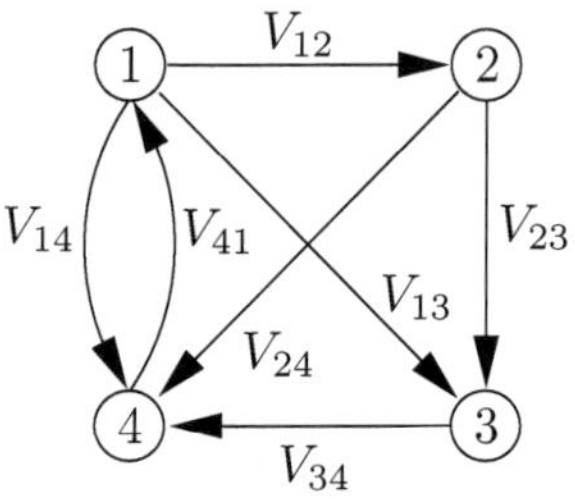

Fig. A.2. V is irreducible $\Leftrightarrow$ the directed graph $\mathcal{G}(V)$ is fully connected, i.e., it consists of $L = 4$ nodes $N_1, \ldots, N_L$. A pair of nodes (N_i, N_j) is connected by a directed edge if $[A_{\mathcal{I}}]_{ij} > 0$. A graph is called *strongly connected* if for each pair of nodes (N_i, N_j) there is a sequence of directed edges leading from N_i to N_j.

A.2 Equivalence of Min-Max and Max-Min Optimization

Lemma A.2. *Consider a continuous real-valued function* $f : \mathbb{R}^n \mapsto \mathbb{R}^n_{++}$. *For any compact set* $\mathcal{X} \subset \mathbb{R}^n$, *we have*

$$\min_{x \in \mathcal{X}} f(x) = \frac{1}{\max_{x \in \mathcal{X}} \frac{1}{f(x)}} \tag{A.1}$$

The same result is obtained by interchanging max *and* min.

Proof. By the extreme value theorem, there is a x_0 such that

$$\min_{x \in \mathcal{X}} f(x) = f(x_0) = \left(\frac{1}{f(x_0)}\right)^{-1} \geq \left(\max_{x \in \mathcal{X}} \frac{1}{f(x)}\right)^{-1} . \tag{A.2}$$

Likewise, there is a x_1 such that

$$\left(\max_{x \in \mathcal{X}} \frac{1}{f(x)}\right)^{-1} = \left(\frac{1}{f(x_1)}\right)^{-1} = f(x_1) \geq \min_{x \in \mathcal{X}} f(x) . \tag{A.3}$$

Comparing (A.2) and (A.3), it follows that these inequalities can only be fulfilled with equality. □

Consider the worst-case SIR, weighted by γ_k.

$$f(p) = \min_{k \in \mathcal{K}} \frac{\mathrm{SIR}_k(p)}{\gamma_k} , \quad \text{on } \mathbb{R}^K_{++} . \tag{A.4}$$

Since $p > 0$ and because of Axiom A1, we have $f(p) > 0$ and $\sup_p f(p) > 0$, thus

$$\inf_p \frac{1}{f(p)} = \frac{1}{\sup_p f(p)} \tag{A.5}$$

Lemma A.2 easily extends to finite sets, thus

$$f(\boldsymbol{p}) = \min_{k \in \mathcal{K}} \frac{\mathrm{SIR}_k(\boldsymbol{p})}{\gamma_k} = \left(\max_{k \in \mathcal{K}} \frac{\gamma_k}{\mathrm{SIR}_k(\boldsymbol{p})} \right)^{-1} . \tag{A.6}$$

Consider some arbitrary $\mathcal{P} \subseteq \mathbb{R}^K_{++}$. If $f(\boldsymbol{p})$ has a finite supremum, then we can write

$$\left(\sup_{\boldsymbol{p} \in \mathcal{P}} \left(\min_{k \in \mathcal{K}} \frac{\mathrm{SIR}_k(\boldsymbol{p})}{\gamma_k} \right) \right)^{-1} = \inf_{\boldsymbol{p} \in \mathcal{P}} \left(\max_{k \in \mathcal{K}} \frac{\gamma_k}{\mathrm{SIR}_k(\boldsymbol{p})} \right) = C(\boldsymbol{\gamma}) . \tag{A.7}$$

Hence, the inf-max indicator function $C(\boldsymbol{\gamma})$, introduced in Subsection 1.4.5, is directly related to the problem of maximizing the worst-case SIR. If the supremum of $f(\boldsymbol{p})$ is not finite, then $C(\boldsymbol{\gamma}) = 0$, which means that arbitrary $\boldsymbol{\gamma} > 0$ can be supported.

A.3 Log-Convex QoS Sets

Consider the QoS region introduced in Subsection 2.6.1 and further discussed in Section 4.4.

Theorem A.3 ([5]). *If the interference functions are log-convex, then $C\big(\boldsymbol{\gamma}(\boldsymbol{q})\big)$ is log-convex on $\mathbb{Q}^K$.*

Proof. Consider two arbitrary points $\hat{\boldsymbol{q}}, \check{\boldsymbol{q}} \in \mathbb{Q}^K$, being connected by a line

$$\boldsymbol{q}(\lambda) = (1-\lambda)\hat{\boldsymbol{q}} + \lambda \check{\boldsymbol{q}} , \quad \lambda \in [0,1] . \tag{A.8}$$

Consider the point $\hat{\boldsymbol{q}}$. The definition (1.22) implies the existence of an $\epsilon > 0$ and a vector $\hat{\boldsymbol{p}} := \hat{\boldsymbol{p}}(\epsilon) > 0$ such that

$$\max_{k \in \mathcal{K}} \log \frac{\gamma_k(\hat{q}_k) \cdot \mathcal{I}_k(\hat{\boldsymbol{p}})}{[\hat{\boldsymbol{p}}]_k} \le \log C\big(\boldsymbol{\gamma}(\hat{\boldsymbol{q}})\big) + \epsilon . \tag{A.9}$$

A similar inequality holds for the point $\check{\boldsymbol{q}}$, with $\check{\boldsymbol{p}} > 0$. Next, we introduce the substitutions $\hat{\boldsymbol{q}} = \mathrm{e}^{\hat{\boldsymbol{s}}}$ and $\check{\boldsymbol{q}} = \mathrm{e}^{\check{\boldsymbol{s}}}$, with

$$\boldsymbol{s}(\lambda) = (1-\lambda)\hat{\boldsymbol{s}} + \lambda \check{\boldsymbol{s}} , \quad \lambda \in [0,1] . \tag{A.10}$$

Now, we can exploit that the functions $\gamma_k(q_k)$ and $\mathcal{I}_k\big(\mathrm{e}^{\boldsymbol{s}}\big)$ are log-convex by assumption. Since e^{s_k} is log-convex and log-concave, and the point-wise product of two log-convex functions is log-convex [23], the function $\mathcal{I}_k\big(\mathrm{e}^{\boldsymbol{s}}\big)/\mathrm{e}^{s_k}$ is log-convex. Thus,

$$
\begin{aligned}
&\log\Big(\gamma_k\big(q_k(\lambda)\big)\cdot\frac{\mathcal{I}_k\big(\mathrm{e}^{\boldsymbol{s}(\lambda)}\big)}{\mathrm{e}^{s_k(\lambda)}}\Big)\\
&\quad=\log\gamma_k\big(q_k(\lambda)\big)+\log\frac{\mathcal{I}_k\big(\mathrm{e}^{\boldsymbol{s}(\lambda)}\big)}{\mathrm{e}^{s_k(\lambda)}}\\
&\quad\le(1-\lambda)\log\gamma_k(\hat{q}_k)+\lambda\log\gamma_k(\check{q}_k)\\
&\qquad\qquad+(1-\lambda)\log\frac{\mathcal{I}_k(\mathrm{e}^{\hat{\boldsymbol{s}}})}{\mathrm{e}^{\hat{s}_k}}+\lambda\log\frac{\mathcal{I}_k(\mathrm{e}^{\check{\boldsymbol{s}}})}{\mathrm{e}^{\check{s}_k}}\\
&\quad=(1-\lambda)\log\frac{\gamma_k(\hat{q}_k)\cdot\mathcal{I}_k(\mathrm{e}^{\hat{\boldsymbol{s}}})}{\mathrm{e}^{\hat{s}_k}}+\lambda\log\frac{\gamma_k(\check{q}_k)\cdot\mathcal{I}_k(\mathrm{e}^{\check{\boldsymbol{s}}})}{\mathrm{e}^{\check{s}_k}}\\
&\quad\le(1-\lambda)\log C\big(\boldsymbol{\gamma}(\hat{\boldsymbol{q}})\big)+\lambda\log C\big(\boldsymbol{\gamma}(\check{\boldsymbol{q}})\big)+2\epsilon\,,
\end{aligned}
$$

where the last inequality follows from (A.9). Consequently,

$$
\begin{aligned}
\log C\big(\boldsymbol{q}(\lambda)\big)&=\inf_{\boldsymbol{s}\in\mathbb{R}^K_+}\Big(\max_{k\in\mathcal{K}}\log\frac{\gamma_k\big(q_k(\lambda)\big)\cdot\mathcal{I}_k(\mathrm{e}^{\boldsymbol{s}})}{\mathrm{e}^{s_k}}\Big)\\
&\le(1-\lambda)\log C\big(\boldsymbol{\gamma}(\hat{\boldsymbol{q}})\big)+\lambda\log C\big(\boldsymbol{\gamma}(\check{\boldsymbol{q}})\big)+2\epsilon. \qquad (A.11)
\end{aligned}
$$

This holds for any $\epsilon>0$. The left-hand side of (A.11) does not depend on ϵ, so letting $\epsilon\to 0$ it can be concluded that $C\big(\boldsymbol{\gamma}(\boldsymbol{q})\big)$ is log-convex on $\mathbb{Q}^K$. □

Next, consider the function $\boldsymbol{p}^{min}(\boldsymbol{q})$, as defined by (2.52) in Subsection 2.7.1.

Theorem A.4 ([5]). *If the interference functions are log-convex, then* $\boldsymbol{p}^{min}(\boldsymbol{q})$ *is log-convex.*

Proof. Consider two arbitrary feasible QoS points $\hat{\boldsymbol{q}},\check{\boldsymbol{q}}\in\operatorname{int}\mathcal{Q}$, connected by a line $\boldsymbol{q}(\lambda)$, as defined by (A.8). Log-convexity implies

$$
\gamma_k\big(q_k(\lambda)\big)\le\gamma_k(\hat{q}_k)^{1-\lambda}\cdot\gamma_k(\check{q}_k)^{\lambda}\,,\quad\forall k\in\mathcal{K}\,. \qquad (A.12)
$$

By $\hat{\boldsymbol{p}}:=\boldsymbol{p}^{min}(\hat{\boldsymbol{q}})$ and $\check{\boldsymbol{p}}:=\boldsymbol{p}^{min}(\check{\boldsymbol{q}})$ we denote the power vectors solving the power minimization problem (2.52) for given targets $\hat{\boldsymbol{q}}$ and $\check{\boldsymbol{q}}$, respectively. It was shown in [1] that these vectors are characterized by fixed point equations

$$
\gamma_k(\hat{q}_k)\cdot\mathcal{I}_k(\hat{\boldsymbol{p}})=\hat{p}_k\,,\quad\forall k\in\mathcal{K}\,, \qquad (A.13)
$$

$$
\gamma_k(\check{q}_k)\cdot\mathcal{I}_k(\check{\boldsymbol{p}})=\check{p}_k\,,\quad\forall k\in\mathcal{K}\,. \qquad (A.14)
$$

Now, we introduce substitutions $\hat{\boldsymbol{p}}=\exp\hat{\boldsymbol{s}}$ (component-wise) and $\check{\boldsymbol{p}}=\exp\check{\boldsymbol{s}}$. The points $\hat{\boldsymbol{s}}$ and $\check{\boldsymbol{s}}$ are connected by a line $\boldsymbol{s}(\lambda)$, as defined by (A.10). Because $\mathcal{I}_k(\mathrm{e}^{\boldsymbol{s}})$ is log-convex on $\mathbb{R}^K$ by assumption,

$$
\mathcal{I}_k(\exp\boldsymbol{s}(\lambda))\le\mathcal{I}_k(\exp\hat{\boldsymbol{s}})^{1-\lambda}\cdot\mathcal{I}_k(\exp\check{\boldsymbol{s}})^{\lambda}\,,\quad\forall k\in\mathcal{K}\,. \qquad (A.15)
$$

Defining

$$
\boldsymbol{p}(\lambda):=\exp\boldsymbol{s}(\lambda)=\hat{\boldsymbol{p}}^{1-\lambda}\cdot\check{\boldsymbol{p}}^{\lambda}\,, \qquad (A.16)
$$

inequality (A.15) can be rewritten as

$$\mathcal{I}_k\big(\boldsymbol{p}(\lambda)\big) \leq \mathcal{I}_k(\hat{\boldsymbol{p}})^{1-\lambda} \cdot \mathcal{I}_k(\check{\boldsymbol{p}})^{\lambda} . \tag{A.17}$$

With (A.12), (A.16), and (A.17), we have

$$\begin{aligned} &\frac{\gamma_k\big(q_k(\lambda)\big) \cdot \mathcal{I}_k\big(\boldsymbol{p}(\lambda)\big)}{p_k(\lambda)} \\ &\quad \leq \frac{\gamma_k(\hat{q}_k)^{1-\lambda} \cdot \gamma_k(\check{q}_k)^{\lambda} \cdot \mathcal{I}_k(\hat{\boldsymbol{p}})^{1-\lambda} \cdot \mathcal{I}_k(\check{\boldsymbol{p}})^{\lambda}}{p_k(\lambda)} \\ &\quad = \left(\frac{\gamma_k(\hat{q}_k) \cdot \mathcal{I}_k(\hat{\boldsymbol{p}})}{(\hat{p}_k)}\right)^{1-\lambda} \cdot \left(\frac{\gamma_k(\check{q}_k) \cdot \mathcal{I}_k(\check{\boldsymbol{p}})}{(\check{p}_k)}\right)^{\lambda} . \end{aligned} \tag{A.18}$$

Exploiting (A.13) and (A.14), inequality (A.18) can be rewritten as

$$\frac{p_k(\lambda)}{\mathcal{I}_k\big(\boldsymbol{p}(\lambda)\big)} \geq \gamma_k\big(q_k(\lambda)\big) , \quad \forall k \in \mathcal{K} .$$

That is, for any $\lambda \in [0,1]$, the power vector $\boldsymbol{p}(\lambda)$ achieves the QoS targets $\boldsymbol{q}(\lambda)$. We know that $\boldsymbol{p}^{min}\big(\boldsymbol{q}(\lambda)\big)$, as defined by (2.52), achieves $\boldsymbol{q}(\lambda)$ with component-wise minimal power [1], thus

$$p_k^{min}\big(\boldsymbol{q}(\lambda)\big) \leq p_k(\lambda) , \quad \forall k \in \mathcal{K} . \tag{A.19}$$

With (A.16) it can be concluded that

$$\begin{aligned} p_k^{min}\big(\boldsymbol{q}(\lambda)\big) &\leq \big(\hat{p}_k\big)^{1-\lambda} \cdot \big(\check{p}_k\big)^{\lambda} \\ &= \big(p_k^{min}(\hat{\boldsymbol{q}})\big)^{1-\lambda} \cdot \big(p_k^{min}(\check{\boldsymbol{q}})\big)^{\lambda} , \quad \forall \lambda \in [0,1] . \end{aligned}$$

This shows that $p_k^{min}(\boldsymbol{q})$ is log-convex on int $\mathcal{Q}$ for all $k \in \mathcal{K}$. □

A.4 Derivatives of Interference Functions

Assume that $\mathcal{I}(\boldsymbol{p})$ on $\mathbb{R}^K_{++}$ is smooth, i.e., all partial derivatives exist. With $\mathcal{I}(\lambda\boldsymbol{p}) = \lambda\mathcal{I}(\boldsymbol{p})$ (Axiom A2), and the generalized chain rule, we have

$$\frac{d\mathcal{I}_k(\lambda\boldsymbol{p})}{d\lambda} = \mathcal{I}_k(\boldsymbol{p}) = \sum_{k=1}^{K} \frac{\partial\mathcal{I}_k(\lambda\boldsymbol{p})}{\partial(\lambda p_k)} p_k = \sum_{k=1}^{K} \frac{\partial\mathcal{I}_k(\boldsymbol{p})}{\partial p_k} p_k . \tag{A.20}$$

Introducing the gradient $\nabla\mathcal{I}_k(\boldsymbol{p}) = \left[\frac{\partial\mathcal{I}_k(\boldsymbol{p})}{\partial p_1}, \ldots, \frac{\partial\mathcal{I}_k(\boldsymbol{p})}{\partial p_K}\right]$, this can be rewritten as

$$\mathcal{I}_k(\boldsymbol{p}) = \nabla\mathcal{I}_k(\boldsymbol{p}) \cdot \boldsymbol{p} . \tag{A.21}$$

A component (k,l) of the local dependency matrix (2.9) equals one if there exists a $\delta_l(\boldsymbol{p}) > 0$ such that $\mathcal{I}_k(\boldsymbol{p} - \delta\boldsymbol{e}_l)$ is strictly monotone decreasing for

$0 \leq \delta \leq \delta_l(\boldsymbol{p})$. This corresponds to the case where the lth component of $\nabla \mathcal{I}_k(\boldsymbol{p})$ is non-zero.

If the function is concave in addition, then representation (3.23) holds, and

$$\mathcal{I}_k(\boldsymbol{p}) = \min_{\boldsymbol{w}_k \in \mathcal{N}_0(\mathcal{I}_k)} \boldsymbol{w}_k^T \boldsymbol{p} \,.$$

For any choice of $\boldsymbol{p}$, we obtain a coupling matrix $\boldsymbol{W} = [\boldsymbol{w}_1, \ldots, \boldsymbol{w}_K]^T$. For each non-zero entry the corresponding entry of the local dependency matrix equals one.

A.5 Non-Smooth Analysis

Definition A.5 (directional derivative). *The directional derivative $F'(\boldsymbol{x}, \boldsymbol{h})$ of the function $F : \mathbb{R}^n \mapsto \mathbb{R}^m$ at a point $\boldsymbol{x}$ in the direction $\boldsymbol{h}$ (unity vector) exists if the following limit exists:*

$$F'(\boldsymbol{x}, \boldsymbol{h}) = \lim_{t \to 0} \frac{F(\boldsymbol{x} + t \cdot \boldsymbol{h}) - F(\boldsymbol{x})}{t} \,.$$

If this holds for all $\boldsymbol{h} \in \mathbb{R}^n$, then F is called directionally differentiable *at the point $\boldsymbol{x}$.*

Definition A.6 (B-derivative). *[129]: A function $F : \mathbb{R}^n \mapsto \mathbb{R}^m$ is said to be* B-differentiable *at a point $\boldsymbol{x}$ if it is directionally differentiable at $\boldsymbol{x}$ and the following limit exists and is finite.*

$$\lim_{\|\boldsymbol{h}\| \to 0} \frac{\|F(\boldsymbol{x} + \boldsymbol{h}) - F(\boldsymbol{x}) - F'(\boldsymbol{x}, \boldsymbol{h})\|}{\|\boldsymbol{h}\|} = 0 \,.$$

This can be rewritten in the equivalent form

$$F(\boldsymbol{x} + \boldsymbol{h}) = F(\boldsymbol{x}) + F'(\boldsymbol{x}, \boldsymbol{h}) + o(\|\boldsymbol{h}\|) \quad \text{as } \|\boldsymbol{h}\| \to 0.$$

In a finite-dimensional Euclidean space, Shapiro (see e.g. [129]) showed that a locally Lipschitz continuous function F is B-differentiable at $\boldsymbol{x}$ if and only if it is directionally differentiable at $\boldsymbol{x}$.

Suppose that F is B-differentiable at $\boldsymbol{x}$. We say that F is B-differentiable of degree 2 at $\boldsymbol{x}$ if

$$F(\boldsymbol{x} + \boldsymbol{h}) = F(\boldsymbol{x}) + F'(\boldsymbol{x}, \boldsymbol{h}) + O(\|\boldsymbol{h}\|^2) \quad \text{as } \|\boldsymbol{h}\| \to 0.$$

Definition A.7 (semi-continuous). *[129]: Let $F : \mathbb{R}^n \mapsto \mathbb{R}^m$ be B-differentiable at a neighborhood of $\boldsymbol{x}$. The directional derivative F' is* semi-continuous *at $\boldsymbol{x}$ if, for every $\epsilon > 0$, there exists a neighborhood $\mathcal{N}$ of $\boldsymbol{x}$ such that, for all $\boldsymbol{h}$ with $\boldsymbol{x} + \boldsymbol{h} \in \mathcal{N}$,*

$$\|F'(\boldsymbol{x}+\boldsymbol{h},\boldsymbol{h}) - F'(\boldsymbol{x},\boldsymbol{h})\| \le \epsilon \cdot \|\boldsymbol{h}\| \, .$$

The directional derivative F' is semi-continuous of degree 2 *at $\boldsymbol{x}$, if there exists a constant L and a neighborhood $\mathcal{N}$ of $\boldsymbol{x}$ such that, for all $\boldsymbol{h}$ with $\boldsymbol{x}+\boldsymbol{h} \in \mathcal{N}$,*

$$\|F'(\boldsymbol{x}+\boldsymbol{h},\boldsymbol{h}) - F'(\boldsymbol{x},\boldsymbol{h})\| \le L \cdot \|\boldsymbol{h}\|^2 \, .$$

Definition A.8 (semi-smooth). *F is* semi-smooth *at $\boldsymbol{x}$ if F is B-differentiable at $\boldsymbol{x}$ and F' is semi-continuous at $\boldsymbol{x}$. If a locally Lipschitz-continuous function F is also convex, then it is semi-smooth for all $\boldsymbol{x}$ [129]. The same holds for concave functions.*

Definition A.9 (linear convergence). *Let the sequence $\{\boldsymbol{p}^{(n)}\}$ converge to $\boldsymbol{p}^*$ such that*

$$\limsup_{n\to\infty} \frac{\|\boldsymbol{p}^{(n+1)} - \boldsymbol{p}^*\|}{\|\boldsymbol{p}^{(n)} - \boldsymbol{p}^*\|} = C < 1 \tag{A.22}$$

Then the sequence is said to converge linearly to $\boldsymbol{p}^*$ with convergence ratio C.

Lemma A.10. *Let F be locally Lipschitz-continuous and semi-smooth at $\boldsymbol{x}$ then for any $\boldsymbol{V} \in \partial F(\boldsymbol{x}+\boldsymbol{h})$,*

$$\boldsymbol{V}\boldsymbol{h} - F'(\boldsymbol{x},\boldsymbol{h}) = o(\|\boldsymbol{h}\|) \quad \text{as } \boldsymbol{h} \to \boldsymbol{0}.$$

If in addition F' is semi-continuous of degree 2, then

$$\boldsymbol{V}\boldsymbol{h} - F'(\boldsymbol{x},\boldsymbol{h}) = O(\|\boldsymbol{h}\|^2) \quad \text{as } \boldsymbol{h} \to \boldsymbol{0}.$$

Proof. See [129]. □

A.6 Ratio of Sequences

Lemma A.11. *Consider sequences $a_n, b_n \in \mathbb{R}$, with limits $a = \lim_{n\to\infty} a_n$ and $b = \lim_{n\to\infty} b_n$. Assuming $b_n \neq 0$ and $b \neq 0$, we have*

$$\lim_{n\to\infty} \frac{a_n}{b_n} = \frac{a}{b} \, . \tag{A.23}$$

Proof. With

$$\frac{a_n}{b_n} - \frac{a}{b} = \frac{a_n(b-b_n) + (a_n - a)b_n}{b_n b}$$

we obtain the inequality

$$\left|\frac{a_n}{b_n} - \frac{a}{b}\right| \le \left|\frac{a_n}{b_n \cdot b}\right| \cdot |b - b_n| + |a_n - a| \cdot \frac{1}{|b|}$$

There exists an n_1 such that $|a_n - a| \leq |a/2|$ for all $n \geq n_1$, and there exists an n_2 such that $|b_n - b| \leq |b/2|$ for all $n \geq n_2$. Consequently, $|a_n| \leq \frac{3}{2}|a|$ and $|b_n| \geq \frac{|b|}{2}$ for all $n \geq \max(n_1, n_2)$. Thus,

$$\left|\frac{a_n}{b_n} - \frac{a}{b}\right| \leq \left|3\frac{|a|}{|b|^2} \cdot |b - b_n| + |a_n - a| \cdot \frac{1}{|b|}\right. .$$

Letting $n \to \infty$ the results follows. □

A.7 Optimizing a Ratio of Linear Functions

Lemma A.12. *Let $\boldsymbol{x}$, $\boldsymbol{b}$, and $\boldsymbol{c}$ be positive, K-dimensional vectors, then the following two equalities hold:*

$$\max_{\boldsymbol{x}} \frac{\boldsymbol{x}^T \boldsymbol{b}}{\boldsymbol{x}^T \boldsymbol{c}} = \max_{1 \leq k \leq K} \frac{[\boldsymbol{b}]_k}{[\boldsymbol{c}]_k} \tag{A.24}$$

$$\min_{\boldsymbol{x}} \frac{\boldsymbol{x}^T \boldsymbol{b}}{\boldsymbol{x}^T \boldsymbol{c}} = \min_{1 \leq k \leq K} \frac{[\boldsymbol{b}]_k}{[\boldsymbol{c}]_k} . \tag{A.25}$$

Proof. We have

$$\max_{\boldsymbol{x}} \frac{\boldsymbol{x}^T \boldsymbol{b}}{\boldsymbol{x}^T \boldsymbol{c}} = \max_{\boldsymbol{x}} \frac{\sum_{k=1}^K [\boldsymbol{x}]_k [\boldsymbol{b}]_k}{\sum_{k=1}^K [\boldsymbol{x}]_k [\boldsymbol{c}]_k} ,$$

where $\boldsymbol{x} \in \mathbb{R}_+^K$, $\boldsymbol{b} \in \mathbb{R}_+^K$, and $\boldsymbol{c} \in \mathbb{R}_+^K$. Defining $[\boldsymbol{x}_1]_k = [\boldsymbol{x}]_k \cdot [\boldsymbol{c}]_k$, $1 \leq k \leq K$, this can be rewritten as

$$\max_{\boldsymbol{x}} \frac{\boldsymbol{x}^T \boldsymbol{b}}{\boldsymbol{x}^T \boldsymbol{c}} = \max_{\|\boldsymbol{x}_1\|=1} \sum_{k=1}^K [\boldsymbol{x}_1]_k \frac{[\boldsymbol{b}]_k}{[\boldsymbol{c}]_k} .$$

Next, we have

$$\sum_{k=1}^K [\boldsymbol{x}_1]_k \frac{[\boldsymbol{b}]_k}{[\boldsymbol{c}]_k} \leq \max_{1 \leq k \leq K} \frac{[\boldsymbol{b}]_k}{[\boldsymbol{c}]_k} \sum_{k=1}^K [\boldsymbol{x}_1]_k = \max_{1 \leq k \leq K} \frac{[\boldsymbol{b}]_k}{[\boldsymbol{c}]_k} . \tag{A.26}$$

There exists a certain index k_0 such that

$$\frac{[\boldsymbol{b}]_{k_0}}{[\boldsymbol{c}]_{k_0}} = \max_{1 \leq k \leq K} \frac{[\boldsymbol{b}]_k}{[\boldsymbol{c}]_k}$$

Now, we use $\tilde{\boldsymbol{x}}$, with

$$[\tilde{\boldsymbol{x}}]_k = \begin{cases} 1 \, , \, k = k_0 \\ 0 \, , \, k \neq k_0 \end{cases} ,$$

which satisfies $\|\tilde{\boldsymbol{x}}\|_1 = 1$. Thus

$$\sum_{k=1}^K [\tilde{x}]_k \frac{[\boldsymbol{b}]_k}{[\boldsymbol{c}]_k} = \max_{1 \leq k \leq K} \frac{[\boldsymbol{b}]_k}{[\boldsymbol{c}]_k} . \tag{A.27}$$

From (A.26) and (A.27) follows (A.24), which concludes the proof. □

A special case is $\boldsymbol{c} = [1, \ldots, 1]^T$. Then problem (A.24) is reduced to maximizing the dot product $\boldsymbol{x}^T\boldsymbol{b}$ under the constraint $\|\boldsymbol{x}\|_1 = 1$.

A.8 Continuations of Interference Functions

The following results are needed in Section 2.5.1. We begin by considering an arbitrary vector

$$\boldsymbol{p} = [p_1, \ldots, p_r, p_{r+1}, \ldots, p_K]^T \tag{A.28}$$

where $p_l = 0$ for $1 \le l \le r$ and $p_l > 0$ for $r+1 \le l \le K$. We introduce an arbitrary sequence $\{\boldsymbol{\epsilon}^{(n)}\}_{n\in\mathbb{N}}$, with

$$\boldsymbol{\epsilon}^{(n)} = [\epsilon_1^{(n)}, \ldots, \epsilon_r^{(n)}]^T > 0$$

and $\lim_{n\to\infty} \boldsymbol{\epsilon}^{(n)} = [0, \ldots, 0]^T$. With the non-zero components of $\boldsymbol{p}$, we define

$$\boldsymbol{p}^{(n)} = [\epsilon_1^{(n)}, \ldots, \epsilon_r^{(n)}, p_{r+1}, \ldots, p_K]^T \in \mathbb{R}^K_{++} \,. \tag{A.29}$$

Note, that there are many possible choices of null sequences $\boldsymbol{\epsilon}^{(n)}$. They all converge to the same limit $\boldsymbol{p} = \lim_{n\to\infty} \boldsymbol{p}^{(n)}$. The first lemma shows that the limit of the resulting interference function is always the same, irrespective of the choice of $\boldsymbol{p}^{(n)}$.

Lemma A.13. *Consider an arbitrary interference function $\mathcal{I}$ defined on $\mathbb{R}^K_{++}$. For any $\boldsymbol{p} \in \mathbb{R}^K_+$ there is a value $\mathcal{I}^c(\boldsymbol{p}) = \mathcal{I}^c(p_{r+1}, \ldots, p_K)$ such that for all possible positive sequences $\{\epsilon_1^{(n)}\}, \ldots, \{\epsilon_r^{(n)}\}$, $n \in \mathbb{N}$, with $\boldsymbol{p} = \lim_{n\to\infty} \boldsymbol{p}^{(n)}$, we have*

$$\lim_{n\to\infty} \mathcal{I}(\boldsymbol{p}^{(n)}) = \mathcal{I}^c(\boldsymbol{p}) \,. \tag{A.30}$$

Proof. We define

$$\begin{aligned} \overline{\epsilon}^{(n)} &= \max_{1\le k\le K} \epsilon_k^{(n)} \\ \underline{\epsilon}^{(n)} &= \min_{1\le k\le K} \epsilon_k^{(n)} \,. \end{aligned}$$

For all $n \in \mathbb{N}$ we have $\overline{\epsilon}^{(n)} \ge \underline{\epsilon}^{(n)} > 0$. With

$$\begin{aligned} \overline{\boldsymbol{p}}^{(n)} &= [\overline{\epsilon}^{(n)}, \ldots, \overline{\epsilon}^{(n)}, p_{r+1}, \ldots, p_K]^T \\ \underline{\boldsymbol{p}}^{(n)} &= [\underline{\epsilon}^{(n)}, \ldots, \underline{\epsilon}^{(n)}, p_{r+1}, \ldots, p_K]^T \end{aligned}$$

we have $\underline{\boldsymbol{p}}^{(n)} \le \boldsymbol{p}^{(n)} \le \overline{\boldsymbol{p}}^{(n)}$, thus $\mathcal{I}(\underline{\boldsymbol{p}}^{(n)}) \le \mathcal{I}(\boldsymbol{p}^{(n)}) \le \mathcal{I}(\overline{\boldsymbol{p}}^{(n)})$. There exist limits

$$\begin{aligned} \overline{C}_1 &= \limsup_{n\to\infty} \mathcal{I}(\overline{\boldsymbol{p}}^{(n)}) \\ \underline{C}_1 &= \liminf_{n\to\infty} \mathcal{I}(\underline{\boldsymbol{p}}^{(n)}) \end{aligned}$$

We have

$$\underline{C}_1 \leq \liminf_{n\to\infty} \mathcal{I}(\boldsymbol{p}^{(n)}) \leq \limsup_{n\to\infty} \mathcal{I}(\boldsymbol{p}^{(n)}) \leq \overline{C}_1 . \tag{A.31}$$

Next, we show that this is fulfilled with equality. Consider an arbitrary $v \in \mathbb{N}$. By assumption we have $\lim_{n\to\infty} \overline{\epsilon}^{(n)} = \lim_{n\to\infty} \underline{\epsilon}^{(n)} = 0$, thus there exists an $n_0 = n_0(v)$ such that $\overline{\epsilon}^{(n)} \leq \underline{\epsilon}^{(v)}$ for all $n \geq n_0$. Thus, $\overline{\boldsymbol{p}}^{(n)} \leq \underline{\boldsymbol{p}}^{(v)}$, and with A3 we have $\mathcal{I}(\overline{\boldsymbol{p}}^{(n)}) \leq \mathcal{I}(\underline{\boldsymbol{p}}^{(v)})$, implying $\overline{C}_1 \leq \mathcal{I}(\underline{\boldsymbol{p}}^{(v)})$. This inequality holds for any $v \in \mathbb{N}$, thus

$$\overline{C}_1 \leq \liminf_{v\to\infty} \mathcal{I}(\underline{\boldsymbol{p}}^{(v)}) = \underline{C}_1 . \tag{A.32}$$

Combining (A.32) with (A.31) we have $\overline{C}_1 = \underline{C}_1$. From (A.31) we know that there exists $\mathcal{I}^c = \lim_{n\to\infty} \mathcal{I}(\boldsymbol{p}^{(n)})$. This limit does not depend on the choice of the null sequences. □

Based on Lemma A.13 we can show scale invariance (A2) on the boundary:

Lemma A.14. *Consider an arbitrary interference function $\mathcal{I}$ defined on $\mathbb{R}^K_{++}$. Let $\boldsymbol{p} \in \mathbb{R}^K_+$. For all $\lambda > 0$ we have*

$$\mathcal{I}^c(\lambda \boldsymbol{p}) = \lambda \mathcal{I}^c(\boldsymbol{p}) . \tag{A.33}$$

Proof. For any $\boldsymbol{p}^{(n)} > 0$ we have $\mathcal{I}(\lambda \boldsymbol{p}^{(n)}) = \lambda \mathcal{I}(\boldsymbol{p}^{(n)})$. The result follows from (A.30) and $\lim_{n\to\infty} \mathcal{I}(\alpha \boldsymbol{p}^{(n)}) = \mathcal{I}^c(\alpha \boldsymbol{p})$. □

Also based on Lemma A.13 we can prove the following Lemma A.15, which shows monotonicity under the restrictive assumption that the compared vectors have zero entries at the same positions.

Lemma A.15. *Consider an arbitrary interference function $\mathcal{I}$ defined on $\mathbb{R}^K_{++}$. Let $\hat{\boldsymbol{p}}$ and $\check{\boldsymbol{p}}$ be two arbitrary vectors from $\mathbb{R}^K_+$ with $\hat{p}_l = \check{p}_l = 0$ for $1 \leq l \leq r$ and $\hat{p}_l \geq \check{p}_l > 0$ for $r+1 \leq l \leq K$. Then*

$$\mathcal{I}^c(\hat{\boldsymbol{p}}) \geq \mathcal{I}^c(\check{\boldsymbol{p}}) . \tag{A.34}$$

Proof. Let $\epsilon^{(n)} > 0$ be an arbitrary null sequence, and

$$\begin{aligned} \hat{\boldsymbol{p}}^{(n)} &= [\epsilon^{(n)}, \ldots, \epsilon^{(n)}, \hat{p}_{r+1}, \ldots, \hat{p}_K]^T \\ \check{\boldsymbol{p}}^{(n)} &= [\epsilon^{(n)}, \ldots, \epsilon^{(n)}, \check{p}_{r+1}, \ldots, \check{p}_K]^T . \end{aligned}$$

From Lemma A.13 we know that $\lim_{n\to\infty} \mathcal{I}(\hat{\boldsymbol{p}}^{(n)}) = \mathcal{I}^c(\hat{\boldsymbol{p}})$ and $\lim_{n\to\infty} \mathcal{I}(\check{\boldsymbol{p}}^{(n)}) = \mathcal{I}^c(\check{\boldsymbol{p}})$. Inequality $\mathcal{I}(\hat{\boldsymbol{p}}^{(n)}) \geq \mathcal{I}(\check{\boldsymbol{p}}^{(n)})$ implies (A.34). □

Note, that Lemma A.15 does not show monotonicity for arbitrary $\hat{\boldsymbol{p}} \geq \check{\boldsymbol{p}}$. This is because Lemma A.13 and Lemma A.15 assume that the positions of the non-zero entries are fixed. So in order to show A3 we need to extend the results to the case of an *arbitrary* sequence $\{\boldsymbol{p}^{(n)}\} > 0$. Based on the previous Lemmas A.13, A.14, and A.15, we show the following result. It extends Lemma A.13 to the case of arbitrary sequences from $\mathbb{R}^K_{++}$. This provides a basis for Theorem 2.17, where general continuity of $\mathcal{I}^c$ is shown.

Lemma A.16. *Let* $\boldsymbol{p} \in \mathbb{R}_+^K$ *be arbitrary. For an arbitrary sequence* $\boldsymbol{p}^{(n)} = [\epsilon_1^{(n)}, \dots, \epsilon_K^{(n)}]^T$, *with* $\epsilon_k^{(n)} \in \mathbb{R}_{++}$, *and* $\lim_{n\to\infty} \boldsymbol{p}^{(n)} = \boldsymbol{p}$, *we have*

$$\lim_{n\to\infty} \mathcal{I}(\boldsymbol{p}^{(n)}) = \mathcal{I}^c(\boldsymbol{p}) . \tag{A.35}$$

Proof. Consider $\overline{\delta}^{(n)} = \max_k |p_k^{(n)} - p_k|$ and $\overline{\boldsymbol{p}}^{(n)} = \boldsymbol{p} + \overline{\delta}^{(n)}\mathbf{1}$, where $\mathbf{1}$ is the all-ones vector. With Lemma A.15 we have

$$\mathcal{I}(\overline{\boldsymbol{p}}^{(n)}) \geq \mathcal{I}^c(\boldsymbol{p}) \tag{A.36}$$

$$\mathcal{I}(\overline{\boldsymbol{p}}^{(n)}) \geq \mathcal{I}(\boldsymbol{p}^{(n)}) . \tag{A.37}$$

Thus,

$$\liminf_{n\to\infty} \mathcal{I}(\overline{\boldsymbol{p}}^{(n)}) \geq \mathcal{I}^c(\boldsymbol{p}) \tag{A.38}$$

$$\liminf_{n\to\infty} \mathcal{I}(\overline{\boldsymbol{p}}^{(n)}) \geq \liminf_{n\to\infty} \mathcal{I}(\boldsymbol{p}^{(n)}) \tag{A.39}$$

$$\limsup_{n\to\infty} \mathcal{I}(\overline{\boldsymbol{p}}^{(n)}) \geq \limsup_{n\to\infty} \mathcal{I}(\boldsymbol{p}^{(n)}) . \tag{A.40}$$

Consider an arbitrary $\epsilon > 0$ and $\mathcal{K}_+ = \{k \in \{1, 2, \dots, K\} : p_k > 0\}$. There exists a $n_0 = n_0(\epsilon)$ such that for all $n \geq n_0$ we have $\overline{\delta}^{(n)} \leq \epsilon$. We define

$$[\mathbf{1}_{\mathcal{K}_+}]_k = \begin{cases} 1 & k \in \mathcal{K}_+ \\ 0 & k \notin \mathcal{K}_+ . \end{cases}$$

The complement of $\mathcal{K}_+$ is $\mathcal{K}_+^c = \mathcal{K} \backslash \mathcal{K}_+$. For all $n \geq n_0$ we have

$$\boldsymbol{p} \leq \boldsymbol{p} + \overline{\delta}^{(n)}\mathbf{1} \leq \boldsymbol{p} + \epsilon\mathbf{1}_{\mathcal{K}_+} + \overline{\delta}^{(n)}\mathbf{1}_{\mathcal{K}_+^c} , \tag{A.41}$$

and thus

$$\mathcal{I}^c(\boldsymbol{p}) \leq \mathcal{I}(\boldsymbol{p} + \overline{\delta}^{(n)}\mathbf{1}) \leq \mathcal{I}(\boldsymbol{p} + \epsilon\mathbf{1}_{\mathcal{K}_+} + \overline{\delta}^{(n)}\mathbf{1}_{\mathcal{K}_+^c}) . \tag{A.42}$$

With Lemma A.13 we have

$$\lim_{n\to\infty} \mathcal{I}(\boldsymbol{p} + \epsilon\mathbf{1}_{\mathcal{K}_+} + \overline{\delta}^{(n)}\mathbf{1}_{\mathcal{K}_+^c}) = \mathcal{I}(\boldsymbol{p} + \epsilon\mathbf{1}_{\mathcal{K}_+}) . \tag{A.43}$$

Combining (A.42) and (A.43) yields

$$\mathcal{I}^c(\boldsymbol{p}) \leq \limsup_{n\to\infty} \mathcal{I}(\boldsymbol{p} + \overline{\delta}^{(n)}\mathbf{1}) \leq \mathcal{I}(\boldsymbol{p} + \epsilon\mathbf{1}_{\mathcal{K}_+}) . \tag{A.44}$$

The function $\mathcal{I}^c$ is an interference function (Lemmas A.15 and Lemmas A.14). It is thus continuous with respect to components from $\mathcal{K}_+$. Relation (A.44) holds for all $\epsilon > 0$, thus letting $\epsilon \to 0$, we know that (A.44) can only be fulfilled with equality. We thus have

$$\mathcal{I}^c(\boldsymbol{p}) = \limsup_{n\to\infty} \mathcal{I}(\overline{\boldsymbol{p}}^{(n)}) . \tag{A.45}$$

By definition, $\overline{\boldsymbol{p}}^{(n)} = \boldsymbol{p} + \overline{\delta}^{(n)}\mathbf{1}$, thus $\mathcal{I}^c(\boldsymbol{p}) = \lim_{n\to\infty} \mathcal{I}(\overline{\boldsymbol{p}}^{(n)})$. Combining (A.40) and (A.45) yields

$$\mathcal{I}^c(\boldsymbol{p}) \geq \limsup_{n\to\infty} \mathcal{I}(\boldsymbol{p}^{(n)}) . \tag{A.46}$$

Now, consider the vector $\tilde{\boldsymbol{p}}^{(n)}$, defined as

$$[\tilde{\boldsymbol{p}}^{(n)}]_k = \begin{cases} p_k^{(n)} & k \in \mathcal{K}_+ \\ 0 & k \notin \mathcal{K}_+ . \end{cases} \tag{A.47}$$

We have $\lim_{n\to\infty} \tilde{\boldsymbol{p}}^{(n)} = \boldsymbol{p}$. Again, we exploit that $\mathcal{I}^c$ is an interference function, so it is continuous with respect to components from $\mathcal{K}_+$. Thus, we have $\lim_{n\to\infty} \mathcal{I}^c(\tilde{\boldsymbol{p}}^{(n)}) = \mathcal{I}^c(\boldsymbol{p})$. So with $\tilde{\boldsymbol{p}}^{(n)} \leq \boldsymbol{p}^{(n)}$ and (A.46) we obtain

$$\begin{aligned} \mathcal{I}^c(\boldsymbol{p}) = \lim_{n\to\infty} \mathcal{I}^c(\tilde{\boldsymbol{p}}^{(n)}) &\leq \liminf_{n\to\infty} \mathcal{I}(\boldsymbol{p}^{(n)}) \\ &\leq \limsup_{n\to\infty} \mathcal{I}(\boldsymbol{p}^{(n)}) \leq \mathcal{I}^c(\boldsymbol{p}) . \end{aligned}$$

We have $\liminf_{n\to\infty} \mathcal{I}(\boldsymbol{p}^{(n)}) \leq \lim_{n\to\infty} \mathcal{I}(\boldsymbol{p}^{(n)}) \leq \limsup_{n\to\infty} \mathcal{I}(\boldsymbol{p}^{(n)})$, so the above inequality is fulfilled with equality. That is, $\mathcal{I}^c(\boldsymbol{p}) = \lim_{n\to\infty} \mathcal{I}(\boldsymbol{p}^{(n)})$. □

With Lemma A.16 we can prove that monotonicity (A3) holds on the extended domain $\mathbb{R}_+^K$, which includes the boundary of $\mathbb{R}_{++}^K$.

Lemma A.17. *Consider arbitrary* $\hat{\boldsymbol{p}}, \check{\boldsymbol{p}} \in \mathbb{R}_+^K$, *with* $\hat{\boldsymbol{p}} \geq \check{\boldsymbol{p}}$. *Then*

$$\mathcal{I}(\hat{\boldsymbol{p}}) \geq \mathcal{I}(\check{\boldsymbol{p}}) . \tag{A.48}$$

Proof. Exploiting Lemma A.16, the proof is similar to the proof of Lemma A.15. □

A.9 Proofs

Proof of Theorem 2.6

Consider arbitrary $k, l \in \mathcal{K}$ such that $[\boldsymbol{D}_\mathcal{I}]_{kl} = 1$. Then there exists a $\hat{\boldsymbol{r}} > 0$ and $\hat{\delta} > 0$ such that

$$\mathcal{I}_k(\hat{\boldsymbol{r}}) < \mathcal{I}_k(\hat{\boldsymbol{r}} + \hat{\delta}\boldsymbol{e}_l) . \tag{A.49}$$

Now, consider an arbitrary δ such that $\delta > \hat{\delta}$. We have $\hat{r}_l < \hat{r}_l + \hat{\delta} < \hat{r}_l + \delta$, so there is a $\lambda = \lambda(\delta) \in (0,1)$ such that

$$\log(\hat{r}_l + \hat{\delta}) = (1-\lambda)\log \hat{r}_l + \lambda \log(\hat{r}_l + \delta) . \tag{A.50}$$

That is, we have

$$\hat{r}_l + \hat{\delta} = (\hat{r}_l)^{1-\lambda} \cdot (\hat{r}_l + \delta)^{\lambda} . \tag{A.51}$$

The value λ for which (A.50) holds is given by

$$\frac{1}{\lambda} = \frac{\log\left(1 + \frac{\delta}{\hat{r}_l}\right)}{\log\left(1 + \frac{\hat{\delta}}{\hat{r}_l}\right)} . \tag{A.52}$$

Because $\mathcal{I}_k$ is log-convex (1.2) is fulfilled. With (A.51) we have

$$\mathcal{I}_k\big(\hat{\boldsymbol{r}} + \hat{\delta}\boldsymbol{e}_l\big) \leq \big(\mathcal{I}_k(\hat{\boldsymbol{r}})\big)^{1-\lambda} \cdot \big(\mathcal{I}_k(\hat{\boldsymbol{r}} + \delta\boldsymbol{e}_l)\big)^{\lambda} .$$

This can be rewritten as

$$\frac{\mathcal{I}_k\big(\hat{\boldsymbol{r}} + \hat{\delta}\boldsymbol{e}_l\big)}{\mathcal{I}_k(\hat{\boldsymbol{r}})} \leq \left(\frac{\mathcal{I}_k(\hat{\boldsymbol{r}} + \delta\boldsymbol{e}_l)}{\mathcal{I}_k(\hat{\boldsymbol{r}})}\right)^{\lambda} .$$

Thus, there is a constant $C_1 = \mathcal{I}_k\big(\hat{\boldsymbol{r}} + \hat{\delta}\boldsymbol{e}_l\big)/\mathcal{I}_k(\hat{\boldsymbol{r}}) > 1$ such that

$$\mathcal{I}_k\big(\hat{\boldsymbol{r}} + \delta\boldsymbol{e}_l\big) \geq C_1^{1/\lambda} \cdot \mathcal{I}_k(\hat{\boldsymbol{r}}) . \tag{A.53}$$

Combining (A.52) and (A.53) we can conclude that

$$\lim_{\delta\to\infty} \mathcal{I}_k(\hat{\boldsymbol{r}} + \delta\boldsymbol{e}_l) = +\infty ,$$

which implies $[\boldsymbol{A}_{\mathcal{I}}]_{kl} = 1$. The converse proof follows immediately from the definition.

Proof of Theorem 4.38

Assume that there exist permutation matrices $\boldsymbol{P}^{(1)}$, $\boldsymbol{P}^{(2)}$ such that $\hat{\boldsymbol{D}}_{\mathcal{I}} = \boldsymbol{P}^{(1)}\boldsymbol{D}_{\mathcal{I}}\boldsymbol{P}^{(2)}$ is block-irreducible with a non-zero main diagonal. We show that this implies the existence of an optimizer for problem (4.64). To this end, we first discuss the simpler case where $\hat{\boldsymbol{D}}_{\mathcal{I}}$ is irreducible. Then, this is extended to block-irreducibility.

Since (4.73) is fulfilled by assumption, Theorem 4.34 implies $PF(\mathcal{I}) > -\infty$, so for every $\epsilon > 0$ there exists a vector $\boldsymbol{p}(\epsilon) > 0$ such that

$$\sum_{k\in\mathcal{K}} \log \frac{\mathcal{I}_k\big(\boldsymbol{p}(\epsilon)\big)}{p_k(\epsilon)} \leq PF(\mathcal{I}) + \epsilon . \tag{A.54}$$

Since $PF(\mathcal{I})$ is invariant with respect to a scaling of $\boldsymbol{p}(\epsilon)$, it can be assumed that $\max_k p_k(\epsilon) = 1$. So there exists a null sequence $\{\epsilon_n\}_{n\in\mathbb{N}}$ and a $\boldsymbol{p}^* \geq 0$, with $\max_k p_k^* = 1$, such that

$$\lim_{n\to\infty} \boldsymbol{p}(\epsilon_n) = \boldsymbol{p}^* .$$

We now show by contradiction that $\boldsymbol{p}^* > 0$. Assume that this is not fulfilled, then $\boldsymbol{p}^*$ has r zero components. Without loss of generality, we can assume that the user indices are chosen such that

$$\lim_{n\to\infty} p_l(\epsilon_n) = \begin{cases} 0, & l = 1,\ldots,r \\ p_l^* > 0, & l = r+1,\ldots,K . \end{cases} \tag{A.55}$$

The assumption of such an ordering is justified because for any permutation matrix $\boldsymbol{P}$ the product $\boldsymbol{P}\boldsymbol{D}_{\mathcal{I}}\boldsymbol{P}^T$ still has the properties of interest (irreducibility, existence of a positive main diagonal after row or column permutation). The first r components of $\boldsymbol{p}(\epsilon_n)$ tend to zero, so for any $C > 0$ and $1 \leq k \leq r$, we have that $\log\big(C/p_k(\epsilon_n)\big)$ tends to infinity. Therefore,

$$\sum_{k\in\mathcal{K}} \log \frac{\mathcal{I}_k\big(\boldsymbol{p}(\epsilon_n)\big)}{p_k(\epsilon_n)} \leq PF(\mathcal{I}) + \epsilon_n, \quad \text{for all } n \in \mathbb{N},$$

can only be fulfilled if

$$\lim_{n\to\infty} \mathcal{I}_k\big(\boldsymbol{p}(\epsilon_n)\big) = 0, \quad k = 1,\ldots,r . \tag{A.56}$$

Consider $\boldsymbol{e}_m$, as defined in (2.4). For any $m, k \in \mathcal{K}$ we have

$$\mathcal{I}_k\big(\boldsymbol{p}(\epsilon_n)\big) \geq \mathcal{I}_k\big(\boldsymbol{p}(\epsilon_n) \circ \boldsymbol{e}_m\big) = \mathcal{I}_k(\boldsymbol{e}_m) \cdot p_m(\epsilon_n) . \tag{A.57}$$

Combining (A.55), (A.56), and (A.57) yields

$$0 = \lim_{n\to\infty} \mathcal{I}_k\big(\boldsymbol{p}(\epsilon_n)\big) \geq \mathcal{I}_k(\boldsymbol{e}_m) \cdot p_m^*, \quad k = 1,\ldots,r , \\ m = r+1,\ldots,K .$$

Since $p_m^* > 0$ for $m = r+1,\ldots,K$, and $\mathcal{I}_k(\boldsymbol{e}_m) \geq 0$, it follows that $\mathcal{I}_k(\boldsymbol{e}_m) = 0$ for $m = r+1,\ldots,K$ and $k = 1,\ldots,r$. Consequently, $\mathcal{I}_1,\ldots,\mathcal{I}_r$ do not depend on $p_{r+1},\ldots,p_K$. This means that $\hat{\boldsymbol{D}}_{\mathcal{I}}$ is reducible, which contradicts the assumption, thus proving $\boldsymbol{p}^* > 0$. Since interference functions are continuous on $\mathbb{R}^K_{++}$ [2], we have

$$\begin{aligned} PF(\mathcal{I}) &\leq \sum_{k\in\mathcal{K}} \log \frac{\mathcal{I}_k(\boldsymbol{p}^*)}{p_k^*} \\ &= \lim_{n\to\infty} \sum_{k\in\mathcal{K}} \log \frac{\mathcal{I}_k\big(\boldsymbol{p}(\epsilon_n)\big)}{p_k(\epsilon_n)} \leq PF(\mathcal{I}) . \end{aligned}$$

Hence, the infimum $PF(\mathcal{I})$ is attained by $\boldsymbol{p}^* > 0$.

Next, we extend the proof to the case where $\hat{\boldsymbol{D}}_{\mathcal{I}}$ is *block-irreducible*. The l-th block on the main diagonal has the dimension $K_l \times K_l$, and $\sum_{l=1}^N K_l = K$. By $\mathcal{I}_k^{(l)}$ we denote the kth interference function of the lth block, where $k = 1,\ldots,K_l$. Since the blocks are de-coupled, we have

$$\inf_{\boldsymbol{p}>0} \sum_{k\in\mathcal{K}} \log \frac{\mathcal{I}_k(\boldsymbol{p})}{p_k} = \sum_{l=1}^{N} PF(\mathcal{I}^{(l)}) , \tag{A.58}$$

$$\text{where} \qquad PF(\mathcal{I}^{(l)}) = \inf_{\boldsymbol{p}\in\mathbb{R}^{K_l}_{++}} \sum_{k=1}^{K_l} \log \frac{\mathcal{I}^{(l)}_k(\boldsymbol{p})}{p_k} . \tag{A.59}$$

By assumption, there exists a row or column permutation such that $\boldsymbol{D}_{\mathcal{I}}$ has a positive main diagonal. The same holds for each block $\hat{\boldsymbol{D}}^{(l)}_{\mathcal{I}}$ on the main diagonal. Since $\hat{\boldsymbol{D}}^{(l)}_{\mathcal{I}}$ is also irreducible, we know from the first part of the proof that there exists a $\hat{\boldsymbol{p}}^{(l)} \in \mathbb{R}^{K_l}_{++}$ such that

$$PF(\mathcal{I}^{(l)}) = \sum_{k=1}^{K_l} \log \frac{\mathcal{I}^{(l)}_k(\hat{\boldsymbol{p}}^{(l)})}{\hat{p}^{(l)}_k} .$$

Defining $\hat{\boldsymbol{p}} = [(\hat{\boldsymbol{p}}^{(1)})^T \dots (\hat{\boldsymbol{p}}^{(N)})^T]^T$ we have

$$PF(\mathcal{I}) = \sum_{l=1}^{N} PF(\mathcal{I}^{(l)}) = \sum_{k=1}^{K} \log \frac{\mathcal{I}_k(\hat{\boldsymbol{p}})}{\hat{p}_k} , \tag{A.60}$$

which completes the first part of the proof.

In order to show the converse, assume that there exists an optimizer $\hat{\boldsymbol{p}} > 0$ which attains the infimum $PF(\mathcal{I}) > -\infty$. The proof is by contradiction. Assume that there are no permutation matrices $\boldsymbol{P}^{(1)}$, $\boldsymbol{P}^{(2)}$, such that $\boldsymbol{P}^{(1)}\boldsymbol{D}_{\mathcal{I}}\boldsymbol{P}^{(2)}$ is block-irreducible with strictly positive main diagonal. From Theorem 4.33 we know that there is a permutation matrix $\check{\boldsymbol{P}}$ such that $\check{\boldsymbol{D}}_{\mathcal{I}} = \boldsymbol{D}_{\mathcal{I}}\check{\boldsymbol{P}}$ has a non-zero main diagonal. There exists a permutation matrix $\boldsymbol{P}_1$ such that $\boldsymbol{P}_1\check{\boldsymbol{D}}_{\mathcal{I}}\boldsymbol{P}_1^T$ takes the canonical form (4.85), i.e.,

$$\boldsymbol{P}_1\check{\boldsymbol{D}}_{\mathcal{I}}\boldsymbol{P}_1^T = \begin{bmatrix} \tilde{\boldsymbol{D}}^{(1)}_{\mathcal{I}} & & \boldsymbol{0} \\ \vdots & \ddots & \\ \tilde{\boldsymbol{D}}^{(r,N)}_{\mathcal{I}} & \dots & \tilde{\boldsymbol{D}}^{(N)}_{\mathcal{I}} \end{bmatrix} = \tilde{\boldsymbol{D}}_{\mathcal{I}} .$$

This matrix cannot be block-diagonal since block-irreducibility of $\boldsymbol{P}_1\boldsymbol{D}_{\mathcal{I}}\check{\boldsymbol{P}}\boldsymbol{P}_1^T$ is ruled out by our hypothesis. Since $\check{\boldsymbol{D}}_{\mathcal{I}}$ has a positive main diagonal, also $\tilde{\boldsymbol{D}}_{\mathcal{I}}$ has a positive diagonal. Let $\tilde{\boldsymbol{p}} = \boldsymbol{P}_1\hat{\boldsymbol{p}}$ and $[\tilde{\mathcal{I}}_1(\tilde{\boldsymbol{p}}), \dots, \tilde{\mathcal{I}}_K(\tilde{\boldsymbol{p}})]^T = \boldsymbol{P}_1[\mathcal{I}_1(\boldsymbol{p}), \dots, \mathcal{I}_K(\boldsymbol{p})]^T$, then

$$\inf_{\boldsymbol{p}>0} \sum_{k\in\mathcal{K}} \frac{\mathcal{I}_k(\boldsymbol{p})}{p_k} = \sum_{k\in\mathcal{K}} \log \frac{\tilde{\mathcal{I}}_k(\tilde{\boldsymbol{p}})}{\tilde{p}_k} = PF(\tilde{\mathcal{I}}) = PF(\mathcal{I}) .$$

Consider the first block $\tilde{\boldsymbol{D}}^{(1)}_{\mathcal{I}} \in \mathbb{R}^{K_1\times K_1}_{+}$ with interference functions $\tilde{\mathcal{I}}^{(1)}_1, \dots, \tilde{\mathcal{I}}^{(1)}_{K_1}$, depending on a power vector $\tilde{\boldsymbol{p}}^{(1)}$, given as the first K_1 components of $\tilde{\boldsymbol{p}}$. This block does not receive interference, so

$$\sum_{k=1}^{K_1} \log \frac{\tilde{\mathcal{I}}_k^{(1)}(\tilde{\boldsymbol{p}}^{(1)})}{\tilde{p}_k^{(1)}} = PF(\tilde{\mathcal{I}}^{(1)}) = \inf_{\boldsymbol{p} \in \mathbb{R}_{++}^{K_1}} \sum_{k=1}^{K_1} \log \frac{\tilde{\mathcal{I}}_k^{(1)}(\boldsymbol{p})}{p_k} .$$

Next, consider the second block $\tilde{\boldsymbol{D}}_{\mathcal{I}}^{(2)} \in \mathbb{R}_+^{K_2 \times K_2}$. If $\tilde{\boldsymbol{D}}_{\mathcal{I}}^{(1,2)} = \boldsymbol{0}$, then

$$\sum_{k=1}^{K_2} \log \frac{\tilde{\mathcal{I}}_k^{(2)}(\tilde{\boldsymbol{p}}^{(2)})}{\tilde{p}_k^{(2)}} = PF(\tilde{\mathcal{I}}^{(2)}) = \inf_{\boldsymbol{p} \in \mathbb{R}_{++}^{K_2}} \sum_{k=1}^{K_2} \log \frac{\tilde{\mathcal{I}}_k^{(2)}(\boldsymbol{p})}{p_k} . \tag{A.61}$$

If $\boldsymbol{D}_{\mathcal{I}}^{(1,2)} \neq \boldsymbol{0}$, then at least one of the interference functions $\tilde{\mathcal{I}}_k^{(2)}(\boldsymbol{p})$, $1 \leq k \leq K_2$, depends on at least one $\tilde{p}_l^{(1)}$, $l = 1, \ldots, K_1$. By scaling $\lambda \cdot \tilde{\boldsymbol{p}}^{(1)}$, $0 < \lambda < 1$, the optimum $PF(\tilde{\mathcal{I}}^{(1)})$ remains unaffected. However, the interference to the second block would be reduced because of the assumed strict monotonicity. So it would be possible to construct a new vector $\check{\boldsymbol{p}}$, with $\check{\boldsymbol{p}} \leq \tilde{\boldsymbol{p}}$, which achieves a better value

$$\sum_{k \in \mathcal{K}} \log \frac{\mathcal{I}_k(\check{\boldsymbol{p}})}{\check{p}_k} < \sum_{k \in \mathcal{K}} \log \frac{\mathcal{I}_k(\tilde{\boldsymbol{p}})}{\tilde{p}_k} = PF(\mathcal{I}) .$$

However, this contradicts the assumption that $\tilde{\boldsymbol{p}}$ is an optimizer. It can be concluded that $\hat{\boldsymbol{D}}_{\mathcal{I}}$ is block-irreducible, with a strictly positive main diagonal.

Proof of Theorem 4.49

Assume that $g(\mathrm{e}^x)$ is convex, then for any $\hat{x}, \check{x} \in \mathbb{R}$, with $x(\lambda) = (1-\lambda)\hat{x} + \lambda\check{x}$, we have

$$g(\mathrm{e}^{x(\lambda)}) \leq (1-\lambda) g(\mathrm{e}^{\hat{x}}) + \lambda g(\mathrm{e}^{\check{x}}) , \quad \forall \lambda \in [0,1] . \tag{A.62}$$

The function $c_k(\boldsymbol{s}) = \mathcal{I}_k(\mathrm{e}^{\boldsymbol{s}})/\mathrm{e}^{s_k}$ is log-convex for all k, i.e.,

$$c_k\big(\boldsymbol{s}(\lambda)\big) \leq c_k(\hat{\boldsymbol{s}})^{1-\lambda} \cdot c_k(\check{\boldsymbol{s}})^{\lambda} , \quad \lambda \in [0,1] , \tag{A.63}$$

where $\boldsymbol{s}(\lambda)$ is defined in (A.10). Exploiting (A.62), (A.63), and the monotonicity of g, we obtain

$$\begin{aligned} g\big(\mathrm{e}^{\log c_k(\boldsymbol{s}(\lambda))}\big) &\leq g\Big(\exp\big\{(1-\lambda)\log c_k(\hat{\boldsymbol{s}}) + \lambda \log c_k(\check{\boldsymbol{s}})\big\}\Big) \\ &\leq (1-\lambda) \cdot g\big(c_k(\hat{\boldsymbol{s}})\big) + \lambda \cdot g\big(c_k(\check{\boldsymbol{s}})\big) . \end{aligned}$$

The sum of convex functions is convex, thus the objective function in (4.109) is convex on $\mathbb{R}^K$.

Conversely, assume that (4.109) is convex. We want to show that this implies convexity of $g(\mathrm{e}^x)$. To this end, consider the set $\mathcal{G}$, which is the set of all g such that (4.109) is convex for *all* log-convex interference functions $\mathcal{I}$. Also consider the set $\mathcal{G}_{lin}$, which is the set of all g such that (4.109) is convex for the specific linear interference functions $\mathcal{I}_1(\mathrm{e}^{\boldsymbol{s}}) = e^{s_2}$ and $\mathcal{I}_2(\mathrm{e}^{\boldsymbol{s}}) = e^{s_1}$.

These functions are also log-convex, thus $\mathcal{G} \subseteq \mathcal{G}_{lin}$. We now show that all $g \in \mathcal{G}_{lin}$ are convex. For an arbitrary $g \in \mathcal{G}_{lin}$, the function

$$F(\boldsymbol{s}, \alpha_1, \alpha_2) = \alpha_1 g(\mathrm{e}^{s_2 - s_1}) + \alpha_2 g(\mathrm{e}^{s_1 - s_2}) \tag{A.64}$$

is convex in $\boldsymbol{s}$ by assumption. Convexity is preserved when we set $s_1 = 0$. Let $\alpha_2 = 1 - \alpha_1$. A convergent series of convex functions is a convex function [23], thus

$$\lim_{\alpha_1 \to 1} F(\boldsymbol{s}, \alpha_1) = g(\mathrm{e}^{s_2}) \tag{A.65}$$

is convex, and therefore $g(\mathrm{e}^{\boldsymbol{s}})$ is convex. It can be concluded that all $g \in \mathcal{G}$ are convex.

Proof of Theorem 4.14

For the proof of Theorem 4.14 we will need the following result:

Lemma A.18. *Let $\boldsymbol{q}$ be the principal left-hand eigenvector of an irreducible stochastic $K \times K$ matrix $\boldsymbol{W}$, then the set $\mathcal{O}_q = \{\boldsymbol{z} \in \mathbb{R}^K : \boldsymbol{q}^T \boldsymbol{z} = 0\}$ equals the range of $(\boldsymbol{I} - \boldsymbol{W})$.*

Proof. Every row stochastic $\boldsymbol{W}$ fulfills $\boldsymbol{W}\mathbf{1} = \mathbf{1}$, so $\mathbf{1}$ is an eigenvector of $\boldsymbol{W}$. Since $\boldsymbol{W}$ is irreducible by assumption, it follows from the Perron-Frobenius theorem (see e.g. [56, 57]) that only the maximum eigenvalue, which equals the spectral radius $\rho(\boldsymbol{W})$, can be associated with a non-negative eigenvector. Thus, $\boldsymbol{W}$ has a maximal eigenvalue $\rho(\boldsymbol{W}) = \rho(\boldsymbol{W}^T) = 1$. Because $\boldsymbol{W}^T$ is irreducible as well, the left-hand principal eigenvector $\boldsymbol{q} > 0$, is unique up to a scaling. We can assume $\|\boldsymbol{q}\|_1 = 1$ without loss of generality. We have $\boldsymbol{q}^T \boldsymbol{W} = \boldsymbol{q}^T$, or equivalently $\boldsymbol{q}^T(\boldsymbol{I} - \boldsymbol{W}) = \mathbf{0}^T$. Thus,

$$\boldsymbol{q}^T(\boldsymbol{I} - \boldsymbol{W})\boldsymbol{s} = 0 \,, \quad \text{for all } \boldsymbol{s} \in \mathbb{R}^K \,. \tag{A.66}$$

Consider the range $\mathcal{R}(\boldsymbol{I} - \boldsymbol{W}) = (\boldsymbol{I} - \boldsymbol{W})\mathbb{R}^K$. For all $\boldsymbol{z} \in \mathcal{R}(\boldsymbol{I} - \boldsymbol{W})$, there exists a $\boldsymbol{s} \in \mathbb{R}^K$ with $(\boldsymbol{I} - \boldsymbol{W})\boldsymbol{s} = \boldsymbol{z}$. From (A.66) we know that $\boldsymbol{q}^T \boldsymbol{z} = 0$, thus $\mathcal{R}(\boldsymbol{I} - \boldsymbol{W})$ is a hyperplane lying in the $(K-1)$-dimensional hyperplane $\mathcal{O}_q$. That is,

$$\mathcal{R}(\boldsymbol{I} - \boldsymbol{W}) \subseteq \{\boldsymbol{z} \in \mathbb{R}^K : \boldsymbol{q}^T \boldsymbol{z} = 0\} = \mathcal{O}_q \,. \tag{A.67}$$

For vector spaces $\mathcal{M}$ and $\mathcal{N}$ such that $\mathcal{M} \subseteq \mathcal{N}$, it is known that $\dim \mathcal{M} = \dim \mathcal{N}$ implies $\mathcal{M} = \mathcal{N}$ (see e.g. [58], p. 198). From (A.67) we have $\dim \mathcal{R}(\boldsymbol{I} - \boldsymbol{W}) \leq K - 1$. So in order to prove the lemma, it remains to show $\dim \mathcal{R}(\boldsymbol{I} - \boldsymbol{W}) \geq K - 1$, thus implying $\dim \mathcal{R}(\boldsymbol{I} - \boldsymbol{W}) = K - 1$.

Because $\boldsymbol{W}$ is irreducible and stochastic by assumption, there exists a decomposition $\boldsymbol{W} = \boldsymbol{B} + \mathbf{1}\boldsymbol{q}^T$ such that $\boldsymbol{I} - \boldsymbol{B}$ is non-singular [137]. For any $\boldsymbol{z} \in \mathcal{O}_q$, we have $\boldsymbol{W}\boldsymbol{z} = \boldsymbol{B}\boldsymbol{z} + \mathbf{1}\boldsymbol{q}^T\boldsymbol{z} = \boldsymbol{B}\boldsymbol{z}$. Thus,

$$(\boldsymbol{I} - \boldsymbol{B})\mathcal{O}_q = (\boldsymbol{I} - \boldsymbol{W})\mathcal{O}_q \,. \tag{A.68}$$

The hyperplane $\mathcal{O}_q$ has dimension $K-1$. Since $(\boldsymbol{I}-\boldsymbol{B})$ is non-singular, we have $\dim(\boldsymbol{I}-\boldsymbol{B})\mathcal{O}_q = K-1$, and with (A.68) we have $\dim(\boldsymbol{I}-\boldsymbol{W})\mathcal{O}_q = K-1$. Also, $(\boldsymbol{I}-\boldsymbol{W})\mathbb{R}^K \supset (\boldsymbol{I}-\boldsymbol{W})\mathcal{O}_q$ implies

$$\dim \mathcal{R}(\boldsymbol{I}-\boldsymbol{W}) \geq \dim(\boldsymbol{I}-\boldsymbol{W}) \, , \mathcal{O}_q = K-1 \, ,$$

which concludes the proof. □

We will now use Lemma A.18 and Lemma 4.9 to prove Theorem 4.14:

The matrix $\boldsymbol{W}$ is irreducible, so Lemma A.18 implies $(\boldsymbol{I}-\boldsymbol{W})\mathbb{R}^K = \mathcal{O}_q$, where $\mathcal{O}_q = \{\boldsymbol{z} \in \mathbb{R}^K : \boldsymbol{q}^T\boldsymbol{z} = 0\}$. That is, for every $\boldsymbol{z} \in \mathcal{O}_q$, there exists a $\boldsymbol{s} \in \mathbb{R}^K$, such that $(\boldsymbol{I}-\boldsymbol{W})\boldsymbol{s} = \boldsymbol{z}$. Consider the special choice $\boldsymbol{z}^* = \log \boldsymbol{t} - \mathcal{C}'\boldsymbol{1}$, with $\mathcal{C}' = \boldsymbol{q}^T \log \boldsymbol{t}$. It can be verified that $\boldsymbol{q}^T\boldsymbol{z}^* = 0$, thus, $\boldsymbol{z}^* \in \mathcal{O}_q$. The associated vector $\boldsymbol{s}^*$ solves

$$(\boldsymbol{I}-\boldsymbol{W})\boldsymbol{s}^* = \log \boldsymbol{t} - \mathcal{C}'\boldsymbol{1} \, . \tag{A.69}$$

From Lemma 4.9 we know that with the substitutions $C' = \exp\{\mathcal{C}'\}$ and $\boldsymbol{p}^* = \exp\{\boldsymbol{s}^*\}$, we have

$$C'\boldsymbol{p}^* = \boldsymbol{\Gamma}\boldsymbol{\mathcal{I}}(\boldsymbol{p}^*, \boldsymbol{W}) \, . \tag{A.70}$$

The vector $\boldsymbol{p}^* > 0$ is a fixed point of $\boldsymbol{\Gamma}\boldsymbol{\mathcal{I}}(\boldsymbol{p}, \boldsymbol{W})/C'$. It was shown in [2] (see also Lemma 2.21) that this implies $C' = C(\gamma, \boldsymbol{W})$. Thus, $\boldsymbol{p}^*$ is a solution of the fixed point equation (4.23), for given $\boldsymbol{W}$.

It remains to prove uniqueness. Suppose that there are two vectors $\boldsymbol{p}^{(1)}$ and $\boldsymbol{p}^{(2)}$, with substitute variables $\boldsymbol{s}^{(1)}$ and $\boldsymbol{s}^{(2)}$, respectively, which fulfill

$$(\boldsymbol{I}-\boldsymbol{W})\boldsymbol{s}^{(1)} = \log \boldsymbol{t} - \mathcal{C}\boldsymbol{1} = (\boldsymbol{I}-\boldsymbol{W})\boldsymbol{s}^{(2)} \, .$$

Then,

$$\boldsymbol{W}(\boldsymbol{s}^{(1)} - \boldsymbol{s}^{(2)}) = (\boldsymbol{s}^{(1)} - \boldsymbol{s}^{(2)}) \, .$$

Since the power vectors can be scaled arbitrarily without affecting the optimum (4.22), we can assume $(\boldsymbol{s}^{(1)} - \boldsymbol{s}^{(2)}) > 0$ without loss of generality. Since $\boldsymbol{W}$ is a stochastic irreducible matrix, there is only one possible positive eigenvector $(\boldsymbol{s}^{(1)} - \boldsymbol{s}^{(2)}) = \mu\boldsymbol{1}$, thus

$$\boldsymbol{p}^{(1)} = \mathrm{e}^{\mu} \cdot \boldsymbol{p}^{(2)} \, .$$

This shows uniqueness up to a scaling.

Proof of Lemma 4.15

Consider the isolated blocks $\boldsymbol{W}^{(n)}$, $1 \leq n \leq i$, which are irreducible by definition. We know from Theorem 4.14 that each of these isolated subsystems is characterized by a fixed point equation of the form (4.23), where all quantities are confined to the respective subsystem, with a unique (up to a scaling) power vector $\boldsymbol{p}^{(n)} \in \mathbb{R}^{K_n}_{++}$ and a min-max level $C(\gamma^{(n)}, \boldsymbol{W}^{(n)})$, as defined by

(4.25). Exploiting that the users $\mathcal{K}_n$ do not depend on powers of other blocks, we can use $\mathcal{I}_k(\boldsymbol{p}, \boldsymbol{W})$ instead of $\mathcal{I}_k(\boldsymbol{p}^{(n)}, \boldsymbol{W}^{(n)})$ for all $k \in \mathcal{K}_n$, as in (4.25). So for all isolated blocks n, with $1 \leq n \leq i$, we have

$$\gamma_k \mathcal{I}_k(\boldsymbol{p}, \boldsymbol{W}) = C(\boldsymbol{\gamma}^{(n)}, \boldsymbol{W}^{(n)}) \cdot p_k \,, \quad \forall k \in \mathcal{K}_n \,. \tag{A.71}$$

The K-dimensional power vector of the complete system is

$$\boldsymbol{p} = [(\boldsymbol{p}^{(1)})^T, \ldots, (\boldsymbol{p}^{(i)})^T, (\boldsymbol{p}^{(i+1)})^T, \ldots, (\boldsymbol{p}^{(N)})^T]^T \,. \tag{A.72}$$

With (A.71), the first i vectors $\boldsymbol{p}^{(1)}, \ldots, \boldsymbol{p}^{(i)}$ are determined up to a scaling. For all users belonging to the isolated blocks, we have

$$\frac{\gamma_k \mathcal{I}_k(\boldsymbol{p}, \boldsymbol{W})}{p_k} \leq \max_{1 \leq n \leq i} C(\boldsymbol{\gamma}^{(n)}, \boldsymbol{W}^{(n)}) \,, \quad \forall k \in \cup_{1 \leq n \leq i} \mathcal{K}_n \,. \tag{A.73}$$

Next, consider the first non-isolated block $i+1$. From the structure of the matrix $\boldsymbol{W}$, it is clear that the interference $\mathcal{I}_k(\boldsymbol{p}, \boldsymbol{W})$, for any $k \in \mathcal{K}_{i+1}$, can only depend on the power vectors $\boldsymbol{p}^{(1)}, \ldots, \boldsymbol{p}^{(i+1)}$. The vectors $\boldsymbol{p}^{(1)}, \ldots, \boldsymbol{p}^{(i)}$ have already been determined. It will now be shown that for an arbitrary $\mu_{i+1} \in \mathbb{R}_{++}$ there is a unique power vector $\boldsymbol{p}^{(i+1)}$ such that

$$\gamma_k \mathcal{I}_k(\boldsymbol{p}, \boldsymbol{W}) = \mu_{i+1} \cdot p_k, \quad \forall k \in \mathcal{K}_{i+1} \,. \tag{A.74}$$

Here, $\boldsymbol{p}$ is defined as by (A.72). The last components $i+2, \ldots, N$ can be chosen arbitrarily because (A.74) does not depend on them. They will be constructed later.

Taking the logarithm of both sides of (A.74) and using $\boldsymbol{s}^{(n)} = \log \boldsymbol{p}^{(n)}$, we obtain (see Lemma 4.9)

$$\begin{aligned}(\boldsymbol{I} - \boldsymbol{W}^{(i+1)})\boldsymbol{s}^{(i+1)} &= -\log \mu_{i+1} + \log \boldsymbol{t}^{(i+1)} + \\ &\quad + \sum_{n=1}^{i} \boldsymbol{W}^{(i+1,n)} \boldsymbol{s}^{(n)} \,. \end{aligned} \tag{A.75}$$

Since $\rho(\boldsymbol{W}^{(i+1)}) < 1$, the matrix $(\boldsymbol{I} - \boldsymbol{W}^{(i+1)})$ is invertible, so we can solve (A.75) for $\boldsymbol{s}^{(i+1)}$. For given $\boldsymbol{s}^{(1)}, \ldots, \boldsymbol{s}^{(i)}$ and μ_{i+1}, the power vector $\boldsymbol{p}^{(i+1)} = \exp \boldsymbol{s}^{(i+1)}$ is unique and it achieves the targets $\boldsymbol{\gamma}^{(i+1)}$ with equality.

By induction, it follows that unique vectors $\boldsymbol{s}^{(n)}$ are obtained for all non-isolated blocks $n = i+2, \ldots, n$. This is ensured because $\rho(\boldsymbol{W}^{(n)}) < 1$ for all non-isolated blocks. Arbitrary levels $\mu_{i+1}, \ldots, \mu_N$ can be achieved. We can choose $\mu_{i+1}, \ldots, \mu_N$ such that the resulting vector $\boldsymbol{p} > 0$ fulfills

$$\frac{\gamma_k \mathcal{I}_k(\boldsymbol{p}, \boldsymbol{W})}{p_k} \leq \max_{1 \leq n \leq i} C(\boldsymbol{\gamma}^{(n)}, \boldsymbol{W}^{(n)}) \,, \quad \text{for all } k \in \mathcal{K} \,.$$

Hence,

$$C(\gamma, \boldsymbol{W}) = \inf_{\tilde{\boldsymbol{p}}>0} \Big(\max_{k\in\mathcal{K}} \frac{\gamma_k \mathcal{I}_k(\tilde{\boldsymbol{p}}, \boldsymbol{W})}{\tilde{p}_k} \Big) \le \max_{k\in\mathcal{K}} \frac{\gamma_k \mathcal{I}_k(\boldsymbol{p}, \boldsymbol{W})}{p_k} \le \max_{1\le n\le i} C(\gamma^{(n)}, \boldsymbol{W}^{(n)}) \,. \tag{A.76}$$

With (4.26), we can conclude that this is fulfilled with equality.

Proof of Theorem 4.21

For any $\boldsymbol{W} \in \mathcal{W}$ and $k \in \mathcal{K}$ we have

$$\gamma_k \mathcal{I}_k(\boldsymbol{p}, \boldsymbol{W}) \le \gamma_k \max_{\boldsymbol{W}\in\mathcal{W}} \mathcal{I}_k(\boldsymbol{p}, \boldsymbol{W}) = \gamma_k \mathcal{I}_k(\boldsymbol{p}) \,,$$

thus

$$C(\gamma, \boldsymbol{W}) \le C(\gamma) \,, \quad \text{for all } \boldsymbol{W} \in \mathcal{W}.$$

The set $\mathcal{W}$ is compact by definition and the function $C(\gamma, \boldsymbol{W})$ is continuous with respect to $\boldsymbol{W}$. Thus, there exists a $\hat{\boldsymbol{W}} \in \mathcal{W}$ such that

$$C(\gamma, \hat{\boldsymbol{W}}) = \max_{\boldsymbol{W}\in\mathcal{W}} C(\gamma, \boldsymbol{W}) \,.$$

Because $\hat{\boldsymbol{W}}$ is irreducible by assumption, we know from Theorem 4.14 that there is a $\hat{\boldsymbol{p}} > 0$ such that

$$\boldsymbol{\Gamma}\boldsymbol{\mathcal{I}}(\hat{\boldsymbol{p}}, \hat{\boldsymbol{W}}) = C(\gamma, \hat{\boldsymbol{W}})\hat{\boldsymbol{p}} \,. \tag{A.77}$$

The proof is by contradiction. Suppose $C(\gamma, \hat{\boldsymbol{W}}) < C(\gamma)$. The vector $\hat{\boldsymbol{p}} > 0$ fulfills (A.77). Because of uniqueness (Lemma 2.21, part 2), $\hat{\boldsymbol{p}} > 0$ cannot be a fixed point of $\boldsymbol{\Gamma}\boldsymbol{\mathcal{I}}(\boldsymbol{p}, \hat{\boldsymbol{W}})/C(\gamma)$. There is an index k_0 such that

$$\mathcal{I}_{k_0}(\hat{\boldsymbol{p}}, \hat{\boldsymbol{W}}) < \max_{\boldsymbol{W}\in\mathcal{W}} \mathcal{I}_{k_0}(\hat{\boldsymbol{p}}, \boldsymbol{W}) \,. \tag{A.78}$$

The maximization in (A.78) would lead to another stochastic matrix $\tilde{\boldsymbol{W}} \in \mathcal{W}$ with a balanced level

$$C(\gamma, \tilde{\boldsymbol{W}}) > C(\gamma, \hat{\boldsymbol{W}}) = \max_{\boldsymbol{W}\in\mathcal{W}} C(\gamma, \boldsymbol{W}) \,.$$

This is a contradiction, thus $C(\gamma, \hat{\boldsymbol{W}}) = C(\gamma)$ and $\hat{\boldsymbol{p}}$ fulfills $\boldsymbol{\Gamma}\boldsymbol{\mathcal{I}}(\hat{\boldsymbol{p}}) = C(\gamma)\hat{\boldsymbol{p}}$.

Proof of Theorem 4.19

A simple way to prove this result is based on Theorem 4.21, which shows that there is a $\boldsymbol{p}^* > 0$ such that

$$c(\gamma) = \sup_{\boldsymbol{p}>0} \min_{k\in\mathcal{K}} \frac{\gamma_k \mathcal{I}_k(\boldsymbol{p})}{p_k} \ge \min_{k\in\mathcal{K}} \frac{\gamma_k \mathcal{I}_k(\boldsymbol{p}^*)}{p_k^*} = C(\gamma) \,.$$

With (4.32) we have $c(\gamma) = C(\gamma)$.

Proof of Theorem 4.22

Assume that there is an irreducible $\boldsymbol{W} \in \mathcal{W}_{\mathcal{I}}$ such that (4.44) holds. We need to show that $\boldsymbol{A}_{\mathcal{I}}$ is irreducible. For all $k, l \in \mathcal{K}$ such that $w_{kl} > 0$, we have

$$\lim_{\delta \to \infty} \mathcal{I}_k(\boldsymbol{p} + \delta \boldsymbol{e}_l) = +\infty \,, \quad \forall \boldsymbol{p} > 0 \,. \tag{A.79}$$

Thus, every non-zero entry in $\boldsymbol{W}$ translates to a non-zero entry in $\boldsymbol{A}_{\mathcal{I}}$. Because $\boldsymbol{W}$ is irreducible by assumption, $\boldsymbol{A}_{\mathcal{I}}$ is irreducible as well.

Conversely, assume that $\boldsymbol{A}_{\mathcal{I}}$ is irreducible. For any $k \in \mathcal{K}$ we define an index set

$$\mathcal{A}_k = \{l \in \mathcal{K} : [\boldsymbol{A}_{\mathcal{I}}]_{kl} = 1\} \,.$$

For all $l \in \mathcal{A}_k$ (A.79) is fulfilled. This is a consequence of definition (2.7) and Lemma 2.1. The matrix $\boldsymbol{A}_{\mathcal{I}}$ is irreducible by assumption. Thus, $\mathcal{A}_k$ is non-empty. The set $\mathcal{L}(\mathcal{I}_k)$ is also non-empty because the trivial case $\mathcal{I}_k(\boldsymbol{p}) = 0$, $\forall \boldsymbol{p} > 0$, is ruled out by (A.79) and the assumption of irreducibility.

Next, consider an arbitrary index $k \in \mathcal{K}$. For some arbitrary $l \in \mathcal{A}_k$ we show by contradiction that there is a $\hat{\boldsymbol{w}} \in \mathcal{L}(\mathcal{I}_k)$ with $\hat{w}_{kl} > 0$. Suppose that there is no such vector, then for all $\boldsymbol{p} > 0$ and $\delta > 0$, we would have

$$\begin{aligned} \mathcal{I}_k(\boldsymbol{p} + \delta \boldsymbol{e}_l) &= \max_{\boldsymbol{w}_k \in \mathcal{L}(\mathcal{I}_k)} \Big(\underline{f}_{\mathcal{I}_k}(\boldsymbol{w}_k) \cdot (p_l + \delta)^{w_{kl}} \cdot \prod_{r \neq l} (p_r)^{w_{kr}} \Big) \\ &= \max_{\boldsymbol{w}_k \in \mathcal{L}(\mathcal{I}_k)} \Big(\underline{f}_{\mathcal{I}_k}(\boldsymbol{w}_k) \prod_{r \neq l} (p_r)^{w_{kr}} \Big) = M_1(\boldsymbol{p}) \,, \end{aligned}$$

where $M_1(\boldsymbol{p}) > 0$ is some constant independent of δ. Thus, $\lim_{\delta \to \infty} \mathcal{I}_k(\boldsymbol{p} + \delta \boldsymbol{e}_l)$ would be bounded, which contradicts the assumption $l \in \mathcal{A}_k$. It can be concluded that for all $l \in \mathcal{A}_k$ there is a $\hat{\boldsymbol{w}}_k^{(l)} \in \mathcal{L}(\mathcal{I}_k)$ such that $[\hat{\boldsymbol{w}}_k^{(l)}]_l > 0$. From Lemma 3.51 we know that $\mathcal{L}(\mathcal{I}_k)$ is a convex set, so any convex combination

$$\tilde{\boldsymbol{w}}_k = (1 - \lambda)\hat{\boldsymbol{w}}_k^{(l_1)} + \lambda \hat{\boldsymbol{w}}_k^{(l_2)} \,, \quad l_1, l_2 \in \mathcal{A}_k \,, \quad 1 < \lambda < 1 \,,$$

is also contained in $\mathcal{L}(\mathcal{I}_k)$. This way, we can construct a $\tilde{\boldsymbol{w}}_k \in \mathcal{L}(\mathcal{I}_k)$ such that $\tilde{w}_{kl} > 0$ for all $l \in \mathcal{A}_k$. This holds for any $k \in \mathcal{K}$, so there is a matrix $\tilde{\boldsymbol{W}} = [\tilde{\boldsymbol{w}}_1, \dots, \tilde{\boldsymbol{w}}_K]^T \in \mathcal{W}_{\mathcal{I}}$ having non-zero entries at the same positions as $\boldsymbol{A}_{\mathcal{I}}$. Because $\boldsymbol{A}_{\mathcal{I}}$ is irreducible by assumption, $\tilde{\boldsymbol{W}}$ is irreducible as well. Also,

$$\mathcal{I}_k(\boldsymbol{p}) = \max_{\boldsymbol{w}_k \in \mathcal{L}(\mathcal{I}_k)} \Big(\underline{f}_{\mathcal{I}_k}(\boldsymbol{w}_k) \prod_{l \in \mathcal{K}} (p_l)^{w_{kl}} \Big) \geq \underline{f}_{\mathcal{I}_k}(\tilde{\boldsymbol{w}}_k) \prod_{l \in \mathcal{K}} (p_l)^{\tilde{w}_{kl}}$$

where $\underline{f}_{\mathcal{I}_k}(\tilde{\boldsymbol{w}}_k) > 0$ because $\tilde{\boldsymbol{W}} \in \mathcal{W}_{\mathcal{I}}$. Hence, (4.44) is fulfilled.

Proof of Theorem 4.23

Consider the set

$$\mathcal{S}(M,\boldsymbol{W}) = \big\{\boldsymbol{p} > 0 : \ \|\boldsymbol{p}\|_\infty = 1 \,, \qquad \gamma_k \underline{f}_{\mathcal{I}_k}(\boldsymbol{w}_k) \prod_{l\in\mathcal{K}} (p_l)^{w_{kl}} \leq M \cdot p_k, \ \forall k\big\}. \tag{A.80}$$

For the proof of Theorem 4.23 we will need the following result.

Lemma A.19. *Let $\boldsymbol{W} \in \mathcal{W}_\mathcal{I}$ be a fixed irreducible stochastic matrix, and $M > 0$ a fixed constant. If the set $\mathcal{S} := \mathcal{S}(M,\boldsymbol{W})$ is non-empty, then there exists a constant $\underline{C} := \underline{C}(M,\boldsymbol{W}) > 0$ such that*

$$\min_{k\in\mathcal{K}} p_k \geq \underline{C} > 0 \,, \quad \textit{for all } \boldsymbol{p} \in \mathcal{S} \,. \tag{A.81}$$

Proof. Consider an arbitrary $\boldsymbol{p} \in \mathcal{S}$. Defining $C_k := M/\big(\gamma_k \underline{f}_{\mathcal{I}_k}(\boldsymbol{w}_k)\big)$, we have

$$\prod_{l\in\mathcal{K}} (p_l)^{w_{kl}} \leq C_k p_k, \quad k \in \mathcal{K} \,. \tag{A.82}$$

For an arbitrary fixed $k \in \mathcal{K}$ we define a dependency set

$$L(k) = \{l \in \mathcal{K} : w_{kl} > 0\} \tag{A.83}$$

and bounds

$$\underline{p}(k) = \min_{l\in L(k)} p_l \,, \tag{A.84}$$

$$\overline{p}(k) = \max_{l\in L(k)} p_l \,. \tag{A.85}$$

Consider an index $\bar{l}(k) \in L(k)$, for which $\overline{p}(k) = p_{\bar{l}(k)}$. We have

$$\begin{aligned}\prod_{l\in\mathcal{K}} (p_l)^{w_{kl}} &= \prod_{l\in L(k)} (p_l)^{w_{kl}} \\ &\geq \big(\overline{p}(k)\big)^{w_{k\bar{l}(k)}} \cdot \big(\underline{p}(k)\big)^{\sum_{l\in L(k)\setminus \bar{l}(k)} w_{kl}} \,.\end{aligned}$$

Defining $\alpha_k = w_{k\bar{l}(k)}$ and exploiting $\sum_{l\in L(k)} w_{kl} = 1$ and (A.82), we have

$$\big(\bar{p}(k)\big)^{\alpha_k} \cdot \big(\underline{p}(k)\big)^{1-\alpha_k} \leq C_k \cdot p_k \,, \quad \forall k \in \mathcal{K} \,. \tag{A.86}$$

Because $\boldsymbol{W}$ is irreducible by assumption, every user causes interference to at least one other user, which means that every index is contained in at least one dependency set. Thus,

$$\begin{aligned}\underline{p} &= \min_{k\in\mathcal{K}} \underline{p}(k) = \min_{k\in\mathcal{K}} p_k \\ \bar{p} &= \max_{k\in\mathcal{K}} \bar{p}(k) = \max_{k\in\mathcal{K}} p_k \,.\end{aligned}$$

Let k_1 be an index such that $p_{k_1} = \underline{p}$. Using $(\underline{p})^{1-\alpha_k} \leq \big(\underline{p}(k)\big)^{1-\alpha_k}$, inequality (A.86) leads to

$$\bar{p}(k_1) \leq (C_{k_1})^{1/\alpha_{k_1}} \underline{p} \,. \tag{A.87}$$

We define the set

$$L_1 = \{k \in \mathcal{K} : p_k \leq \bar{p}(k_1)\} \,. \tag{A.88}$$

For all $k \in L_1$ we have

$$\big(\bar{p}(k)\big)^{\alpha_k} \big(\underline{p}(k)\big)^{1-\alpha_k} \leq C_k \cdot \bar{p}(k_1) \leq C_k \cdot (C_{k_1})^{1/\alpha_{k_1}} \underline{p} \,, \tag{A.89}$$

where the first inequality follows from (A.86) and the second from (A.87). Again, using $(\underline{p})^{1-\alpha_k} \leq \big(\underline{p}(k)\big)^{1-\alpha_k}$, inequality (A.89) leads to

$$\bar{p}(k) \leq (C_k)^{1/\alpha_k} \cdot (C_{k_1})^{1/(\alpha_k \alpha_{k_1})} \underline{p} \,, \quad \forall k \in L_1 \,. \tag{A.90}$$

There exists a $k_2 \in L_1$ such that

$$\bar{p}(k_2) = \max_{k \in L_1} \bar{p}(k) \geq \bar{p}(k_1) \,. \tag{A.91}$$

Here we have exploited $k_1 \in L_1$. Inequality (A.91) implies $L_1 \subseteq L_2$. With the index k_2 we define the set

$$L_2 = \{k \in \mathcal{K} : p_k \leq \bar{p}(k_2)\} \,. \tag{A.92}$$

Similar to the derivation of (A.89), we can use (A.86) and (A.90) to show that for all $k \in L_2$,

$$\big(\bar{p}(k)\big)^{\alpha_k} \big(\underline{p}(k)\big)^{1-\alpha_k} \leq C_k \cdot (C_{k_2})^{1/\alpha_{k_2}} \cdot (C_{k_1})^{1/(\alpha_{k_1}\alpha_{k_2})} \cdot \underline{p} \,.$$

Using $(\underline{p})^{1-\alpha_k} \leq \big(\underline{p}(k)\big)^{1-\alpha_k}$ we have for all $k \in L_2$

$$\bar{p}(k) \leq (C_k)^{1/\alpha_k} \cdot (C_{k_2})^{1/\alpha_{k_2}\alpha_k} \cdot (C_{k_1})^{1/(\alpha_{k_1}\alpha_{k_2}\alpha_k)} \cdot \underline{p} \,.$$

If L_2 is non-empty, then there is a $k_3 \in L_2$ such that

$$\bar{p}(k_3) = \max_{k \in L_2} \bar{p}(k) \geq \bar{p}(k_2) \,. \tag{A.93}$$

The inequality in (A.93) follows from $L_1 \subseteq L_2$. With k_3 we define the set

$$L_3 = \{k \in \mathcal{K} : p_k \leq \bar{p}(k_3)\} \,. \tag{A.94}$$

Inequality (A.93) implies $L_2 \subseteq L_3$.

The above steps are repeated until there is an $N \in \mathbb{N}$ such that $L_N = \emptyset$. Then we have

$$L_1 \subseteq L_2 \subseteq L_3 \subseteq \cdots \subseteq L_{N-1} \tag{A.95}$$

and

$$\bar{p}(k_N) \leq (C_{k_N})^{1/\alpha_{k_N}} \cdot (C_{k_{N-1}})^{1/\alpha_{k_N}\alpha_{k_{N-1}}} \times \dots$$
$$\dots \times (C_{k_1})^{1/(\alpha_{k_1}\alpha_{k_2}\,\cdots\,\alpha_{k_N})} \cdot \underline{p} \,. \tag{A.96}$$

By assumption, the powers are upper bounded by $\bar{p} = 1$ so we have $\bar{p}(k_N) \leq \bar{p}$. We now show by contradition that $\bar{p}(k_N) = \bar{p}$. Suppose that this is not true, i.e., $\bar{p}(k_N) < \bar{p}$, then the set L_{N-1} cannot contain all indices $\mathcal{K}$, because otherwise $\bar{p}(k_N) = \max_{k \in L_{N-1}} \bar{p}(k) = \bar{p}$. Thus, there is a non-empty set

$$G_1 = [1, \dots, K] \backslash L_{N-1} \,. \tag{A.97}$$

For any $\bar{k} \in G_1$ and any $k \in L_{N-1}$ we always have

$$p_{\bar{k}} > \bar{p}(k_N) \,, \tag{A.98}$$

because otherwise $p_{\bar{k}} \in L_N$ which would contradict $L_N = \emptyset$. We now show by contradiction that inequality (A.98) implies $[\boldsymbol{W}]_{k\bar{k}} = 0$. Suppose that this is not true, then $\bar{k} \in L(k)$, thus $\bar{p}(k) = \max_{s \in L(k)} p_s \geq p_{\bar{k}}$. With (A.98) we would have

$$\bar{p}(k) > \bar{p}(k_N) = \max_{t \in L_{N-1}} \bar{p}(t) \geq \bar{p}(k) \,.$$

This contradiction shows that $[\boldsymbol{W}]_{k\bar{k}} = 0$ for arbitrary $\bar{k} \in G_1$ and $k \in L_{N-1}$. That is, the directed graph of $\boldsymbol{W}$ has no paths between nodes from the non-intersecting sets G_1 and L_{N-1}. Thus, $\boldsymbol{W}$ would be reducible, which contradicts the assumption that $\boldsymbol{W}$ is irreducible. Hence, $\bar{p}(k_N) = \bar{p}$ holds.

Setting $\bar{p}(k_N) = \bar{p} = 1$ in (A.96) we obtain

$$\min_{k \in \mathcal{K}} p_k = \underline{p} \geq \underline{C} \tag{A.99}$$

with a constant $\underline{C} > 0$. □

The proof of Lemma A.19 characterizes the constant

$$\underline{C}(M, \boldsymbol{W}) = \inf_{\boldsymbol{p} \in \mathcal{S}(M, \boldsymbol{W})} (\min_{k \in \mathcal{K}} p_k) \,.$$

Now, we will use this result to prove Theorem 4.23. To this end, consider an arbitrary $\epsilon > 0$. From (1.22) it can be observed that there always exists a vector $\boldsymbol{p}(\epsilon) > 0$, with $\max_k p_k(\epsilon) = 1$ (because $\boldsymbol{p}$ can be scaled arbitrarily) and

$$\gamma_k \mathcal{I}_k\big(\boldsymbol{p}(\epsilon)\big) \leq M_\epsilon \cdot p_k(\epsilon), \quad \forall k \in \mathcal{K} \,, \tag{A.100}$$

where $M_\epsilon = \big(C(\boldsymbol{\gamma}) + \epsilon\big)$.

For arbitrary $\boldsymbol{W} \in \mathcal{W}_\mathcal{I}$ we define

$$C_\mathcal{I}(\boldsymbol{\gamma}, \boldsymbol{W}) = \inf_{\boldsymbol{p} > 0} \Big(\max_{k \in \mathcal{K}} \frac{\gamma_k \underline{f}_{\mathcal{I}_k}(\boldsymbol{w}_k) \cdot \prod_{l \in \mathcal{K}} (p_l)^{w_{kl}}}{p_k} \Big) \,.$$

We have

$$\begin{aligned}
&\max_{\boldsymbol{W}\in\mathcal{W}_\mathcal{I}} C_\mathcal{I}(\boldsymbol{\gamma},\boldsymbol{W}) \\
&\quad = \max_{\boldsymbol{W}\in\mathcal{W}_\mathcal{I}} \inf_{\boldsymbol{p}>0} \Big(\max_{k\in\mathcal{K}} \frac{\gamma_k \underline{f}_{\mathcal{I}_k}(\boldsymbol{w}_k)\cdot\prod_{l\in\mathcal{K}}(p_l)^{w_{kl}}}{p_k}\Big) \\
&\quad \le \inf_{\boldsymbol{p}>0} \max_{\boldsymbol{W}\in\mathcal{W}_\mathcal{I}} \Big(\max_{k\in\mathcal{K}} \frac{\gamma_k \underline{f}_{\mathcal{I}_k}(\boldsymbol{w}_k)\cdot\prod_{l\in\mathcal{K}}(p_l)^{w_{kl}}}{p_k}\Big) \\
&\quad = \inf_{\boldsymbol{p}>0} \max_{k\in\mathcal{K}} \Big(\max_{\boldsymbol{w}_k\in\mathcal{L}(\mathcal{I}_k)} \frac{\gamma_k \underline{f}_{\mathcal{I}_k}(\boldsymbol{w}_k)\cdot\prod_{l\in\mathcal{K}}(p_l)^{w_{kl}}}{p_k}\Big) \\
&\quad = \inf_{\boldsymbol{p}>0} \max_{k\in\mathcal{K}} \Big(\frac{\gamma_k \mathcal{I}_k(\boldsymbol{p})}{p_k}\Big) = C(\boldsymbol{\gamma}) \,.
\end{aligned}$$

Thus, $C_\mathcal{I}(\boldsymbol{\gamma},\boldsymbol{W}) \le C(\boldsymbol{\gamma})$ for all $\boldsymbol{W}\in\mathcal{W}_\mathcal{I}$. By assumption, there exists an irreducible $\hat{\boldsymbol{W}}\in\mathcal{W}_\mathcal{I}$. We have

$$M_\epsilon = \big(C(\boldsymbol{\gamma})+\epsilon\big) \ge C_\mathcal{I}(\boldsymbol{\gamma},\hat{\boldsymbol{W}}) \,.$$

Consider the set (A.80). We have $\mathcal{S}(M_\epsilon,\hat{\boldsymbol{W}}) \ne \emptyset$. This follows from the irreducibility of $\hat{\boldsymbol{W}}$, which implies the existence of a $\tilde{\boldsymbol{p}} > 0$ such that $\gamma_k \underline{f}_{\mathcal{I}_k}(\hat{\boldsymbol{w}}_k)\prod_l(\tilde{p}_l)^{\hat{w}_{kl}} = C_\mathcal{I}(\boldsymbol{\gamma},\hat{\boldsymbol{W}})\tilde{p}_l$ (see Theorem 4.14). Thus, the set $\mathcal{S}(C_\mathcal{I}(\boldsymbol{\gamma},\hat{\boldsymbol{W}}),\hat{\boldsymbol{W}})$ is nonempty, and because $M_\epsilon \ge C_\mathcal{I}(\boldsymbol{\gamma},\hat{\boldsymbol{W}})$, the set $\mathcal{S}(M_\epsilon,\hat{\boldsymbol{W}})$ is non-empty as well.

Lemma A.19 implies the existence of a constant $\underline{C}(M_\epsilon,\hat{\boldsymbol{W}})$ such that

$$\min_{k\in\mathcal{K}} p_k \ge \underline{C}(M_\epsilon,\hat{\boldsymbol{W}}) > 0 \,, \quad \forall \boldsymbol{p}\in\mathcal{S}(M_\epsilon,\hat{\boldsymbol{W}}) \,. \tag{A.101}$$

The bound $\underline{C}(M_\epsilon,\hat{\boldsymbol{W}})$ is monotone decreasing in M_ϵ because the set $\mathcal{S}(M_\epsilon,\hat{\boldsymbol{W}})$ is enlarged by increasing M_ϵ. Thus,

$$0 < \underline{C}(M_1,\hat{\boldsymbol{W}}) \le \underline{C}(M_\epsilon,\hat{\boldsymbol{W}}) \,, \quad 0<\epsilon\le 1 \,. \tag{A.102}$$

Because of (3.126) (representation theorem), we have $\mathcal{I}_k\big(\boldsymbol{p}(\epsilon)\big) \ge \underline{f}_{\mathcal{I}_k}(\hat{\boldsymbol{w}}_k)\prod_l\big(p_l(\epsilon)\big)^{\hat{w}_{kl}}$ With (A.100) we know that $\boldsymbol{p}(\epsilon)\in\mathcal{S}(M_\epsilon,\hat{\boldsymbol{W}})$. Combining (A.101) and (A.102), we have

$$\min_{k\in\mathcal{K}} p_k(\epsilon) \ge \underline{C}(M_1,\hat{\boldsymbol{W}}) > 0 \,, \quad 0<\epsilon\le 1 \,. \tag{A.103}$$

The family of vectors $\boldsymbol{p}(\epsilon)$ is bounded. There exists a zero sequence $\{\epsilon_n\}$ and a vector $\hat{\boldsymbol{p}}$ from the compact set $\{\boldsymbol{p}>0 : \|\boldsymbol{p}\|_\infty \le 1$ such that $\hat{\boldsymbol{p}} = \lim_{n\to\infty}\boldsymbol{p}(\epsilon_n)$. With (A.103) we have

$$\hat{\boldsymbol{p}} = \lim_{n\to\infty}\boldsymbol{p}(\epsilon_n) \ge \underline{C}(M_1,\hat{\boldsymbol{W}}) > 0 \,.$$

It was shown in [2] that every interference function is continuous on $\mathbb{R}^K_{++}$, so

$$\gamma_k\mathcal{I}_k(\hat{\boldsymbol{p}}) = \lim_{n\to\infty}\gamma_k\mathcal{I}_k\big(\boldsymbol{p}(\epsilon_n)\big) \le C(\boldsymbol{\gamma})\,\hat{p}_k, \;\; \forall k\in\mathcal{K} \,, \tag{A.104}$$

where the inequality follows from (A.100). Defininig $\tilde{\mathcal{I}}_k(\boldsymbol{p}) = \frac{1}{C(\boldsymbol{\gamma})}\mathcal{I}_k(\boldsymbol{p})$, we have

$$\gamma_k \tilde{\mathcal{I}}_k(\hat{\boldsymbol{p}}) \leq \hat{p}_k, \quad \forall k \in \mathcal{K} . \tag{A.105}$$

Next, consider the set

$$E = \{\boldsymbol{p} \in \mathbb{R}^K_{++} \ : \ p_k \geq \gamma_k \tilde{\mathcal{I}}_k(\boldsymbol{p}), \ \forall k \in \mathcal{K}\} . \tag{A.106}$$

With (A.105) we know that E is non-empty. Consider an arbitrary $\boldsymbol{p} \in E$. We define the index set

$$G(\boldsymbol{p}) = \{k \in \mathcal{K} : p_k = \gamma_k \tilde{\mathcal{I}}_k(\boldsymbol{p})\} , \tag{A.107}$$

and its complement

$$U(\boldsymbol{p}) = \mathcal{K} \backslash G(\boldsymbol{p}) . \tag{A.108}$$

The set $G(\boldsymbol{p})$ is non-empty. In order to show this, suppose that there is a $\boldsymbol{p}' \in E$ with $G(\boldsymbol{p}') = \emptyset$, i.e., $p'_k > \gamma_k \tilde{\mathcal{I}}_k(\boldsymbol{p}')$ for all $k \in \mathcal{K}$. This would imply the contradiction

$$1 = C(\boldsymbol{\gamma}, \tilde{\boldsymbol{\mathcal{I}}}) = \inf_{\boldsymbol{p}>0} \left(\max_{k \in \mathcal{K}} \frac{\gamma_k \tilde{\mathcal{I}}_k(\boldsymbol{p})}{p_k} \right) \leq \frac{\gamma_k \tilde{\mathcal{I}}_k(\boldsymbol{p}')}{p'_k} < 1 , \tag{A.109}$$

where $C(\boldsymbol{\gamma}, \tilde{\boldsymbol{\mathcal{I}}})$ is the min-max optimum for the normalized interference functions $\tilde{\mathcal{I}}_1, \ldots, \tilde{\mathcal{I}}_K$.

From (A.105) we know that $\hat{\boldsymbol{p}} \in E$. Let $\hat{\boldsymbol{p}}^{(1)}$ be the vector with components $\hat{p}^{(1)}_k = \gamma_k \tilde{\mathcal{I}}_k(\hat{\boldsymbol{p}}) \leq \hat{p}_k$, $k \in \mathcal{K}$. If $\hat{\boldsymbol{p}}^{(1)} = \hat{\boldsymbol{p}}$, then $\hat{\boldsymbol{p}}$ is a fixed point fulfilling (4.45). In this case the proof is completed. Otherwise, axiom A3 yields $\gamma_k \tilde{\mathcal{I}}_k(\hat{\boldsymbol{p}}^{(1)}) \leq \gamma_k \tilde{\mathcal{I}}_k(\hat{\boldsymbol{p}}) = \hat{p}^{(1)}_k$, thus $\hat{\boldsymbol{p}}^{(1)} \in E$. That is, the set E has at least two elements. In what follows, we will show that there always exists a $\boldsymbol{p} \in E$ such that $G(\boldsymbol{p}) = \mathcal{K}$.

Consider two arbitrary vectors $\hat{\boldsymbol{p}}, \check{\boldsymbol{p}} \in E$ and $\boldsymbol{p}(\lambda) = \hat{\boldsymbol{p}}^{1-\lambda} \cdot \check{\boldsymbol{p}}^{\lambda}$ (componentwise), with $0 < \lambda < 1$. For any $k \in \mathcal{K}$ we have

$$1 \geq \gamma_k^{1-\lambda} \cdot \gamma_k^{\lambda} \cdot \frac{(\tilde{\mathcal{I}}_k(\hat{\boldsymbol{p}}))^{1-\lambda}}{(\hat{p}_k)^{1-\lambda}} \cdot \frac{(\tilde{\mathcal{I}}_k(\check{\boldsymbol{p}}))^{\lambda}}{(\check{p}_k)^{\lambda}} \geq \gamma_k \frac{\tilde{\mathcal{I}}_k(\boldsymbol{p}(\lambda))}{p_k(\lambda)} . \tag{A.110}$$

The first inequality follows from $C(\boldsymbol{\gamma}, \tilde{\boldsymbol{\mathcal{I}}}) = 1$ and $\hat{\boldsymbol{p}}, \check{\boldsymbol{p}} \in E$, similar to (A.109). The second inequality follows because $\tilde{\mathcal{I}}_k$ is log-convex by assumption. From (A.110) we know that $\boldsymbol{p}(\lambda) \in E$. For any $k \in U(\hat{\boldsymbol{p}}) \cup U(\check{\boldsymbol{p}})$, at least one of the factors in (A.110) is strictly less than one, thus $p_k(\lambda) > \gamma_k \tilde{\mathcal{I}}_k(\boldsymbol{p}(\lambda))$, which implies $k \in U(\boldsymbol{p}(\lambda))$. Therefore,

$$U(\hat{\boldsymbol{p}}) \cup U(\check{\boldsymbol{p}}) \subseteq U(\boldsymbol{p}(\lambda)) . \tag{A.111}$$

Note that we have assumed $U(\boldsymbol{p}) \neq \emptyset$ for all vectors $\boldsymbol{p}$ under consideration. Because $U(\boldsymbol{p}) = \emptyset$ would mean that $\boldsymbol{p}$ is a fixed point, in which case the proof would be completed.

Next, let $\overline{U}$ denote the set of all $k \in \mathcal{K}$ such that there is a vector $\boldsymbol{p}^{(k)} \in E$ with $k \in U(\boldsymbol{p}^{(k)})$, that is, $p_k^{(k)} > \gamma_k \tilde{\mathcal{I}}_k(\boldsymbol{p}^{(k)})$. With (A.111) we can construct a vector $\overline{\boldsymbol{p}} \in E$ such that $\overline{U} = U(\overline{\boldsymbol{p}})$. Thus, for all vectors $\boldsymbol{p} \in E$ we have $U(\boldsymbol{p}) \subseteq U(\overline{\boldsymbol{p}})$.

Next, consider the fixed point iteration

$$\overline{p}_k^{(n+1)} = \gamma_k \tilde{\mathcal{I}}_k(\overline{\boldsymbol{p}}^{(n)}), \quad \text{with } \overline{p}_k^{(0)} = \overline{p}_k \,, \quad \forall k \in \mathcal{K} \,, \tag{A.112}$$

where the superscript n, with $n \geq 0$, denotes the nth iteration step. Because $\overline{\boldsymbol{p}} \in E$ we have $\overline{p}_k^{(1)} = \gamma_k \tilde{\mathcal{I}}_k(\overline{\boldsymbol{p}}^{(0)}) \leq \overline{p}_k^{(0)}$ for all $k \in \mathcal{K}$. Exploiting A3, this leads to

$$\overline{p}_k^{(2)} = \gamma_k \tilde{\mathcal{I}}_k(\overline{\boldsymbol{p}}^{(1)}) \leq \gamma_k \tilde{\mathcal{I}}_k(\overline{\boldsymbol{p}}^{(0)}) = \overline{p}_k^{(1)}, \quad \forall k \in \mathcal{K} \,.$$

Thus $\overline{\boldsymbol{p}}^{(1)} \in E$. We also have $U(\overline{\boldsymbol{p}}^{(1)}) \subseteq U(\overline{\boldsymbol{p}})$. This follows by contradiction: suppose that there exists a $k \in U(\overline{\boldsymbol{p}}^{(1)})$ and k is not contained in $U(\overline{\boldsymbol{p}}) = \overline{U}$. This would imply $\overline{p}_k^{(1)} > \gamma_k \tilde{\mathcal{I}}_k(\overline{\boldsymbol{p}}^{(1)})$, thus leading to the contradiction $k \in \overline{U}$. For the complementary sets, this implies

$$G(\overline{\boldsymbol{p}}^{(1)}) \supseteq G(\overline{\boldsymbol{p}}^{(0)}) = G(\overline{\boldsymbol{p}}) \,.$$

For any $k \in G(\overline{\boldsymbol{p}})$ we have

$$\overline{p}_k^{(1)} = \gamma_k \tilde{\mathcal{I}}_k(\overline{\boldsymbol{p}}^{(0)}) = \overline{p}_k^{(0)} \,.$$

Thus, $\overline{p}_k^{(1)} = \overline{p}_k^{(0)}$ for all $k \in G(\overline{\boldsymbol{p}})$.

In a similar way, we show $\overline{\boldsymbol{p}}^{(n)} \in E$, which implies $G(\overline{\boldsymbol{p}}^{(n)}) \supseteq G(\overline{\boldsymbol{p}}^{(0)})$. Thus, any $k \in G(\overline{\boldsymbol{p}})$ is contained in $G(\overline{\boldsymbol{p}}^{(n)})$. This implies $k \in G(\overline{\boldsymbol{p}}^{(n-1)})$. By induction, we have for all $n \in \mathbb{N}$

$$\overline{p}_k^{(n)} = \gamma_k \tilde{\mathcal{I}}_k(\overline{\boldsymbol{p}}^{(n-1)}) = \overline{p}_k^{(n-1)} = \overline{p}_k^{(n-2)} = \cdots = \overline{p}_k^{(0)} \,.$$

Thus, for any $k \in G(\overline{\boldsymbol{p}}^{(0)})$ we have

$$\overline{p}_k^{(n)} = \overline{p}_k^{(0)} \quad \text{for all } n \in \mathbb{N} \,. \tag{A.113}$$

The fixed point iteration (A.112) converges to a limit

$$\overline{\boldsymbol{p}}^* = \lim_{n \to \infty} \overline{\boldsymbol{p}}^{(n)} \,.$$

The finite limit exists because the sequence $\overline{\boldsymbol{p}}^{(n)}$ is monotone decreasing and $\overline{\boldsymbol{p}}^{(n)} > 0$ für all $n \in \mathbb{N}$. Independent of the choice of n, we have

$$\begin{aligned} \|\overline{\boldsymbol{p}}^{(n)}\|_\infty \geq \max_{k \in G(\overline{\boldsymbol{p}}^{(n)})} \overline{p}_k^{(n)} &\geq \max_{k \in G(\overline{\boldsymbol{p}}^{(0)})} \overline{p}_k^{(n)} \\ &= \max_{k \in G(\overline{\boldsymbol{p}}^{(0)})} \overline{p}_k^{(0)} = C_1 > 0 \,, \end{aligned}$$

where C_1 is constant. The sequence $\|\overline{\boldsymbol{p}}^{(n)}\|_\infty$ converges as well, so there is another constant C_2 such that

$$C_2 = \lim_{n\to\infty} \|\overline{\boldsymbol{p}}^{(n)}\|_\infty \geq C_1 > 0 \,. \tag{A.114}$$

Because of the monotone convergence of $\overline{\boldsymbol{p}}^{(n)}$ we have

$$\|\overline{\boldsymbol{p}}^{(n)}\|_\infty \geq C_2 > C_1 > 0 \quad \text{for all } n \in \mathbb{N} \,.$$

The ratio of two convergent sequences is convergent if the denominator sequence has a non-zero limit, so

$$\hat{\boldsymbol{p}}^{(n)} = \frac{1}{\|\overline{\boldsymbol{p}}^{(n)}\|_\infty} \overline{\boldsymbol{p}}^{(n)} \,, \quad n \in \mathbb{N} \,,$$

is convergent as well. We have $\|\hat{\boldsymbol{p}}^{(n)}\|_\infty = 1$. Also, we have shown $\hat{p}_k^{(n)} \geq \gamma_k \tilde{\mathcal{I}}_k(\boldsymbol{p}^{(n)}$ for all n and k. We have

$$\hat{p}_k^{(n)} \geq \frac{\gamma_k}{C(\gamma)} \mathcal{I}_k(\boldsymbol{p}^{(n)} \geq \frac{\gamma_k}{C(\gamma)} \underline{f}_{\mathcal{I}_k}(\hat{\boldsymbol{w}}_k) \prod_{l\in\mathcal{K}} (p_l)^{\hat{w}_{kl}} \,.$$

Thus, there is a constant $M_1 > 0$ such that $\hat{\boldsymbol{p}}^{(n)} \in \mathcal{S}(M_1, \hat{\boldsymbol{W}})$, as defined by (A.80). With Lemma A.19 we know that all $\hat{\boldsymbol{p}}^{(n)}$ fulfill

$$\hat{p}_k^{(n)} \geq \underline{C}(M_1, \hat{\boldsymbol{W}}) > 0 \,, \quad \text{for all } n \in \mathbb{N} \text{ and } k \in \mathcal{K} \,. \tag{A.115}$$

Next, consider the limit

$$\boldsymbol{p}^* = \lim_{n\to\infty} \hat{\boldsymbol{p}}^{(n)} \,.$$

Because of (A.115) we have $\boldsymbol{p}^* > 0$. For all $k \in \mathcal{K}$ we have

$$\begin{aligned}
\gamma_k \tilde{\mathcal{I}}_k(\hat{\boldsymbol{p}}^{(n)}) &= \frac{1}{\|\overline{\boldsymbol{p}}^{(n)}\|_\infty} \cdot \gamma_k \tilde{\mathcal{I}}_k(\overline{\boldsymbol{p}}^{(n)}) \\
&= \frac{1}{\|\overline{\boldsymbol{p}}^{(n)}\|_\infty} \cdot \overline{p}_k^{(n+1)} = \frac{\|\overline{\boldsymbol{p}}^{(n+1)}\|_\infty}{\|\overline{\boldsymbol{p}}^{(n)}\|_\infty} \cdot \hat{p}_k^{(n+1)} \,.
\end{aligned}$$

Because of $\lim_{n\to\infty} \|\overline{\boldsymbol{p}}^{(n+1)}\|_\infty / \|\overline{\boldsymbol{p}}^{(n)}\|_\infty = 1$, we have

$$\gamma_k \tilde{\mathcal{I}}_k(\boldsymbol{p}^*) = \lim_{n\to\infty} \gamma_k \tilde{\mathcal{I}}_k(\hat{\boldsymbol{p}}^{(n)}) = \lim_{n\to\infty} \hat{p}_k^{(n+1)} = p_k^*, \quad \forall k \in \mathcal{K} \,.$$

That is, $\boldsymbol{p}^* > 0$ fulfills $p_k^* C(\gamma) = \gamma_k \mathcal{I}_k(\boldsymbol{p}^*)$ for all $k \in \mathcal{K}$.

Proof of Theorem 4.26

The proof is by contradiction. Suppose that for any $\gamma > 0$ there exists a $\hat{\boldsymbol{p}} > 0$ such that

$$C(\boldsymbol{\gamma})\hat{p}_k = \gamma_k \mathcal{I}_k(\hat{\boldsymbol{p}}) \quad \text{for all } k \in \mathcal{K}\,, \tag{A.116}$$

where $C(\boldsymbol{\gamma})$ is defined as by (1.22).

In order to simplify the discussion, we assume that $\boldsymbol{A}_\mathcal{I}$ has a single isolated block $\boldsymbol{A}_\mathcal{I}^{(1)}$ on its main diagonal. The proof for several isolated blocks is similar. The block $\boldsymbol{A}_\mathcal{I}^{(1)}$ is associated with users $1, \ldots, l_1$. The superscript $(\cdot)^{(1)}$ will be used in the following to indicate that the respective quantity belongs to the first block. The interference functions $\mathcal{I}_1, \ldots, \mathcal{I}_{l_1}$ and powers $p_1, \ldots p_{l_1}$ are collected in vectors $\boldsymbol{\mathcal{I}}^{(1)}(\boldsymbol{p})$ and $\boldsymbol{p}^{(1)}$, respectively.

For arbitrary $\boldsymbol{\gamma} > 0$ we define

$$\begin{aligned} \underline{C}(\boldsymbol{\gamma}) &= \inf_{\boldsymbol{p}^{(1)}>0} \Big(\max_{1\le k \le l_1} \frac{\gamma_k \mathcal{I}_k^{(1)}(\boldsymbol{p}^{(1)})}{p_k^{(1)}} \Big) \\ &= \inf_{\boldsymbol{p}>0} \Big(\max_{1\le k \le l_1} \frac{\gamma_k \mathcal{I}_k(\boldsymbol{p})}{p_k} \Big) \le C(\boldsymbol{\gamma})\,. \end{aligned} \tag{A.117}$$

The last inequality holds because the maximum is restricted to the indices $k \le l_1$. Also, $\mathcal{I}_k^{(1)}(\boldsymbol{p}^{(1)}) = \mathcal{I}_k(\boldsymbol{p})$ because k belongs to an isolated block.

We will now show that $\underline{C}(\boldsymbol{\gamma}) = C(\boldsymbol{\gamma})$. To this end, suppose that $\underline{C}(\boldsymbol{\gamma}) < C(\boldsymbol{\gamma})$. Because $\boldsymbol{A}_\mathcal{I}^{(1)}$ is irreducible, Corollary 4.25 implies the existence of a $\tilde{\boldsymbol{p}}^{(1)} > 0$ such that

$$\underline{C}(\boldsymbol{\gamma})\tilde{p}_k^{(1)} = \gamma_k \mathcal{I}_k^{(1)}(\tilde{\boldsymbol{p}}^{(1)}), \quad 1 \le k \le l_1\,. \tag{A.118}$$

This is compared with (A.116). We focus on the indices $k \le l_1$. These users belong to the isolated block, so $\hat{\boldsymbol{p}}$ can be replaced by the vector $\hat{\boldsymbol{p}}^{(1)}$, which is the subvector of $\hat{\boldsymbol{p}}$ consisting of the first l_1 components. That is,

$$C(\boldsymbol{\gamma})\hat{p}_k^{(1)} = \gamma_k \mathcal{I}_k^{(1)}(\hat{\boldsymbol{p}}^{(1)}), \quad 1 \le k \le l_1\,. \tag{A.119}$$

Comparing (A.118) and (A.119), and using Lemma 2.21 (part 2), it can be concluded that $C(\boldsymbol{\gamma}) = \underline{C}(\boldsymbol{\gamma})$. The same can be shown for any isolated block.

For arbitrary $\boldsymbol{\gamma} > 0$, we define SIR targets

$$\gamma_k(\lambda) = \begin{cases} \lambda \cdot \gamma_k\,, & k \le l_1 \\ \gamma_k\,, & k > l_1 \end{cases} \qquad \lambda > 0\,, \tag{A.120}$$

which are collected in a vector $\boldsymbol{\gamma}(\lambda) = [\gamma_1(\lambda), \ldots, \gamma_K(\lambda)]^T$. The l_1-dimensional vector $\boldsymbol{\gamma}^{(1)}(\lambda) > 0$ contains the targets associated with the users of the first block $\boldsymbol{A}_\mathcal{I}^{(1)}$.

From (A.116) we know that for any $\boldsymbol{\gamma}(\lambda) > 0$ there is a $\boldsymbol{p}(\lambda) > 0$ such that

$$C\big(\boldsymbol{\gamma}(\lambda)\big)\, p_k(\lambda) = \gamma_k(\lambda)\mathcal{I}_k\big(\boldsymbol{p}(\lambda)\big) \quad \text{for all } k \in \mathcal{K}\,. \tag{A.121}$$

Introducing a sub-vector $\boldsymbol{p}^{(1)}(\lambda)$, defined by

$$p_k^{(1)}(\lambda) = p_k(\lambda), \quad 1 \le k \le l_1 ,$$

the first l_1 components of (A.121) can be written as

$$C(\gamma(\lambda)) \cdot p_k^{(1)}(\lambda) = \lambda \cdot \gamma_k \cdot \mathcal{I}_k^{(1)}(\boldsymbol{p}^{(1)}(\lambda)), \quad 1 \le k \le l_1 .$$

For arbitrary $\lambda > 0$, we have

$$\begin{aligned} C(\gamma(\lambda)) = \underline{C}(\gamma(\lambda)) &= \inf_{\boldsymbol{p}^{(1)}>0} \max_{1\le k\le l_1} \frac{\gamma_k(\lambda) \cdot \mathcal{I}_k^{(1)}(\boldsymbol{p}^{(1)})}{p_k^{(1)}} \\ &= \lambda \cdot \inf_{\boldsymbol{p}^{(1)}>0} \max_{1\le k\le l_1} \frac{\gamma_k \cdot \mathcal{I}_k^{(1)}(\boldsymbol{p}^{(1)})}{p_k^{(1)}} = \lambda \cdot \underline{C}(\gamma) \\ &= \lambda \cdot C(\gamma) , \end{aligned} \tag{A.122}$$

By assumption (4.48), we have $\underline{C}_1(\gamma) > 0$, so

$$\begin{aligned} C(\gamma(\lambda)) &= \inf_{\boldsymbol{p}>0} \max_{k\in\mathcal{K}} \frac{\gamma_k(\lambda) \cdot \mathcal{I}_k(\boldsymbol{p})}{p_k} \\ &\ge \inf_{\boldsymbol{p}>0} \max_{k>l_1} \frac{\gamma_k \cdot \mathcal{I}_k(\boldsymbol{p})}{p_k} = \underline{C}_1(\gamma) > 0 . \end{aligned} \tag{A.123}$$

Here we have exploited that $\gamma_k(\lambda) = \gamma_k$ for $k > l_1$. Combining (A.122) and (A.123) we obtain

$$\lambda \cdot C(\gamma) \ge \underline{C}_1(\gamma) > 0 .$$

This inequality holds for all $\lambda > 0$. By letting $\lambda \to 0$, we obtain a contradiction, thus concluding the proof.

References

1. R. D. Yates, "A framework for uplink power control in cellular radio systems," *IEEE J. Select. Areas Commun.*, vol. 13, no. 7, pp. 1341–1348, Sept. 1995.
2. M. Schubert and H. Boche, *QoS-Based Resource Allocation and Transceiver Optimization.* Foundations and Trends in Communications and Information Theory, 2005/2006, vol. 2, no. 6.
3. H. Boche and M. Schubert, "The structure of general interference functions and applications," *IEEE Trans. Inform. Theory*, vol. 54, no. 11, pp. 4980 - 4990, Nov. 2008.
4. ——, "Concave and convex interference functions – general characterizations and applications," *IEEE Trans. Signal Processing*, vol. 56, no. 10, pp. 4951–4965, Oct. 2008.
5. ——, "A calculus for log-convex interference functions," *IEEE Trans. Inform. Theory*, vol. 54, no. 12, pp. 5469–5490, Dec. 2008.
6. ——, "Nash bargaining and proportional fairness for wireless systems," *IEEE/ACM Trans. on Networking*, vol. 17, no. 5, pp. 1453–1466, Oct. 2009.
7. ——, "A superlinearly and globally convergent algorithm for power control and resource allocation with general interference functions," *IEEE/ACM Trans. on Networking*, vol. 16, no. 2, pp. 383–395, Apr. 2008.
8. ——, "Perron-root minimization for interference-coupled systems with adaptive receive strategies," *IEEE Trans. Commun.*, vol. 57, no. 10, pp. 3173–3164, Oct. 2009.
9. M. Schubert and H. Boche, "Robust resource allocation," in *Proc. IEEE Information Theory Workshop (ITW), Chengdu, China*, Oct. 2006, pp. 556–560.
10. H. Boche and M. Schubert, "A generalization of Nash bargaining and proportional fairness to log-convex utility sets with power constraints," *IEEE Trans. Inform. Theory*, vol. 57, no. 6, June 2011.
11. ——, "A unifying approach to interference modeling for wireless networks," *IEEE Trans. Signal Processing*, vol. 58, no. 6, pp. 3282–8297, June 2010.
12. M. Schubert and H. Boche, "Dynamic resource allocation and interference avoidance for wireless networks," Tutorial at IEEE Internat. Conf. on Acoustics, Speech, and Signal Proc. (ICASSP), Dallas, USA, Mar. 2010.
13. D. Gesbert, M. Kountouris, R. W. Heath Jr., C.-B. Chae, and T. Sälzer, "Shifting the MIMO paradigm," *IEEE Signal Processing Magazine*, pp. 36–46, Sept. 2007.

14. H. Boche, S. Naik, and T. Alpcan, "Characterization of non-manipulable and pareto optimal resource allocation strategies for interference coupled wireless systems," in *IEEE Infocom*, 2010.
15. S. U. Pillai, T. Suel, and S. Cha, "The perron-frobenius theorem: Some of its applications," *IEEE Signal Processing Magazine*, vol. 22, no. 2, pp. 62–75, Mar. 2005.
16. J. F. Nash Jr., "The bargaining problem," *Econometrica, Econometric Soc.*, vol. 18, pp. 155–162, 1950.
17. ——, "Two-person cooperative games," *Econometrica, Econometric Soc.*, vol. 21, no. 1, pp. 128–140, 1953.
18. H. J. M. Peters, *Axiomatic Bargaining Game Theory.* Kluwer Academic Publishers, Dordrecht, 1992.
19. W. Thomson, *Handbook of Game Theory, Vol. 2.* Elsevier Science, 1994, ch. Cooperative Models of Bargaining.
20. A. I. Khinchin, "The entropy concept in probability theory," *Usp. Mat. Nauk.*, vol. 8, no. 55, pp. 3–20, 1953, (in Russian), English translation in Mathematical Foundations of Information Theory. New York: Dover, 1957.
21. D. K. Faddeev, "On the concept of entropy of a finite probabilistic scheme," *Usp. Mat. Nauk.*, vol. 11, no. 67, pp. 227–231, 1956, (in Russian), German translation in "Arbeiten zur Informationsteorie I". VEB Dt. Verlag der Wissenschaften, Berlin 1961.
22. R. Ash, *Information Theory.* Dover Pub. Inc., New York, 1965.
23. S. Boyd and L. Vandenberghe, *Convex Optimization.* Cambridge University Press, 2004.
24. Z.-Q. Luo and W. Yu, "An introduction to convex optimization for communications and signal processing," *IEEE J. Select. Areas Commun.*, vol. 24, no. 8, pp. 1426–1438, Aug. 2006.
25. Z.-Q. Luo, T. N. Davidson, G. B. Giannakis, and K. Wong, "Transceiver optimization for block-based multiple access through ISI channels," *IEEE Trans. Signal Proc.*, vol. 52, no. 4, pp. 1037–1052, Apr. 2004.
26. M. Bengtsson and B. Ottersten, *Handbook of Antennas in Wireless Communications.* CRC press, Aug. 2001, ch. 18: Optimal and Suboptimal Transmit Beamforming.
27. D. Hammarwall, M. Bengtsson, and B. Ottersten, "On downlink beamforming with indefinite shaping constraints," *IEEE Trans. Signal Proc.*, vol. 54, no. 9, pp. 3566–3580, Sept. 2006.
28. A. Wiesel, Y. C. Eldar, and S. Shamai (Shitz), "Linear precoding via conic optimization for fixed MIMO receivers," *IEEE Trans. Signal Proc.*, vol. 54, no. 1, pp. 161–176, 2006.
29. M. Biguesh, S. Shahbazpanahi, and A. B. Gershman, "Robust downlink power control in wireless cellular systems," *EURASIP Journal on Wireless Communications and Networking*, no. 2, pp. 261–272, 2004.
30. M. Payaró, A. Pascual-Iserte, and M. A. Lagunas, "Robust power allocation designs for multiuser and multiantenna downlink communication systems through convex optimization," *IEEE Selec. Areas in Commun.*, vol. 25, no. 7, pp. 1390 – 1401, Sep. 2007.
31. C. Farsakh and J. A. Nossek, "Spatial covariance based downlink beamforming in an SDMA mobile radio system," *IEEE Trans. Commun.*, vol. 46, no. 11, pp. 1497–1506, Nov. 1998.

32. F. Rashid-Farrokhi, K. J. Liu, and L. Tassiulas, "Transmit beamforming and power control for cellular wireless systems," *IEEE J. Select. Areas Commun.*, vol. 16, no. 8, pp. 1437–1449, Oct. 1998.
33. M. Schubert and H. Boche, "Solution of the multi-user downlink beamforming problem with individual SINR constraints," *IEEE Trans. Veh. Technol.*, vol. 53, no. 1, pp. 18–28, Jan. 2004.
34. P. Kumar, R. Yates, and J. Holtzman, "Power control based on bit error (BER) measurements," in *Proc. IEEE Military Communications Conference MILCOM 95*, McLean, VA, Nov. 1995, pp. 617–620.
35. S. Ulukus and R. Yates, "Adaptive power control and MMSE interference suppression," *ACM Wireless Networks*, vol. 4, no. 6, pp. 489–496, 1998.
36. S. Hanly, "An algorithm for combined cell-site selection and power control to maximize cellular spread spectrum capacity," *IEEE Journal on Selected Areas in Communications*, vol. 13, no. 7, pp. 1332–1340, Sept. 1995.
37. R. Yates and H. Ching-Yao, "Integrated power control and base station assignment," *IEEE Trans. on Vehicular Technology*, vol. 44, no. 3, pp. 638 – 644, Aug. 1995.
38. C. W. Sung, "Log-convexity property of the feasible SIR region in power-controlled cellular systems," *IEEE Communications Letters*, vol. 6, no. 6, pp. 248–249, June 2002.
39. D. Catrein, L. Imhof, and R. Mathar, "Power control, capacity, and duality of up- and downlink in cellular CDMA systems," *IEEE Trans. Commun.*, vol. 52, no. 10, pp. 1777–1785, 2004.
40. H. Boche and S. Stańczak, "Log-convexity of the minimum total power in CDMA systems with certain quality-of-service guaranteed," *IEEE Trans. Inform. Theory*, vol. 51, no. 1, pp. 374–381, Jan. 2005.
41. ——, "Convexity of some feasible QoS regions and asymptotic behavior of the minimum total power in CDMA systems," *IEEE Trans. Commun.*, vol. 52, no. 12, pp. 2190 – 2197, Dec. 2004.
42. S. Stanczak, M. Wiczanowski, and H. Boche, *Theory and Algorithms for Resource Allocation in Wireless Networks*, ser. Lecture Notes in Computer Science (LNCS). Springer-Verlag, 2006.
43. M. Chiang, C. W. Tan, D. Palomar, D. O'Neill, and D. Julian, "Power control by geometric programming," *IEEE Trans. Wireless Commun.*, vol. 6, no. 7, pp. 2640–2651, July 2007.
44. C. W. Tan, D. P. Palomar, and M. Chiang, "Exploiting hidden convexity for flexible and robust resource allocation in cellular networks," in *IEEE Infocom*, May 2007, pp. 964–972.
45. J. Zander and S.-L. Kim, *Radio Resource Management for Wireless Networks*. Artech House, Boston, London, 2001.
46. S. Koskie and Z. Gajic, "SIR-based power control algorithms for wireless CDMA networks: An overview," *Dynamics of Continuous Discrete and Impulsive Systems B: Applications and Algorithms*, vol. 10-S, pp. 286–293, 2003.
47. S. Stanczak, M. Wiczanowski, and H. Boche, *Fundamentals of Resource Allocation in Wireless Networks: Theory and Algorithms*, ser. Foundations in Signal Processing, Communications and Networking, R. M. W. Utschick, H. Boche, Ed. Springer, 2008, vol. 3.
48. J. M. Aein, "Power balancing in systems employing frequency reuse," *COMSAT Tech. Rev.*, vol. 3, no. 2, pp. 277–300, 1973.

49. H. J. Meyerhoff, "Method for computing the optimum power balance in multibeam satellites," *COMSAT Tech. Rev.*, vol. 4, no. 1, pp. 139–146, 1974.
50. J. Zander, "Performance of optimum transmitter power control in cellular radio systems," *IEEE Trans. on Vehicular Technology*, vol. 41, no. 1, pp. 57–62, Feb. 1992.
51. G. J. Foschini and Z. Miljanic, "A simple distributed autonomous power control algorithm and its convergence," *IEEE Trans. on Vehicular Technology*, vol. 42, no. 4, pp. 541–646, nov 1993.
52. N. Bambos, S. C. Chen, and G. J. Pottie, "Radio link admission algorithms for wireless networks with power control and active link quality protection," in *INFOCOM (1)*, 1995, pp. 97–104.
53. R. A. Monzingo and T. W. Miller, *Introduction to Adaptive Arrays.* Wiley, New York, 1980.
54. H. Alavi and R. Nettleton, "Downstream power control for a spread spectrum cellular mobile radio system," in *Proc. IEEE Globecom*, 1982, pp. 84–88.
55. H. Wielandt, "Unzerlegbare, nicht negative Matrizen," *Math. Z.*, no. 52, pp. 642 – 648, 1950, and Mathematische Werke/Mathematical Works, vol. 2, 100-106 de Gruyter, Berlin, 1996.
56. R. A. Horn and C. R. Johnson, *Matrix Analysis.* Cambridge University Press, MA, 1985.
57. F. R. Gantmacher, *The Theory of Matrices, Vol. 2.* New York: Chelsea Publishing Comp., 1959.
58. C. D. Meyer, *Matrix Analysis and Applied Linear Algebra.* SIAM, 2000.
59. D. Gerlach and A. Paulraj, "Base station transmitting antenna arrays for multipath environments," *Signal Processing (Elsevier Science)*, vol. 54, pp. 59–73, 1996.
60. G. Montalbano and D. T. M. Slock, "Matched filter bound optimization for multiuser downlink transmit beamforming," in *Proc. IEEE Internat. Conf. on Universal Personal Communications (ICUPC), Florence, Italy*, Oct. 1998.
61. J. Zander and M. Frodigh, "Comment on performance of optimum transmitter power control in cellular radio systems," *IEEE Trans. on Vehicular Technology*, vol. 43, no. 3, p. 636, Aug. 1994.
62. M. Schubert and H. Boche, "SIR balancing for multiuser downlink beamforming – a convergence analysis," in *Proc. IEEE Int. Conf. on Comm. (ICC). New York, USA*, Apr. 2002, pp. 841 – 845 vol.2.
63. H. Boche and M. Schubert, "A general duality theory for uplink and downlink beamforming," in *Proc. IEEE Vehicular Techn. Conf. (VTC) fall, Vancouver, Canada*, Sept. 2002, pp. 87 – 91 vol.1.
64. E. Visotsky and U. Madhow, "Optimum beamforming using transmit antenna arrays," in *Proc. IEEE Vehicular Techn. Conf. (VTC) spring, Houston, Texas*, vol. 1, May 1999, pp. 851–856.
65. P. Viswanath and D. Tse, "Sum capacity of the vector Gaussian broadcast channel and uplink-downlink duality," *IEEE Trans. Inform. Theory*, vol. 49, no. 8, pp. 1912–1921, Aug. 2003.
66. S. Vishwanath, N. Jindal, and A. Goldsmith, "Duality, achievable rates, and sum-rate capacity of Gaussian MIMO broadcast channels," *IEEE Trans. Inform. Theory*, vol. 49, no. 10, pp. 2658– 2668, Oct. 2003.
67. W. Yu and T. Lan, "Transmitter optimization for the multi-antenna downlink with per-antenna power constraints," *IEEE Trans. Signal Proc.*, vol. 55, no. 6, pp. 2646–2660, June 2007.

68. H. Weingarten, Y. Steinberg, and S. Shamai (Shitz), "The capacity region of the Gaussian MIMO broadcast channel," *IEEE Trans. Inform. Theory*, pp. 3936–3964, Sept. 2006.
69. W. Yu and J. M. Cioffi, "Sum capacity of Gaussian vector broadcast channels," *IEEE Trans. Inform. Theory*, vol. 50, no. 9, pp. 1875–1892, 2004.
70. S. Vorobyov, A. Gershman, and Z.-Q. Luo, "Robust adaptive beamforming using worst-case performance optimization: a solution to the signal mismatch problem," *IEEE Trans. Signal Proc.*, vol. 51, no. 2, pp. 313 – 324, Feb. 2003.
71. R. G. Lorenz and S. Boyd, "Robust minimum variance beamforming," *IEEE Trans. Signal Proc.*, vol. 53, no. 5, pp. 1684–1696, 2005.
72. A. M. Rubinov, "Towards monotonic analysis," *Nonlinear Analysis and Related Topics, Proceedings of Institute of Mathematics, Minsk*, pp. 147–154, 1999.
73. A. M. Rubinov and B. M. Glover, "Duality for increasing positively homogeneous functions and normal sets," *Recherche Opérationelle/Operations Research*, vol. 32, no. 1, pp. 105–123, 1999.
74. C. Huang and R. Yates, "Rate of convergence for minimum power assignment algorithms in cellular radio systems," *Baltzer/ACM Wireless Networks*, vol. 4, pp. 223–231, 1998.
75. K. K. Leung, C. W. Sung, W. S. Wong, and T. Lok, "Convergence theorem for a general class of power-control algorithms," *IEEE Trans. Commun.*, vol. 52, no. 9, pp. 1566 – 1574, Sept. 2004.
76. N. Bambos, S. Chen, and G. Pottie, "Channel access algorithms with active link protection for wireless communication networks with power control," *IEEE/ACM Trans. on Networking*, vol. 8, no. 5, pp. 583 – 597, Oct. 2000.
77. M. Xiao, N. Shroff, and E. Chong, "A utility-based power-control scheme in wireless cellular systems," *IEEE/ACM Trans. on Networking*, vol. 11, no. 2, pp. 210–221, Apr. 2003.
78. A. Koskie and Z. Gajic, "A Nash game algorithm for SIR-based power control for 3G wireless CDMA networks," *IEEE/ACM Trans. on Networking*, vol. 13, no. 5, Oct. 2005.
79. J.-W. Lee, R. Mazumdar, and N. Shroff, "Joint resource allocation and base-station assignment for the downlink in CDMA networks," *IEEE/ACM Trans. on Networking*, vol. 14, no. 1, pp. 1–14, Feb. 2006.
80. H. Tuy, "Monotonic optimization: Problems and solution approaches," *SIAM Journal on Optimization archive*, vol. 11, no. 2, pp. 464 – 494, 2000.
81. H. Boche and M. Schubert, "On the structure of the multiuser QoS region," *IEEE Trans. Signal Processing*, vol. 55, no. 7, pp. 3484–3495, July 2007.
82. M. Schubert and H. Boche, "A unifying theory for uplink and downlink multiuser beamforming," in *Proc. IEEE Intern. Zurich Seminar, Switzerland*, Feb. 2002.
83. F. Rashid-Farrokhi, L. Tassiulas, and K. J. Liu, "Joint optimal power control and beamforming in wireless networks using antenna arrays," *IEEE Trans. Commun.*, vol. 46, no. 10, pp. 1313–1323, Oct. 1998.
84. J.-B. Hiriart-Urruty and C. Lemaréchal, *Fundamentals of Convex Analysis*. Springer Berlin, Heidelberg, New York, 2001.
85. H. Boche and S. Stanczak, "Strict convexity of the feasible log-SIR region," *IEEE Trans. Commun.*, vol. 56, no. 9, pp. 1511–1518, Sept. 2008.
86. ——, "The Kullback-Leibler divergence and nonnegative matrices," *IEEE Trans. Inform. Theory*, vol. 52, no. 12, pp. 5539–5545, Dec. 2006.

87. S. Stanczak and H. Boche, "On the convexity of feasible QoS regions," *IEEE Trans. Inform. Theory*, vol. 53, no. 2, pp. 779–783, Feb. 2007.
88. ——, "The infeasible SIR region is not a convex set," *IEEE Trans. on Communications*, vol. 54, no. 11, Nov. 2006.
89. F. Meshkati, H. Poor, and S. Schwartz, "Energy-efficient resource allocation in wireless networks: An overview of game-theoretic approaches," *IEEE Signal Processing Magazine*, vol. 24, pp. 58–68, 2007.
90. F. Meshkati, H. V. Poor, S. C. Schwartz, and N. B. Mandayam, "A utility-based approach to power control and receiver design in wireless data networks," *IEEE Trans. Commun.*, 2003, submitted.
91. A. Leshem and E. Zehavi, "Cooperative game theory and the frequency selective gaussian interference channel," *IEEE Journal of Selected Topic in Communications, Special issue on Game theory in communication and networking*, vol. 26, no. 7, pp. 1078 – 1088, Sept. 2008.
92. E. Larsson, E. Jorswieck, J. Lindblom, and R. Mochaourab, "Game Theory and the Flat-Fading Gaussian Interference Channel," *IEEE Signal Processing Magazine*, vol. 26, no. 5, pp. 18–27, Feb. 2009.
93. A. Leshem and E. Zehavi, "Game theory and the frequency selective interference channel," *IEEE Signal Processing Magazine*, vol. 26, no. 5, pp. 28 – 40, 2009.
94. S. Lasaulce, M. Debbah, and E. Altman, "Methodologies for analyzing equilibria in wireless games," *IEEE Signal Processing Magazine*, vol. 26, no. 5, pp. 41 – 52, 2009.
95. R. Hunger and M. Joham, "A complete description of the QoS feasibility region in the vector broadcast channel," *IEEE Trans. Signal Proc.*, vol. 58, no. 7, pp. 3870–3877, July 2010.
96. E. Larsson and E. Jorswieck, "Competition versus Collaboration on the MISO Interference Channel," *IEEE Journal on Selected Areas in Communications*, vol. 26, no. 7, pp. 1059 – 1069, Sept. 2008.
97. E. Jorswieck and E. Larsson, "Complete Characterization of the Pareto Boundary for the MISO Interference Channel," *IEEE Trans. on Signal Processing*, vol. 56, no. 10, pp. 5292 – 5296, Oct. 2008.
98. F. Kelly, A. Maulloo, and D. Tan, "Rate control for communication networks: Shadow prices, proportional fairness and stability," *Journal of Operations Research Society*, vol. 49, no. 3, pp. 237–252, Mar. 1998.
99. J. Brehmer and W. Utschick, "On proportional fairness in nonconvex wireless systems," in *Proc. Int. ITG Workshop on Smart Antennas (WSA) 2009, Berlin, Germany*, Feb. 2009.
100. A. Leshem and E. Zehavi, "Bargaining over the interference channel," *IEEE J. Select. Areas Commun.*, vol. 26, no. 7, 2008.
101. E. G. Larsson and E. A. Jorswieck, "The MISO interference channel: Competition versus collaboration," in *Proc. Allerton Conf. on Commun., Control and Computing, Monticello, USA*, 2007.
102. R. Mazumdar, L. G. Mason, and C. Douligeris, "Fairness in network optimal flow control: Optimality of product forms," *IEEE Trans. Commun.*, vol. 39, no. 5, pp. 775–782, May 1991.
103. H. Yaïche, R. Mazumdar, and C. Rosenberg, "A game theoretic framework for bandwidth allocation and pricing in broadband networks," *IEEE/ACM Trans. on Networking*, vol. 8, no. 5, pp. 667–678, 2000.

104. C. Touati, E. Altman, and J. Galtier, "Utility based fair bandwidth allocation," in *Proc. of the IASTED International Conference on Networks, Parallel and Distributed Processing and Applications (NPDPA)*, Tsukuba, Japan, Oct. 2002.
105. Z. Han, Z. Ji, and K. J. R. Liu, "Fair multiuser channel allocation for OFDMA networks using Nash bargaining solutions and coalitions," *IEEE Trans. Commun.*, vol. 53, no. 8, pp. 1366– 1376, Aug. 2005.
106. C. Touati, E. Altman, and J. Galtier, "Generalized Nash bargaining solution for bandwidth allocation," *Computer Networks*, vol. 50, no. 17, pp. 3242–3263, 2006.
107. J. E. Suris, L. A. DaSilva, Z. Han, and A. B. MacKenzie, "Cooperative game theory for distributed spectrum sharing," in *Proc. IEEE Int. Conf. on Comm. (ICC), Glasgow, Scotland*, June 2007.
108. G. de Clippel, "An axiomatization of the Nash bargaining solution," *Social Choice and Welfare, Springer*, vol. 29, no. 2, pp. 201–210, Sept. 2007.
109. J. P. Conley and S. Wilkie, "An extension of the Nash bargaining solution to nonconvex problems," *Games and Economic Behavior*, vol. 13, no. 1, pp. 26–38, March 1996.
110. Y. Xu and N. Yoshihara, "Alternative characterizations of three bargaining solutions for nonconvex problems," *Games and Economic Behavior Technology*, vol. 57, pp. 86–92, 2006.
111. C. Touati, H. Kameda, and A. Inoie, "Fairness in non-convex systems," University of Tsukuba, Tech. Rep. CS-TR-05-4, Sept. 2005.
112. M. Kaneko, "An extension of Nash bargaining problem and the Nash social welfare function," *Theory and Decision*, vol. 12, no. 2, pp. 135–148, June 1980.
113. M. Herrero, "The Nash program: Non-convex bargaining problems," *Journal of Economic Theory*, vol. 49, pp. 266–277, 1989.
114. L. Zhou, "The Nash bargaining theory with non-convex problems," *Econometrica, Econometric Soc.*, vol. 3, no. 65, pp. 681–686, May 1997.
115. V. Denicolo and M. Mariotti, "Nash bargaining theory, nonconvex problems and social welfare orderings," *Theory and Decisions (Springer Netherlands)*, vol. 48, no. 4, pp. 351–358, June 2000.
116. E. A. Ok and L. Zhou, "Revealed group preferences on non-convex choice problems," *Economic Theory, Springer*, vol. 13, no. 3, pp. 671–687, 1999.
117. J.-Y. Le Boudec, "Rate adaptation, congestion control and fairness: A tutorial," Tutorial, Ecole Polytechnique Federale de Lausanne (EPFL), Tech. Rep., 2003.
118. H. Boche, M. Schubert, N. Vucic, and S. Naik, "Non-symmetric Nash bargaining solution for resource allocation in wireless networks and connection to interference function calculus," in *Proc. European Signal Processing Conference (EUSIPCO)*, Poznań, Poland, Sept. 2007, (invited).
119. H. Boche, M. Wiczanowski, and S. Stańczak, "Unifying view on min-max fairness, max-min fairness, and utility optimization in cellular networks," in *EURASIP Journal on Wireless Communications and Networking*, 2007, ID 34869.
120. H. Boche, M. Wiczanowski, and S. Stanczak, "Characterization of the fairness gap in resource allocation for wireless cellular networks," in *Proc. IEEE Int. Symp. on Inf. Theory and Applications (ISITA) , Seoul, Korea*, Oct. 2006.
121. H. Boche and S. Naik, "Revisiting proportional fairness: Anonymity among users in interference coupled wireless systems," *IEEE Trans. Commun.*, vol. 58, no. 10, pp. 2995–3000, Oct. 2010.

122. X. Qiu and K. Chawla, "On the performance of adaptive modulation in cellular systems," *IEEE Trans. Commun.*, vol. 47, no. 6, pp. 884–895, June 1999.
123. S. Hayashi and Z.-Q. Luo, "Spectrum management for interference-limited multiuser communication systems," *IEEE Trans. Inform. Theory*, vol. 55, no. 3, pp. 1153–1175, Mar. 2009.
124. H. Boche, S. Naik, and E. Jorswieck, "Detecting misbehavior in distributed wireless interference networks," 2011, submitted for possible publication.
125. E. Altman and Z. Altman, "S-modular games and power control inwireless networks," *IEEE Trans. on Automatic Control*, vol. 48, no. 5, pp. 839–842, May 2003.
126. M. Schubert and H. Boche, "A generic approach to QoS-based transceiver optimization," *IEEE Trans. Commun.*, vol. 55, no. 8, pp. 1557 – 1566, Aug. 2007.
127. L. Qi and J. Sun, "A nonsmooth version of newton's method," *Mathematical Programming*, vol. 58, pp. 353–367, 1993.
128. F. H. Clarke, *Optimization and Non-Smooth Analysis.* Wiley, New York, 1983.
129. L. Qi, "Convergence analysis of some algorithms for solving nonsmooth equations," *Mathematics of Operations Research*, vol. 18, no. 1, pp. 227–244, Feb. 1993.
130. S. Hanly, "Congestion measures in DS-CDMA networks," *IEEE Trans. Commun.*, vol. 47, no. 3, Mar. 1999.
131. H. Boche and M. Schubert, "A general theory for SIR-balancing," *EURASIP Journal on Wireless Communications and Networking*, 2006, http://www.hindawi.com/journals/wcn/, Article ID 60681, 18 pages.
132. B. Han, M. Overton, and T. P.-Y. Yu, "Design of hermite subdivision schemes aided by spectral radius optimization," *submitted to SIAM J. Scient. Comp.*, 2002.
133. R. Nettleton and H. Alavi, "Power control for a spread spectrum cellular mobile radio system," in *Proc. IEEE Vehicular Techn. Conf. (VTC)*, 1983, pp. 242–246.
134. J. Zander, "Distributed cochannel interference control in cellular radio systems," *IEEE Trans. on Vehicular Technology*, vol. 41, no. 3, pp. 305–311, Aug. 1992.
135. H. Boche and M. Schubert, "Resource allocation in multi-antenna systems – achieving max-min fairness by optimizing a sum of inverse SIR," *IEEE Trans. Signal Processing*, vol. 54, no. 6, pp. 1990–1997, June 2006.
136. N. Vucic and M. Schubert, "Fixed point iteration for max-min sir balancing with general interference functions," in *Proc. IEEE Internat. Conf. on Acoustics, Speech, and Signal Proc. (ICASSP)*, Prague, Czech Republic, May 2011.
137. E. Deutsch and M. Neumann, "Derivatives of the Perron root at an essentially nonnegative matrix and the group inverse of an M-matrix," *J. Math. Anal. Appl.*, vol. I-29, no. 102, pp. 1–29, 1984.

Index

analysis
  concave interference functions 50
  convex interference functions 61
  general interference functions 40
  log-convex interference functions 75
asymptotic coupling matrix 19
axioms
  general interference function 4
  standard interference function 23

beamforming 9, 174, 185

comprehensive hull 48
comprehensive set 30
comprehensiveness 30
concave interference function 6, 50, 51
continuation 27
continuity
  of general interference function 27
  of standard interference function 28
convex interference function 6, 61
convex optimization 5
coupling coefficient 8
coupling matrix 8, 19

dependency matrix 20
  global 20
  local 20
dependency set 20
duality 13

extended power vector 9, 33

fairness gap 119
feasibility 12, 13
fixed point 32
  characterization of Boundary Points 29
  characterization of boundary points 31
  iteration 142, 155, 171, 192

greatest convex minorant 66, 67, 91
greatest log-convex minorant 80

hidden convexity 6

indicator function 12
individual power constraints 34
interference coupling 17
interference function 4
  concave 6
  continuation 205
  convex 6
  derivatives 201
  general 4
  linear 8
  log-concave 6
  log-convex 6
  non-smooth 202
  standard 23
  strictly log-convex 21
  strictly monotone 21, 22, 49
  strictly monotone w.r.t. noise 24
  weakly standard 24
irreducible 9, 11, 197

Jacobi matrix 164

Kullback-Leibler distance 78

least concave majorant 57, 91
least log-concave majorant 82
least log-concave minorant 82
linear convergence 203
linear interference function 8
link gain matrix 8
Lipschitz continuous 161
locally Lipschitz continuous 161
log-convex interference function 6, 72
log-convex sets
  achievability of the boundary 138
  bounded 104
  unbounded 103
logarithmic convexity 6

majorant 41
  least concave 57
  least log-concave 82
max-min balancing *see* min-max balancing
max-min fairness *see* min-max balancing
Max-min SIR balancing *see* min-max balancing
min-max balancing 12, 183
minorant 41
  greatest convex 66, 67
  greatest log-convex 80
monotone optmization 16

Nash bargaining 99, 100
Nash bargaining solution 100
Nash equilibrium 100
Nash product 102
NBS *see* Nash bargaining solution
Newton's method 167
Non-symmetric Nash bargaining 108
normal set *see* comprehensive set

Pareto boundary 103, 140
Pareto optimal 103
Pareto optimality 140
Perron root 12, 184
Perron-Frobenius theory 8, 110, 119, 185
PEV iteration *see* principal eigenvector iteration

power constraint 32
  individual 34
  sum 33
power constraints
  individual 140
power control 7, 23
power minimization 36, 155
problem
  min-max balancing balancing 183
  Nash bargaining 99
  power minimization 155
  QoS balancing 35
  utility maximization 138
  weighted sum utility 15, 137
proportional fairness 99, 122

QoS *see* Quality of Service
QoS balancing 36
QoS region 32
  individual power constraints 34
  sum-power constraint 33
quadratic convergence 170
quality-of-service 29

receive strategy 10, 156
reducible 197
relatively closed 42
robustness 177

semi smooth 203
semi-continuous 75, 202
signal-to-interference ratio 8
signal-to-interference-plus-noise ratio 9
SINR *see* signal-to-interference-plus-noise ratio, 33
SINR region 33
SIR *see* signal-to-interference ratio
SIR region 12, 13, 29
solution outcome 100
spectral radius 12
standard interference function 23
  continuity 28
  structure 87
strict log-convexity 21
strict monotonicity 21, 149
strongly coupled system 147
sublevel set 12, 30, 42
super-linear convergence 155

superlevel set 42
synthesis
  of concave interference functions 55
  of interference functions 46
  of log-convex interference functions 77

transmit strategy 13, 156

utility maximization 138
utility region 99
utility set 99

weighted sum utility 15, 137
weighted utility 137

Printed by Publishers' Graphics LLC USA
SO20120330-050
2012